BIOCHEMICAL SOCIETY SYMPOSIA

No. 65

CELL BEHAVIOUR: CONTROL AND MECHANISM OF MOTILITY

BIOCHEMICAL SOCIETY SYMPOSIUM No. 65

4th Abercrombie Conference on Cell Behaviour held at St Catherine's College, Oxford, 28 September–1 October 1997

Cell Behaviour: Control and Mechanism of Motility

ORGANIZED AND EDITED BY

J.M. LACKIE, G.A. DUNN AND G.E. JONES

PORTLAND PRESS

Published in the United Kingdom by Portland Press,
59 Portland Place, London W1N 3AJ, U.K.
on behalf of The Biochemical Society
Tel: (+44) 171 580 5530; e-mail: edit@portlandpress.co.uk

Published in North America by Princeton University Press,
41 William Street, Princeton, New Jersey 08540, U.S.A.

© 1999 The Biochemical Society, London

ISBN 1 85578 124 7   ISSN 0067-8964

British Library Cataloguing in Publication Data
A catalogue record for this book is available from the British Library

Typeset by Portland Press Ltd
Printed in Great Britain by Information Press Ltd, Eynsham, U.K.

Contents

Section 1: Introduction

Section 2: Motile Responses

Section 3: Signal Transduction

Dedication

This volume commemorates the 70th birthday, in November 1998, of Juri Vasiliev, whose laboratory in Moscow maintained strong and affectionate links with Michael Abercrombie and colleagues.

Acknowledgements

The Editors would like to thank the many people involved in the preparations for the meeting, especially Christine Moorhouse from Yamanouchi, who provided so much professional assistance with a thousand details of organization. In addition to the financial support of the British Society for Cell Biology and the Society for Experimental Biology, we thank the Company of Biologists, the Yamanouchi Research Institute, SmithKline Beecham, Zeneca Pharmaceuticals and Portland Press for supporting the meeting. Our special thanks are due to the commissioning editor at Portland Press, Sarah Harrison, who provided tireless support during the preparation of the book.

Preface

Towards a molecular view of cell behaviour

In 1982, when the first symposium to commemorate Michael Abercrombie's pioneering work in cell behaviour was published, we little thought that the time would come for publishing a fourth Abercrombie volume. It already seemed that the study of cell behaviour would soon be elbowed off stage by the dazzling success of molecular cell biology. Nevertheless, cell-behaviour studies continued to be required for investigating how cells co-operate to build and maintain whole organisms and how this co-operation can break down in certain disease conditions. Moreover, they soon took on a valuable new role, as documented in the second and third Abercrombie volumes (see below), of supporting molecular research into the mechanisms of cell motility and motile responses. This new role is rapidly assuming much greater importance and is placing cell behaviour studies at the forefront of cell biology research.

During the five years since the last Abercrombie symposium, two factors have emerged that presage a much closer integration between molecular biology and cell behaviour. First, the phase of documenting the molecular structure of the cell's motile machinery is drawing to a close and the molecular dynamics of many individual components of the cytoskeleton are beginning to be understood. The stage is now set for embarking on the more formidable task of unravelling how the functioning of these components is integrated and controlled. Secondly, the tools required for this new phase are already being assembled. Genetic studies of the motile machinery are already well advanced in *Dictyostelium* and, in the *Introduction* to this latest Abercrombie symposium, Günther Gerisch and colleagues outline the remarkable progress that has been made by coupling these studies with the analysis of whole-cell dynamics. Rapid progress in molecular genetics will soon permit the expression of proteins to be manipulated as readily in vertebrate tissue cells.

While *Dictyostelium* provides a valuable model of co-operative cell behaviour, the control of *Cell Motility* assumes a much greater importance in vertebrate cells, and this is dealt with in the first part of the book. The first section concentrates on the motile responses of vertebrate cells and outlines some of the challenges. One outstanding challenge for the future is to understand the environmental factors that direct cells so that we can begin to regulate cell movement for our own purposes. In this section, Curtis and Wilkinson explore the reactions of cells to microfabricated topographical features of the substratum, whereas Bob Tranquillo examines the complexities of the mechanical interactions between cells and fibrillar extracellular matrices. Thus considerable effort is being put into manipulating the physical and chemical properties of artificial environments so as to persuade cells to align themselves with electrode arrays (the man–machine interface perhaps) or to construct tissue equivalents than can be used in prosthetic surgery. If we could construct tubes with cellular walls that resembled blood vessels then coronary bypass surgery might be made easier and, if we could direct nerves to make appropriate reconnections, we might revolutionize repair following traumatic injury to nerve tracts or even within the brain. In other situations, as Marc Mareel's group reminds us, the ability to recognize anarchic behaviour by metastatic tumour cells will allow appropriate intervention, especially if we knew what sort of intervention was required.

Among the early successes of the new approach to studying the control of cell locomotion, the most striking have been concerned with the mechanisms of *Signal Transduction*, and we have devoted the next five chapters to this topic. Many signals come through G-protein-coupled receptors and eventually work through pathways that involve small G-proteins. Other signals cross the membrane in other ways from integrins or through receptor tyrosine kinases, but again act later through the small G-proteins. Martin Humphries discusses integrin activation and ligand binding at a molecular/atomic level and how they subsequently trigger intracellular signals. The relevance for biology stems not only from the role that integrins play in cell translocation, but also in providing structural support for cells in tissues, as proven for epithelia in genetic diseases. Once initiated, signal transduction from activated integrins follows ever-diversifying pathways and David Critchley's team takes us through the latest evidence pointing to a role for the cytoskeletal proteins talin, α-actinin and filamin as linkers between the β-subunits of integrins and F-actin. Another pathway involving microtubules is important for axonal remodelling. Patricia Salinas describes the consequences of Wnt binding to its surface receptor, which leads to an eventual decrease in the level of stable microtubules in cerebellar neurons.

The Rho family of GTPases features in many of the articles in this volume, but chapters 8 and 9 emphasize their critical importance to cell behaviour. Anne Ridley and colleagues consider the roles of Rho, Rac and Cdc42 in the regulation of cellular protrusion, adhesion and migration on a substratum and demonstrate that different cell types are likely to use Rho proteins in subtly different ways. This message is reinforced by Michiels and Collard, who show that the expression of Tiam-1 protein, an activator of Rac, has very different effects on the migration of different cell types. As we learn more about the dynamics of the machinery itself it is clear that the opportunities for fine control of the location of motor activity and the rate and extent of its action are extensive.

The second part of the book deals with the mechanism of cell motility, and the next five chapters take us into the realm of *Cytoskeletal Dynamics*. The first contribution by Sasha Bershadsky and colleagues provides a link between the previous topic of signal transduction and cytoskeletal dynamics. Their chapter stresses the significance of the observation that microtubule disruption induces cell contractility and stimulates assembly of focal adhesions. Thus microtubule dynamics may generate local fluctuations in contractility that could be used to regulate signal transduction from nearby focal contacts. The next two chapters concern the organization and dynamics of filamentous actin and myosin in two very different types of motile vertebrate cells. Both of these contributions challenge the widespread belief that the contractile machinery of non-muscle cells has a sarcomeric organization, looser but similar in principle to that of muscle cells. Louise Cramer focuses on the organization of filamentous actin in fibroblasts and describes a curious gradation in the polarity of ventral bundles that span the whole length of the cell and comprise the majority of the cell's actin bundles. Gary Borisy's group treats us to a radically new insight into the organization of myosin II and actin in the leading lamellae of fish keratocytes. Clusters of myosin II bipolar filaments embedded in an actin meshwork are formed at the leading edge as the cell moves forwards. This arrangement remains quite stable until it is close to the encroaching cell body when it contracts into a bundle parallel to the leading edge. This contraction is postulated to drive the forward translocation of the cell body. Returning to the microtubule array, Manfred Schliwa's team confronts another long-standing problem: the enigmatic role of the centrosome in determining the direction of cell migration. In a detailed analysis of the time course of turning events, they demonstrate that centrosome position is probably an effect rather than the cause of the cell's direction of travel. The challenge now is to gather together the strands of evidence represented in these chapters and to explain the mechanism of motility in terms of cytoskeletal organization. In the final chapter of this section, Mike Sheetz and his associates describe how the various dynamical processes involved in receptor recycling and assembly and the interaction of cytoskeletal components can become co-ordinated into the five-step cycle of the motile process itself.

Explaining the *Dynamics of Motility* of the whole cell is, of course, one of the ultimate goals of cell motility research. We need no longer speculate about the functions of the different components of the motor without data to restrict our fancies. Deletion studies in *Dictyostelium*, coupled with newer and ever more sophisticated methods for analysing the ensuing motile dynamics, have begun to reveal quite unexpected effects. The chapter by Igor Weber and collaborators concentrates on the various classes of actin-binding proteins. They report the astonishing observation that multiple deletions of what were thought to be essential regulators of actin dynamics can have effects that are so subtle that they can only be revealed under special conditions. Tom Stossel's group take up the story of the regulation of actin dynamics in vertebrate cells and find a similar picture, that cells can maintain adequate locomotion in the absence of several seemingly important components. Nevertheless, effects are observed and the time has come to analyse function rather than to speculate about function on the basis of known molecular interactions. The old concept of "one protein, one function" is no longer adequate and it may soon be necessary to conceive these regulatory proteins simply as elements of a complex parallel computer that has evolved to use many different pathways, either simultaneously or interchangeably, to implement each single action. Michelle Peckham and her colleagues adopt a different genetic approach, one of transfecting cells in order to express foreign or mutant myosins, but again the effects are subtle and variable and require large databases for their detection. Finally, we should not forget that cell motility is a mechanical process and, as with muscle research, an essential line of enquiry is to examine the pattern of forces generated by the motile system. Elliot Elson and others consider the different forces that could arise as a result of various molecular interactions and discuss the results of measuring these forces in both wild-type and genetically modified *Dictyostelium* amoebae.

In the longer term, these new lines of enquiry may well lead to a molecular explanation of the control of cell protrusion. It is one aspect of this control, contact inhibition of cell locomotion, which Abercrombie and colleagues postulated becomes defective during the invasion of malignant cells into normal cell populations. Albert Harris, in the last chapter of this volume, sees an understanding of the molecular mechanism of contact inhibition, and how it fails in malignant cells, as the Holy Grail of cell biology. These are but two problems in a list he has formulated of twelve unanswered questions about how tissue cells move. We can only hope that Albert's list will serve to stimulate cell motility research during the next century in the same way that Hilbert's list of great unsolved problems stimulated the mathematics of this century,

It is perhaps worth saying that, as editors, we asked our contributors to try to write in an approachable way, aiming at a wider audience than the committed specialists. We wanted the book to be a source book for those who work in the field, an introduction for those who are thinking of getting into the field — perhaps to seek a function for their recently cloned protein — and a reference book for those few students who are taught about the marvels of cell behaviour at a level above the cataloguing of components. We hoped that this latest volume in the Abercrombie series might be a source of insight and ideas, a collection of essays rather than a dry review, and above all that it might capture some of the current excitement in the field. Whether our hopes have been realized we leave the reader to determine: the meeting itself certainly bubbled with ideas and lively discussion and that in itself bodes well for the future.

John M. Lackie
Graham A. Dunn
Gareth E. Jones

Previous Abercrombie Conferences

1. Bellairs, R., Curtis, A. and Dunn, G. (eds.) (1982) *Cell Behaviour: A Tribute to Michael Abercrombie*, Cambridge University Press, Cambridge
2. Heaysman, J. E. M., Middleton, C. A. and Watt, F. M. (eds.) (1987) *Cell Behaviour: Shape, Adhesion and Motility*, J. Cell Sci., Suppl. 8
3. Jones, G., Wigley, C. and Warn, R. (eds.) (1993) *Cell Behaviour: Adhesion and Motility*, S.E.B. Symposium XLVII, pp. 91–106, Company of Biologists, Cambridge

Abbreviations

ABP-120 (-280)	actin-binding protein of 120 kDa (or 280 kDa)
AFM	atomic-force microscopy
ANOVA	analysis of variance
APC	adenomatous polyposis coli
ARF	ADP-ribosylation factor
BDM	butanedione monoxime
ConA	concanavalin A
CSF	colony-stimulating factor
DH domain	Dbl homology domain
DRIMAPS	digitally recorded interference microscopy with automatic phase shifting
EC1 domain	extracellular domain 1
ECM	extracellular matrix
EGF	epidermal growth factor
ERM proteins	ezrin/radixin/moesin proteins
ES cells	embryonic stem cells
FcγR	receptor that recognizes the Fc portion of immunoglobulins
FPCL	fibroblast-populated collagen lattice
GAP	GTPase-activating protein
GDI	guanine nucleotide dissociation inhibitor
GEF	guanine nucleotide exchange factor
GF	120 kDa gelation factor
GFP	green fluorescent protein
gp80 (etc.)	glycoprotein of 80 kDa (etc.)
GSK	glycogen synthase kinase
GTP[S]	guanosine 5′-[γ-thio]triphosphate
HGF	hepatocyte growth factor
ICAM	intercellular adhesion molecule
JNK	Jun N-terminal kinase
LC_{20}	myosin regulatory light chain of 20 kDa
LPA	lysophosphatidic acid
mAb	monoclonal antibody
MAP	microtubule-associated protein
MAPK	mitogen-activated protein kinase
MDCK cells	Madin–Darby canine kidney cells
MHC	β-cardiac myosin II heavy chain
MLK	mixed-lineage kinase
PAF	platelet-activating factor

PAK	p21-activated kinase
PDGF	platelet-derived growth factor
PH domain	pleckstrin homology domain
PI 3-kinase	phosphoinositide 3-kinase
pp125^{FAK}	focal adhesion kinase
PtdIns $(4,5)P_2$	phosphatidylinositol 4,5-bisphosphate
RICM	reflection interference contrast microscopy
ROK	Rho-associated kinase
SF	scatter factor
Tiam	T-lymphoma invasion and metastasis
t-MHC	truncated MHC
t-MIα	truncated myosin Iα head fragment
VASP	vasodilator-stimulated phosphoprotein
WASP	Wiskott–Aldrich syndrome protein

Biochem. Soc. Symp. **65**, 1–14
Printed in Great Britain

1

Patterns of cellular activities based on protein sorting in cell motility, endocytosis and cytokinesis

Günther Gerisch[1], Markus Maniak and Ralph Neujahr

Max-Planck-Institut für Biochemie, D-82152 Martinsried, Germany

Abstract

In order to move persistently, a cell has to harmonize its protrusion and retraction with attachment and detachment from the substrate. Time-series analyses based on fluctuations in these activities are being used in combination with advanced imaging techniques to unravel the network of protein–protein interactions that tune the activities in a motile cell and co-ordinate them in space and time. Fusions with the green fluorescent protein have helped to visualize the recruitment of cytoskeletal proteins from a soluble pool and their transient assembly into supramolecular structures. Using a series of mutants deficient in specific cytoskeletal proteins has revealed common themes and interrelationships between cell motility, endocytosis and cytokinesis. For instance, a phagocytic cup competes with leading-edge formation, and recruits the same actin-associated proteins. Cytokinesis is based on the fine tuning of activities in the microtubule system and the actin network in the cell cortex. Cells dividing on a substrate apply tension to the surface on which they adhere, as determined by the silicone rubber technique. Actin-associated proteins are sorted during cytokinesis either to the extensions formed at the poles of a dividing cell or to the cleavage furrow. A major effort will be required to elucidate the mechanisms that dictate the pattern of local activities and drive the translocation of proteins in cell motility, endocytosis and cytokinesis.

[1]To whom correspondence should be addressed.

Introduction

As an introduction to the 4th Abercrombie Conference, it seems appropriate to refer to a classical paper by Michael Abercrombie, Joan E.M. Heaysman and Susan M. Pegrum [1], which marked the beginning of the quantitative analysis of cell motility. One statement made in this paper is that the activities in the leading lamella of a fibroblast fluctuate between periods of protrusion, standstill and withdrawal. Another principal observation is that, when these fluctuations are measured at points separated by some 6 μm, they appear to be largely independent of each other. As pointed out at the end of the paper, although the leading lamella is not firmly attached to a substrate, its fluctuations are linked with the mechanism of displacement of the whole cell. In the decades following the publication of this paper, the integration of local activities into a pattern that allows a cell to move persistently on a substrate has become a major issue in the field of cell motility, and a number of powerful techniques have been developed to allow quantification of activities underlying the movement of a cell and its interaction with a substrate.

In this overview, we wish to broaden the scope from the regulation of cell motility to the interrelationship of the diverse functions mediated by the actin system. The actin system not only mediates cell motility, but is also involved in phagocytosis and endocytic trafficking, and plays a crucial role in cytokinesis, the division of a mitotic cell. The question is: what is common to these different functions of the actin system, which peculiarities distinguish them, and how are these divergent activities co-ordinated? We will place emphasis on the fact that a common feature of these cellular activities is the transient assembly of actin together with other proteins recruited from a cytosolic pool. Most of the examples will be taken from work on *Dictyostelium*, the organism most familiar to us, assuming that major conclusions will also apply to other systems.

Regulation of force generation and cell attachment in motile cells

Three major technical advancements exemplify progress in the quantitative analysis of cell motility. One is the technique of reflection interference contrast microscopy (RICM), designed by Adam Curtis [2] and used extensively for the analysis of cell–substrate interactions in a series of papers by David Gingell and his colleagues [3–5]; the second is correlation analysis of local activities of moving cells, as introduced by Graham Dunn and his colleagues [6,7]; and the third is the silicone rubber technique, pioneered by Albert Harris et al. [8] and refined by Ken Jacobson and his colleagues [9,10], which has provided an elegant way to visualize and quantify the forces applied by a moving cell on to the substrate surface.

Despite the independence of local activities in a lamellipod, as demonstrated in the reference paper by Abercrombie et al. [1], there must be an overall co-ordination of local activities within a cell in order to permit persistent movement. This co-ordination has been studied using time-series analysis in fibroblasts and in the fast-moving cells of *Dictyostelium*. Cross-correlation analysis of fluctuations at the front and tail regions in moving fibroblasts showed that retraction of the tail is the initial event and that expansion of the front lags behind [7]. A similar analysis applied to *Dictyostelium* indicated, in contrast, a pioneering role for the front of a cell. Phases of rapid front expansion preceded by about 10 s phases of increased rates of retraction at the tail of a cell. RICM imaging revealed that not only speed of movement but also the size and shape of the area of cell–substrate contact fluctuates in *Dictyostelium* cells [11]. In accordance with a leading role for the front region, gain of contact area near to the front preceded the loss of contact area near to the retracting tail [12]. It seems, according to these results, that cells have developed two different modes of co-ordinating activities in the actin cytoskeleton in order to make movement persistent. In fibroblasts, a signal is elicited by tail retraction which stimulates the leading edge to propagate, whereas in *Dictyostelium* (and probably other fast-moving cells), a signal is generated by an active front which causes other regions to retract.

Vectorial cross-correlation analysis indicated that, in wild-type cells of *Dictyostelium*, retraction occurred preferentially in the same direction as expansion. This is a prerequisite for persistence of the direction of movement. This persistence proved to be impaired in mutant cells deficient in a number of proteins that regulate the cross-linkage and polymerization state of actin, indicating that the spatial co-ordination of activities in a motile cell depends on the integrity of the actin network [12].

These results indicate that activities in a motile cell are integrated into a spatio–temporal pattern of changes in the network of actin filaments, which are densely cross-linked to each other in the cell cortex and are connected to the plasma membrane. Dynamic properties of the system controlling activities in such a network have been elucidated by time-series analysis in leucocytes. The shape and movement of these cells often fluctuates in an oscillatory manner, suggesting non-linear interactions between control elements responsible for actin-based cell motility [13,14]. Comparable results obtained in *Dictyostelium* [15] showed a similar phenomenon in the control system of cell motility in this organism.

Protein assembly from a common cytosolic pool into a variety of supramolecular structures

The local assembly of actin and associated proteins occurs transiently at sites where specific activities are evoked either by processes inherent to the cell or in response to an external signal. For instance, in chemotactically responsive neutrophils or *Dictyostelium* cells, a new leading edge can be formed in a uniform environment and at any site of the cell surface as an autonomous process. In a gradient of chemoattractant, however, the formation of a leading edge is biased towards the source of chemoattractant.

We would like to single out one protein, coronin, as a prototype to illustrate the involvement of a basically cytosolic protein in multiple functions of the actin system (Fig. 1). Coronin is designed for participating in multiple protein–protein interactions by the presence of five WD repeats and a leucine zipper domain [16]. Fusions of coronin with green fluorescent protein (GFP)

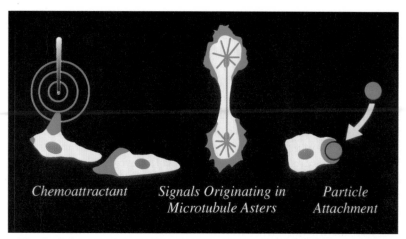

*Chemoattractant   Signals Originating in   Particle
                         Microtubule Asters   Attachment*

**Fig. 1. Scheme of the redistribution of coronin reflecting re-organi-
zation of the actin system in response to chemotactic stimulation,
particle attachment and signals emanating from the mitotic apparatus.** Sources of stimuli are shown in red, and areas of strong coronin
accumulation in the cell cortex in green.

have shown that this protein assembles dynamically in actin-rich protrusions
of the cell, e.g. in crown-shaped extensions of the cell surface (hence the name),
at leading edges that are spontaneously formed or induced by a gradient of
chemoattractant [17], and in phagocytic cups induced by the attachment of particles [18]. During cytokinesis, coronin accumulates in association with actin
filaments at the two poles of a cell. This local assembly of coronin in cytokinesis is determined by the asters of microtubules. Evidence for this
microtubule-directed accumulation of coronin is provided by its peculiar localization in multinucleate myosin II null cells. During mitosis, coronin
accumulates in the cortex of these cells on top of the centrosomes from which
the microtubule asters arise (Fig. 2).

The analysis of knockout mutants has indicated that coronin enhances re-
organization of the actin network [17,19]. Coronin thus accelerates cell
motility at the sites of its accumulation, contributes to endocytosis, and by its
presence at the poles assists in cytokinesis of cells anchored to a substrate. A
human coronin homologue in leucocytes is, like the *Dictyostelium* protein,
associated with the actin cortex. In addition, human coronin is part of the
NADPH oxidase complex that produces the respiratory burst in activated neutrophils [20].

Involvement of the actin system in particle uptake and in multiple steps of endocytic traffic

As has been demonstrated for coronin, a single protein can participate in
different functions of the actin system. What, then, are the common features of
these functions, and what distinguishes, for instance, a phagocytic cup from a
leading edge?

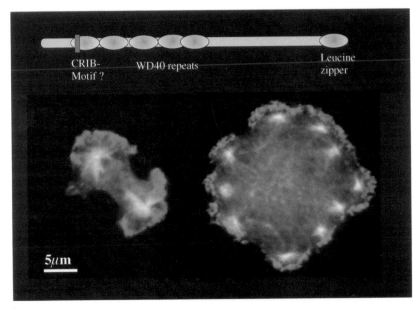

Fig. 2. Domain organization and microtubule-induced accumulation of coronin during mitosis in uninucleate and multinucleate cells of *Dictyostelium discoideum*. Coronin, with its five WD repeats, belongs to a family of proteins that undergo multiple interactions with other proteins. In addition, coronin contains a leucine zipper motif at the C-terminal end and a putative Rac interaction site (CRIB). The fluorescence images show myosin II null cells in mitosis labelled for coronin (green) and for tubulin (yellow) to visualize spindles and asters of the mitotic apparatus. In the uninucleate cell (left), the cell cortex in the cleavage furrow is devoid of coronin and strong accumulation is seen in ruffles of the polar regions, which are sites of highly dynamic actin re-organization. In the multinucleate cell (right), coronin is concentrated at the cortical docking sites of the microtubule asters. Two bent spindles connecting asters are recognizable in this image.

The first steps in the response to particle attachment are comparable in *Dictyostelium* to the chemotactic response: in the same way that a cell can turn towards the source of attractant by the formation of a new leading edge, it can also form an extension around the surface of a particle. This formation of a phagocytic cup is initiated by the assembly of F-actin and coronin at the site of contact with the particle [18]. More importantly, progression of a phagocytic cup competes with the protrusion of a leading edge. At any stage before final closure of the phagosome by membrane fusion, phagocytosis can be reversed when an active leading edge is formed at another site. In turn, a leading edge is retracted when phagocytosis proceeds to completion. From these findings, illustrated in Fig. 3, two conclusions can be drawn. The first is that phagocytosis is subject to continuous control in a zipper-like fashion; it does not proceed automatically to completion after the induction of a phagocytic cup. Secondly, there is a common control element in the formation of a leading edge and a phagocytic cup. The interdependence of the two processes may be due to the

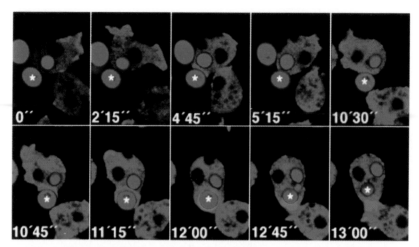

Fig. 3. Coronin–GFP redistribution reflects competition between leading-edge activity and particle uptake. The confocal sections show heat-killed, tetramethylrhodamine isothiocyanate (TRITC)-labelled yeast particles false coloured in red, and coronin–GFP with increasing fluorescence intensity from dark to light green. Times are in minutes (') and seconds ("). The sequence shows an unsuccessful and then successful attempt to take up the particle labelled with an asterisk. At 4 min 45 s, coronin accumulation was equally strong at a phagocytic cup and a leading edge. Subsequently the leading edge dominated and the particle was released, until a newly initiated phagocytic cup led to its engulfment at 12 min 00 s. Finally, the coronin dissociated from the phagosome and took part in the formation of a new leading edge. Adapted from [18] with permission. © 1995 Cell Press.

common requirement for a limiting factor, for instance a regulatory protein. Alternatively, a negative signal may emanate from a site of local actin assembly, thus suppressing assembly at a second site.

Phagocytosis is distinguished from cell-to-substrate adhesion of a motile cell by the eventual engulfment of the particle. How is this final step controlled? In macrophages, phagocytosis of opsonized particles is mediated by receptors that recognize either the Fc portion of immunoglobulins (FcγRs) or a complement protein (CR3). These receptors are linked to signal transduction pathways that ultimately lead to the local accumulation of actin [21]. FcγRs activate tyrosine kinases and, further downstream in the signalling cascade, phosphoinositide 3-kinase. The phosphoinositide 3-kinase appears to be the clue to the final step in particle uptake, i.e. the closure of phagosomes into intracellular organelles (Fig. 4). In the presence of the phosphoinositide 3-kinase inhibitor wortmannin, macrophages still produce membrane ruffles forming a cup in response to FcγR-activating particles, but these ruffles recede into the cytoplasm instead of proceeding to engulf the particle [22].

The actin system is involved not only in the uptake of a particle, but also in later steps that accompany the pathway of endocytic vesicles through the cell. In *Dictyostelium*, endocytic vesicles are re-decorated with actin and coronin after they have passed the phase of acidification (Fig. 5). Finally, coronin is

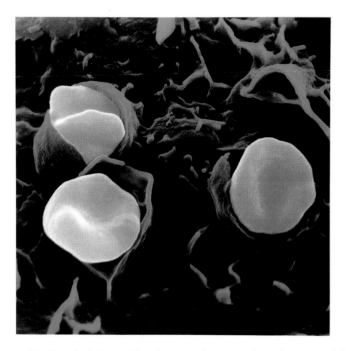

Fig. 4. Ruffling is induced by the attachment of erythrocytes (pink) to the surface of a macrophage (blue), but phagocytosis is not completed in the presence of wortmannin, an inhibitor of phosphoinositide 3-kinase. Magnification ×4500. Scanning electron micrograph courtesy N. Araki. Reproduced from The Journal of Cell Biology (1996) J. Cell Biol. **135** (journal cover) by copyright permission of The Rockefeller University Press.

replaced by vacuolin, a third protein recruited from the cytoplasm, which marks the late post-lysosomal phase and is thought to programme the vesicles for exocytosis [23]. These data are in accordance with previous findings indicating a role for actin in endocytic trafficking in mammalian cells [24]. One of these functions has been specified by the discovery of RhoD, a protein which regulates the actin cytoskeleton and controls endosome motility by an interplay with Rab5 [25]. It has been inferred from these results that actin-based endosome motility critically influences the balance between the fusion and fission of vesicles along the endocytic pathway.

So-called redundancy: a phenomenon related to the function of proteins in a multi-purpose network

Quite a few actin-binding proteins, which had been considered to be important because of their abundance in the cell and their strong activity *in vitro*, have been eliminated genetically from *Dictyostelium*. However, phenotypic analysis of the mutants has often led to disappointing results. Even the most active actin cross-linkers, such as α-actinin, could be eliminated without

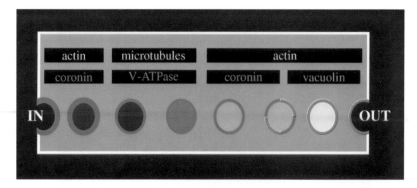

Fig. 5. Diagram of the early and late decoration of vesicle membranes with actin and coronin in the endocytic pathway of Dictyostelium. Nascent endosomes are coated with actin and coronin (green). Subsequently the lumen of the vesicles becomes acidified by vacuolar H^+-ATPase (blue), the actin coat disappears and the vesicles move bidirectionally along microtubules. After neutralization of their contents, the vesicles associate again with actin. Coronin also re-decorates the vesicles, but is finally replaced by vacuolin (red), another protein recruited from the cytoplasm. The vacuolin- and actin-coated vesicles are ready for exocytosis. This is a summary of data from Hacker et al. [49] and Jenne et al. [23], and unpublished work (M. Maniak).

any conspicuous alteration in cell behaviour or multicellular development. In some cases, however, the elimination of two or more proteins with related functions led to more salient deficiencies [26,27]. These findings have raised the issue of 'redundancy', i.e. the notion that one cellular function is guaranteed by multiple proteins that can replace each other. This phenomenon may have one of the following explanations: (i) two or more proteins may act in an additive manner, and elimination of either one of them is not critical so long as the sum of activities is beyond a certain threshold; (ii) a member of a protein family may have functions in common with other members and, on top of that, a unique activity that is only required for specific purposes or under special conditions.

We wish to illustrate these possibilities by referring to the α-actinin/spectrin superfamily of actin-cross-linking and -bundling proteins (Fig. 6). These proteins share a strongly conserved type of actin-binding domain of about 220 amino acid residues in size [28,29]. Some of these proteins have EF-hand motifs, which make the actin-binding activity sensitive to Ca^{2+}. Members of the fimbrin/plastin class contain two of the actin-binding domains in tandem in a single polypeptide chain. Other members of the superfamily are di- or tetrameric. They are distinguished from each other by the repeat structure of their rod domains, which are responsible for the mode of subunit interaction. The rod domains consist of triple α-helical repeats, of repeats forming an IgG-like fold [30], or of heptads that form a dimeric parallel coiled-coil [31].

All five members of the superfamily identified in Dictyostelium have been eliminated by mutation, separately or in pairs. An additive effect of α-actinin and the 120 kDa gelation factor (ABP-120), i.e. two of the actin-cross-linking proteins, is indicated by the finding that elimination of these proteins sepa-

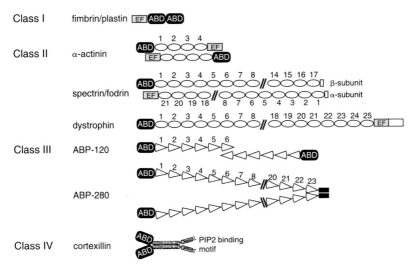

Fig. 6. Representative members of the α-actinin/spectrin superfamily of actin-binding proteins. Class I proteins are distinguished by the presence of two actin-binding domains (ABDs) in a monomer. The rod domains in class II proteins consist of triple α-helical repeats, in class III proteins of repeats forming an IgG-like fold [30], and in class IV proteins of heptad repeats forming a parallel coiled-coil [31]. EF-hand and phosphatidylinositol 4,5-bisphosphate (PIP2)-binding motifs point to regulatory Ca^{2+}- and lipid-binding sites. For spectrin, only one α- and one β-subunit are shown, and for dystrophin only a single subunit is shown. The gelation factor is modelled according to Fucini et al. ([30]; and P. Fucini, personal communication). Five members of the superfamily have been identified in cells of *Dictyostelium*: plastin, α-actinin, ABP-120 and two isoforms of cortexillin. Courtesy of Jan Faix, MPI für Biochemie, Martinsried, Germany.

rately had no marked effect on multicellular development, whereas a double knockout prevented, at least under certain conditions, the formation of tipped aggregates [26]. This means that development is blocked before the beginning of the 'finger' stage, at which the cell aggregate lengthens by pushing its tip towards the water/air interphase. To produce the finger shape, forces need to be generated by the cell mass that surpass the surface tension at the interphase. Apparently, with one of the two actin-cross-linking proteins alone, the forces generated in the aggregate remain below this threshold.

Evidence for the requirement of a single member of the superfamily for a specific purpose has been provided by the finding that ABP-120 is essential for the precise phototactic and thermotactic orientation of *Dictyostelium* slugs. This protein cannot be replaced, with respect to this particular function, by any of the related proteins tested [32]. The rod of this protein consists of IgG-like repeats [30] and resembles in its sequence the rod of ABP-280 (non-muscle filamin), an actin-cross-linking protein of mammalian cells known to be involved in intracellular signal transduction [33]. It is the phosphorylatable rod portion that is responsible for the binding of ABP-280 to signal-regulated kinases [34].

The cortexillins (Fig. 6) provide another example of a unique function of certain members of the α-actinin/spectrin superfamily. The two cortexillin isoforms of *Dictyostelium* are the only family members whose elimination strongly impairs cytokinesis [31]. It is most probably the ability of these proteins to bundle actin filaments in an anti-parallel fashion which, in combination with the regulatory properties of the C-terminal portion of cortexillins, is responsible for their important role in mitotic cleavage [31].

Cytokinesis: the microtubule and actin systems in concert

During mitotic cell division, the cortical actin system mediates the formation of a cleavage furrow and finally causes the daughter cells to separate. These activities are linked to the organization of the microtubule system into spindle and asters, and they are connected to the segregation of chromosomes along the spindle. One question concerning cytokinesis is whether the changes in cell shape, in particular the constriction of the cleavage furrow, require special proteins or the re-programming of a set of actin-associated proteins that are involved in the motility of interphase cells. A second question concerns the signals accommodating activities in the actin system to the organization and dynamics of the microtubule system in the mitotic apparatus. Where are these signals generated, and how are they transmitted?

Coronin and myosin II, the conventional double-headed myosin, exemplify two different actin-based activities in cytokinesis. One is the activity at the midline of the cell, which leads to the formation of the cleavage furrow and finally to complete separation of the daughter cells. The other activity occurs at the poles of the cell, where ruffling areas of the cell surface act in opposing directions as leading edges and help to draw daughter cells away from each other. As shown in Fig. 2, coronin is localized specifically to the actin-rich regions at the poles of a dividing *Dictyostelium* cell [19], at sites where actin is also highly concentrated. Myosin II, under the conditions used by us [35], accumulates near to, but not strictly within, the cleavage furrow, suggesting that it stabilizes the position of the furrow and limits its expansion towards the polar regions of the dividing cell. Myosin II is dispensable for cytokinesis if the cells are attached to a substrate surface [36]. Similarly, myosin II is not essential for interphase cells to move [37]. However, myosin II plays a supporting role in both cytokinesis and cell motility. It increases the fidelity and speed of cytokinesis [36] and makes this process independent of cell–substrate interaction [38,39]. For motility, myosin II activity is particularly important when cells are plated on to a strongly adhesive substrate such as polylysine-coated glass, from which it is difficult for myosin II null cells to detach and retract [40].

The dual activities of the actin system in cytokinesis, i.e. in the middle region and at the poles of the cell, are manifested in the forces that a dividing cell applies to the substrate. By increasing the compliance of a silicone layer, Burton and Taylor [41] have adapted the silicone rubber technique to the determination of forces exerted by a dividing fibroblast. On the silicone substrate, the forces are visualized by the wrinkles on its surface (Fig. 7). The strongest forces are produced by the constricting furrow, but tension is also generated at

Fig. 7. Wrinkles on a silicone rubber sheet visualize forces applied by a dividing fibroblast. The length of the wrinkles radiating out from the area of cell attachment is proportional to the forces that the cell applies to the compliant substrate surface. Traction is strongest at the cell equator, but is also generated at the cell poles. From Burton and Taylor [41]. Courtesy K. Burton, with the permission of *Nature*.

the two poles of the cell, nicely demonstrating the relative contributions of different surface regions to the division of a substrate-anchored cell.

An excellent system in which to study the coupling of activities in the cell cortex to the microtubule system is provided by myosin II null cells of *Dictyostelium* grown up into multinucleate bodies. After these large cells are transferred from a suspension culture on to an adhesive surface, they are capable of dividing when their nuclei undergo synchronous mitoses (Fig. 8). Cytokinesis is primed in these cells by anaphase movement of the centrosomes towards the cell surface and their firm docking to the cell cortex through the ends of microtubules, which form an aster around each centrosome. Subsequently, the cell surface starts to ruffle in the vicinity of the asters, in a similar manner to the two polar regions of a normal, uninucleate cell. The ruffling regions are rich in filamentous actin and in coronin. The smooth areas between the asters become depleted of coronin and predisposed to initiate furrows, which expand and eventually cleave large cells into portions containing various numbers of nuclei. The cleavage furrows are initiated independently of the localization of the spindles. Thus in *Dictyostelium* a cleavage furrow is initiated at the space between asters, following the same rule as found in sea urchin blastomeres by Rappaport [42,43].

Major goals of current research in cytokinesis are (1) to identify proteins which, in response to signals from the microtubule system, are recruited to the cleavage furrow, (2) to elucidate the mechanism by which this recruitment occurs, and (3) to specify the function of these proteins in the initiation and progression of a furrow. Candidate proteins for transmitting signals from the microtubule asters to the actin cortex are monomeric G-proteins of the Ras

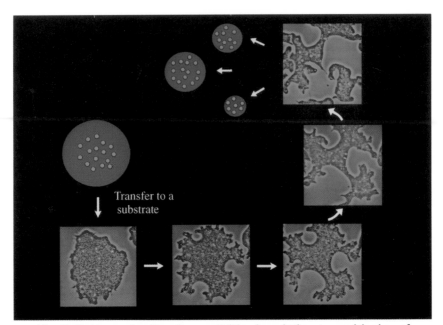

Fig. 8. Pattern of cell-surface activities in relation to positioning of the mitotic apparatus. To study the influence of the microtubule system on the cortical actin system during mitosis, multinucleate cells are obtained by cultivating myosin II null cells of *Dictyostelium discoideum* in suspension and transferring them on to a glass surface. The cells produce GFP–α-tubulin (orange label in the square panels), which makes it possible to relate microtubule dynamics and centrosome translocation to organization of the cell surface (phase-contrast images in green). Cleavage furrows are formed between the docking sites of microtubule asters, whereas surface protrusions are linked to asters of microtubules emanating from the centrosomes. At the end of cytokinesis, the multinucleate cells give rise to daughter cells with irregular numbers of nuclei.

superfamily [44,45] and GTPase-activating proteins that regulate the activity of these small GTPases [46,47]. Candidate proteins for providing the cleavage furrow with the mechanical properties required for its constriction are, in addition to myosin II, the cortexillins responsible for normal cleavage in *Dictyostelium* cells [31]. These actin-bundling proteins are discussed in Chapter 15 by Weber, Niewöhner and Faix. It will be of interest to find out whether similar proteins are involved in cytokinesis in higher eukaryotes, and which more specialized proteins may have evolved to adapt the mechanics of cytokinesis to the various conditions of cleavage in embryogenesis and tissue formation of animals.

Concluding remarks

In this overview, only a few examples could be presented to illustrate the versatility of the actin system and the variety of proteins that associate with filamentous actin to fulfil different tasks. Exogenous signals and internal control mechanisms determine the spatial and temporal order of actin assembly. Obvious regulator proteins are small GTPases of the Ras superfamily that determine the type of actin organization [46], programme the endocytic pathway [48] and control the changes in cell shape during cytokinesis. To find out how proteins are recruited within seconds from a common cytoplasmic pool in order to build up multiprotein complexes of different structures and functions will be a formidable challenge for the future, not only for molecular cell biologists but also for theoreticians familiar with the study of complicated and complex systems.

References

1. Abercrombie, M., Heaysman, J.E. and Pegrum, S.M. (1970) Exp. Cell Res. **59**, 393–398
2. Curtis, A.S.G. (1964) J. Cell Biol. **20**, 199–215
3. Gingell, D. and Todd, I. (1979) Biophys. J. **26**, 507–526
4. Bailey, J. and Gingell, D. (1988) J. Cell Sci. **90**, 215–224
5. Gingell, D. and Owens, N. (1992) J. Cell Sci. **101**, 255–266
6. Dunn, G.A. and Brown, A.F. (1987) J. Cell Sci. Suppl. **8**, 81–102
7. Dunn, G.A. and Zicha, D. (1995) J. Cell Sci. **108**, 1239–1249
8. Harris, A.K., Wild, P. and Stopak, D. (1980) Science **208**, 177–179
9. Oliver, T., Dembo, M. and Jacobson, K. (1995) Cell Motil. Cytoskeleton **31**, 225–240
10. Dembo, M., Oliver, T., Ishihara, A. and Jacobson, K. (1996) Biophys. J. **70**, 2008–2022
11. Schindl, M., Wallraff, E., Deubzer, B., Witke, W., Gerisch, G. and Sackmann, E. (1995) Biophys. J. **68**, 1177–1190
12. Weber, I., Wallraff, E., Albrecht, R. and Gerisch, G. (1995) J. Cell Sci. **108**, 1519–1530
13. Wymann, M.P., Kernen, P., Bengtsson, T., Andersson, T., Baggiolini, M. and Deranleau, D.A. (1990) J. Biol. Chem. **265**, 619–622
14. Hartmann, R.S., Lau, K., Chou, W. and Coates, T.D. (1994) Biophys. J. **67**, 2535–2545
15. Weber, I. (1995) Ph.D. Thesis, TU München
16. de Hostos, E.L., Bradtke, B., Lottspeich, F., Guggenheim, R. and Gerisch, G. (1991) EMBO J. **10**, 4097–4104
17. Gerisch, G., Albrecht, R., Heizer, C., Hodgkinson, S. and Maniak, M. (1995) Curr. Biol. **5**, 1280–1285
18. Maniak, M., Rauchenberger, R., Albrecht, R., Murphy, J. and Gerisch, G. (1995) Cell **83**, 915–924
19. de Hostos, E.L., Rehfuess, C., Bradtke, B., Waddell, D.R., Albrecht, R., Murphy, J. and Gerisch, G. (1993) J. Cell Biol. **120**, 163–173
20. Grogan, A., Reeves, E., Keep, N., Wientjes, F., Trotty, N., Burlingame, A.L., Hsuan, J.J. and Segal, A.W. (1997) J. Cell Sci. **110**, 3071–3081
21. Greenberg, S. (1995) Trends Cell Biol. **5**, 93–99
22. Araki, N., Johnson, M.T. and Swanson, J.A. (1996) J. Cell Biol. **135**, 1249–1260
23. Jenne, N., Rauchenberger, R., Hacker, U., Kast, T. and Maniak, M. (1998) J. Cell Sci. **111**, 61–70
24. Durrbach, A., Louvard, D. and Coudrier, E. (1996) J. Cell Sci. **109**, 457–465
25. Murphy, C., Saffrich, R., Grummt, M., Gournier, H., Rybin, V., Rubino, M., Auvinen, P., Lütcke, A., Parton, R.G. and Zerial, M. (1996) Nature (London) **384**, 427–432
26. Witke, W., Schleicher, M. and Noegel, A.A. (1992) Cell **68**, 53–62
27. Schleicher, M., Andre, B., Andreoli, C., Eichinger, L., Haugwitz, M., Hofmann, A., Karakesisoglou, J., Stöckelhuber, M. and Noegel, A.A. (1995) FEBS Lett. **369**, 38–42
28. Hartwig, J.H. and Kwiatkowski, D.J. (1991) Curr. Opin. Cell Biol. **3**, 87–97
29. Matsudaira, P. (1991) Trends Biochem. Sci. **16**, 87–92

30. Fucini, P., Renner, C., Herberhold, C., Noegel, A.A. and Holak, T.A. (1997) Nature Struct. Biol. **4**, 223–230

31. Faix, J., Steinmetz, M., Boves, H., Kammerer, R.A., Lottspeich, F., Mintert, U., Murphy, J., Stock, A., Aebi, U. and Gerisch, G. (1996) Cell **86**, 631–642

32. Fisher, P.R., Noegel, A.A., Fechheimer, M., Rivero, F., Prassler, J. and Gerisch, G. (1997) Curr. Biol. **7**, 889–892

33. Ohta, Y., Stossel, T.P. and Hartwig, J.H. (1991) Cell **67**, 275–282

34. Marti, A., Luo, Z., Cunningham, C., Ohta, Y., Hartwig, J., Stossel, T.P., Kyriakis, J.M. and Avruch, J. (1997) J. Biol. Chem. **272**, 2620–2628

35. Neujahr, R., Heizer, C., Albrecht, R., Ecke, M., Schwartz, J.M., Weber, I. and Gerisch, G. (1997) J. Cell Biol. **139**, 1793–1804

36. Neujahr, R., Heizer, C. and Gerisch, G. (1997) J. Cell Sci. **110**, 123–137

37. Wessels, D., Soll, D.R., Knecht, D., Loomis, W.F., De Lozanne, A. and Spudich, J. (1988) Dev. Biol. **128**, 164–177

38. De Lozanne, A. and Spudich, J.A. (1987) Science **236**, 1086–1091

39. Knecht, D.A. and Loomis, W.F. (1987) Science **236**, 1081–1086

40. Jay, P.Y., Pham, P.A., Wong, S.A. and Elson, E.L. (1995) J. Cell Sci. **108**, 387–393

41. Burton, K. and Taylor, D.L. (1997) Nature (London) **385**, 450–454

42. Rappaport, R. (1961) J. Exp. Zool. **148**, 81–89

43. Rappaport, R. (1996) Cytokinesis in Animal Cells, Cambridge University Press, Cambridge

44. Larochelle, D.A., Vithalani, K.K. and De Lozanne, A. (1997) Mol. Biol. Cell **8**, 935–944

45. Tuxworth, R.I., Cheetham, J.L., Machesky, L.M., Spiegelmann, G.B., Weeks, G. and Insall, R.H. (1997) J. Cell Biol. **138**, 605–614

46. Hall, A. (1998) Science **279**, 509–514

47. Faix, J. and Dittrich, W. (1996) FEBS Lett. **394**, 251–257

48. Novick, P. and Zerial, M. (1997) Curr. Opin. Cell Biol. **9**, 496–504

49. Hacker, U., Albrecht, R. and Maniak, M. (1997) J. Cell Sci. **110**, 105–112

Biochem. Soc. Symp. **65**, 15–26
Printed in Great Britain

2

New depths in cell behaviour: reactions of cells to nanotopography

Adam Curtis and Chris Wilkinson

Centre for Cell Engineering, University of Glasgow, Glasgow G12 8QQ, Scotland, U.K.

Abstract

The physical and molecular biological bases of the reactions of cells to features of the topography of the substratum or environment on which and in which cells live, both in culture and in the embryo, are discussed. The fact that most, if not all, cells react to micrometric and nanometric topography is stressed. Some cell types will react to steps as shallow as 11 nm. Methods of fabricating such topographies in a variety of materials are outlined. Types of topography and the reactions of cells to these are described. It is emphasized that different cell types are sensitive to fairly specific ranges of size of topography. Reactions to topography include cell orientation, changes in cell motility, cell adhesion and cell shape. The term 'contact guidance' has been used in this field, but the term 'topographic reaction' is more appropriate, since it covers the wide range of reactions that are reported. In addition, the reactions involve activation of tyrosine kinases, cytoskeletal condensation and further downstream activation and inactivation of gene expression. The reactions to topography are probably due to stretch reactions of the cells to the substratum and not to chemical details of the substratum. The reasons for this are that a given cell type reacts in much the same way to the same topography made with different materials and that, when both chemical patterns and topographic ones are offered to cells, topography tends to have a greater effect than chemical patterns.

Introduction

Tissue cells normally live in environments with marked and often organized micro- and even nano-topography. This may, for instance, be in the form of oriented molecules such as collagen, or simply in the shapes of adjoining cells. The conventional tissue culture environment is supposedly planar, although nanotopography can be found even in this system. A wide variety of animal cells and some fungal cells react to microtopography and to nanotopography [1,2]. In using these terms, we mean topography whose main features are respectively in the region of 0.11–100 μm and 1–100 nm in scale. On this basis, nanotopography refers to those objects of a size similar to that of large protein molecules and small assemblies of molecules.

The classic demonstration of the reaction of cells to microtopography has been the alignment of cells to fibres [3] and to groove/ridge structures [4]. This alignment is often termed 'contact guidance'. Although contact guidance has been known since the very early days of cell culture, little quantitative examination of the reaction of cells to topography has been carried out. Exceptions to this subjective approach are notably the papers of Clark et al. [5–7]. Yet if we are to understand how cells react and if we are to be able to design ideal substrata for particular cells, to be used especially in tissue repair, we need this information. Precision and quantification are required, both of the methods of examination of the cells and in the definition of that topography. Recently it has become clear that these reactions occur on a nanometric scale.

My colleagues and I have demonstrated that cells will react to steps in the range 50–100 nm, and that some cells will detect 10 nm steps [8]. Still more recently, we have demonstrated that cells will also react to small columnar structures 15 nm high at 50 nm spacing. These reactions demonstrate a previously almost unsuspected capability of cells. However, the early work of Rosenberg [9] on the reaction of cells to steps in multilayers of behenic acid deserves mention and praise.

In this chapter we analyse the significance of the phenomenon and discuss the possible mechanisms behind some of the reactions. It is important to appreciate that the dimensions of topography that lead to the maximal reaction of a cell may vary very appreciably from one cell type to another. It is also important to appreciate that some of these surfaces that can now be fabricated resemble those of the real cellular environment far more closely than the planar cell culture dish or the randomly patterned surface of a metal prosthesis.

Types of topography and their fabrication

Although random and quasi-random topography may be imposed on a surface by very many mechanisms, including simple abrasion, this approach is, in our view, inadequate. Such topographies tend to be highly variable and in effect indescribable because of their randomness. Fig. 1 shows an atomic-force microscopic (AFM) view of the surface of a freshly unpacked polystyrene culture dish with ridges of about 10 nm high, probably representing tool marks from the pressing tool.

Photolithography followed by reactive-ion etching is capable of imposing very precise topography on suitable substrates with a resolution of about 0.2 μm laterally and vertical resolutions of 5 nm. Electron-beam lithography, using an electron beam to write on to the resist before etching, allows resolutions of about 5 nm, laterally or vertically, to be attained. Examples are shown in Fig. 2.

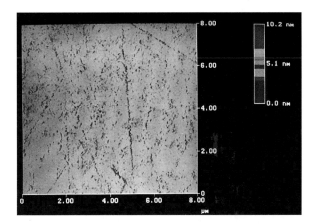

Fig. 1. Surface of a polystyrene tissue culture dish. An AFM view is shown (Nanoscope III in contact mode).

If these methods are used to fabricate masters, they may in turn be used to prepare multiple copies in various polymers by embossing or casting.

Three-dimensional structures are difficult to fabricate in a controlled manner, and much reliance has been placed on mats or tangles of fibres. These are clearly irreproducible. Collagen and fibrin can be oriented to a degree by stretching their gels as they form, and more effectively by applying very large magnetic fields as they gel from solution (see [10,11]). In principle, fibres might be oriented in the form of a knitted mat. Cima and colleagues [12] have developed ink-jet printing methods for fabrication. The droplets can be of about 50 μm in diameter and, by multiple passes with a printer, a three-dimensional structure can be built up. At the moment the system does not appear to be able to produce a reproducibility better than 70 μm, but it has the great advantage that layers, and even positions of specific chemistry, can be built up. We our-

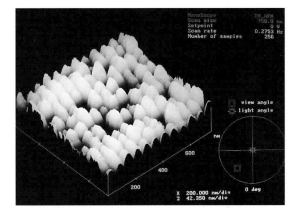

Fig. 2. Electron-beam lithography: an example of nano-fabrication. Shown is an array of 60 nm dots on 125 nm centres etched into silica to a depth of 50 nm (AFM image; Nanoscope III). Courtesy of Dr. B. Casey.

selves have preferred to use methods based on photolithography and/or embossing, in which structures formed on a surface are built into three-dimensional systems by glueing and folding methods. One might, however, apply the methods of electrolytically etching doped silicone to form pillars and pores [13]. Extrusion methods should, in principle, allow the manufacture of tubes and groups of tubes to give three-dimensional highly oriented structures. Clearly, precision and reproducibility are the requirements of the investigator, and ease of mass production is the requirement of groups manufacturing prosthetic devices.

The range of phenomena

We will start by commenting that there is an enormous literature that describes subjectively the effects of various topographies, usually poorly defined, on cell behaviour, but we consider that such work, although often interesting and inspiring, lacks the precision that would allow firm conclusions to be drawn about cell behaviour. Consequently, we do not review such work.

Although most investigations have concentrated on groove/ridge topography, it is probably accurate to state that oriented topography tends to orient cells and thus to alter their morphology, sometimes to a very elongated form (see Fig. 3). This has been thoroughly reviewed [1]. In addition, adhesion and motility of the cells may be enhanced by contact with the topography. Contact inhibition of movement may become more marked on convex surfaces [4]. Within the cell, actin and perhaps microtubule orientation and architecture are changed, with marked orientation of actin bundles to the groove edges [8]. Early events include stimulation of tyrosine phosphorylation in regions of the cell over the groove edges. Very recently we have demonstrated that changes in gene expression result from the reaction to this type of topography (see below).

Less is known about the effects of other types of topography, but orientation and behavioural reactions are apparently frequent. Deep pits can lead to cell trapping and to the activation of cells such as macrophages (Fig. 4). Spikes at an appropriate spacing lead to the formation of layers of cells bridging between the spikes (see [14]).

Topography or chemistry?

One important controversy thrown up by these observations is whether cells react to the topography or to the chemistry of the substrate in these reactions. On one side of the argument, it can be pointed out that, in all situations where a specific solid or semi-solid chemical signal has been placed on a surface, there exists a topography. For instance, the very thin layers of derivatized laminin attached to a silica surface by Britland et al. [2] are probably less than three molecules in thickness, but have a clear edge topography and a different surface texture from the silica, even after culture of cells in a protein-containing medium upon it. In this experiment, as in others by Letourneau and Cypher [15] and Thiery et al. [16], the chemical tracks also had a clear topography. It is

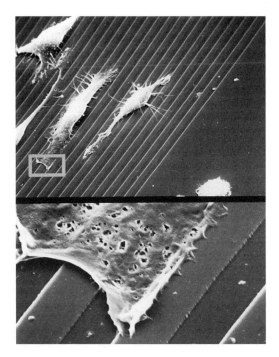

Fig. 3. Orientation of cells to a groove/ridge topography. P388D1 macrophages are shown after 20 min on a silica topography with grooves 60 nm deep and 5 μm wide. Note the alignment of cells on the grooves, with attachment at the groove edges.

(a)   (b)   (c)

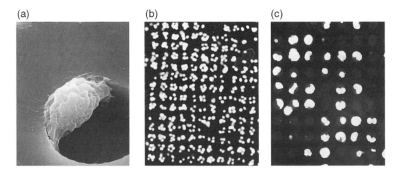

Fig. 4. Reaction of cells to pits. P388D1 macrophages were grown on a silica surface with pits 6 μm deep. The cells were stained with fluorescent phalloidin. Note that all of the cells are in the pits. (a) Scanning electron micrograph of one cell located on entry to a pit of 5 μm diameter. (b) Cells stained with rhodamine–phalloidin for actin and viewed by fluorescence microscopy. Groups of three or four cells per pit are present in pits of 10 μm diameter. (c) Rhodamine–phalloidin staining of macrophages in arrays of pits of 5 μm diameter.

noteworthy that the cells often aligned to the edges of these chemical tracks. Of course, in the final molecular analysis, chemistry and shape come together. However, for the present discussion we take the viewpoint that if specific chemistry has a clearly stronger effect on attachment, orientation and activation than does topography, then the argument is resolved in favour of chemistry.

The first line of evidence comes from the observations that groove/ridge topography and fibre topography have been fabricated in a very wide range of materials (Table 1) and have elicited much the same set of reactions from a given cell type. It would not be expected that these materials would show highly similar adsorption of proteins from the medium. It is, of course, possible that the topography would induce specific adsorption which then influenced cell reaction. Might a sharp edge adsorb fibronectin preferentially from a serum medium? Would this then lead chemically to the cell orienting itself?

However, this line of argument would be destroyed if the adsorptions were sufficiently similar. The next stage of the argument is that the silica topographies and their polymer replicas do not show any chemical differences between different parts of the topography. Fused silica is an isotropic material, so that the reactive-ion etching step should not induce chemical patterns. There is the possibility that some fluorine might be left on the exposed surface but, since it has been our practice to 'blanket-etch' any sample after the fabrication of grooves, this second total exposure to the reactive plasma should induce identical chemistry over the entire surface.

Similarly, casts or embossings are unlikely to have chemical differences, but it must be agreed that local strains induced in the process of parting the master and the polymer could produce local chemical changes.

The direct experimental approach is more rewarding. Britland et al. [2] combined topography and chemical patterning of laminin bound to the surface and examined the orientation of neurites to these devices. In one set of devices, Britland and his colleagues aligned the chemical patterns to the bottom of grooves, and in the other they set the chemical tracks, which were 25 μm wide, at 90° to the groove orientation. They used a range of depths of grooves. The results from devices with orthogonally arranged chemical and topographical patterns are most interesting. The neurites aligned to chemical tracks when the grooves were shallower than 0.5 μm, although alignment tended steadily

Table 1. Substratum materials for which topographic reactions have been demonstrated.

Topography	Materials
Groove/ridge topography	Silica; polydioxanone; cellulose acetates; polyurethane; polymethyl methacrylate; titanium dioxide (on titanium); glasses (an unspecifiable set of materials)
Tunnels and pits	Silica; polystyrene (with oxidized surfaces)
Fibres	Silica; platinum; nylons; fibronectin; cellulose acetates; fibrin; polylactides; polyglycollides

towards the grooves as depth increased. At depths greater than 2 μm, alignment to topography was much stronger than that to the chemical patterns (Fig. 5). It is clear that topography is a stronger factor than chemistry when the two are competing but, when both are aligned as in the other set of devices made by Britland et al. [2], the two factors act together.

However, this type of experiment needs to be repeated for other cell types and other substrata. Britland et al. [17] carried out such an experiment with BHK cells and fibronectin substrates and found a less clear distinction, but did not use very deep grooves.

Reactions

Although the orientation of cells to fibres and grooves was discovered at an early date (see Weiss [3]) and is perhaps the most obvious reaction, other important reactions take place when a cell contacts a non-planar surface which would not occur at all, or occur so obviously, on a planar surface.

Adhesion

It has long been known that random roughness and micro-roughness can affect cell adhesion. Most papers report increases in adhesion, but one interesting report [18] described positive reactions (rugophilia) of some cell types to roughness and negative reactions of other cell types. In the earlier work on attachment to groove/ridge topography, we noted the tendency of cells to attach to discontinuities, but the area of the vertical walls was too great and of questionable accessibility to cells for an estimate to be made of the true surface area of the devices.

Consequently we were reluctant to attempt measurement of any change in cell attachment resulting from the topography. With the advent of the pro-

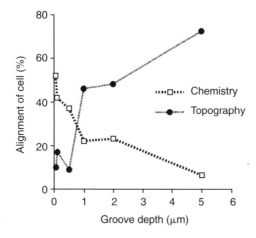

Fig. 5. Reaction of cells to topography rather than to chemistry as groove depth increases. Grooves were orthogonal to chemical tracks of laminin. Redrawn from Britland et al. [2].

duction of very shallow grooves, where the walls were 100 nm or less in height, the increased area of the device might amount to less than 2% of the plan-view area, so measurements of cell attachment to the plan area would be a reasonable indicator of any change in adhesion. Such changes have been found.

Orientation

On the groove/ridge substrata, cells take up the orientation of the topography over 24 h or so, although some cells, such as macrophages, react in some tens of minutes, while others, such as chondrocytes, react very slowly. Very shallow grooves do not orient the cells well, and several studies have shown that orientation responses vary from cell type to cell type. There are also more gradations of reaction that depend on groove width and pitch dimension. In most cases, orientation is parallel to the groove length, but Nakata et al. [19] have reported situations where neurites are produced orthogonally to the length of the grooves.

Movement

Surprisingly few studies have been made of movement on topography, but Curtis and Varde [4] have studied movement on silica fibres and Curtis et al. [20] have described movement on grooves. Acceleration of cell movement was observed in the second of these papers but, unfortunately, the first paper inexplicably does not give quantitative data on cell movement. Tranquillo et al. [10] have shown that movement of fibroblasts in oriented collagen gels is preferentially aligned to the orientation of the collagen.

It is interesting to note that it is unclear as to which method should be used for analysing cell movement on various topographies. It has been conventional to use measurements that define the walk, random or otherwise, of the cells in two dimensions (see Gail and Boone [21]), and to look for the degree of departure from randomness and the persistence of direction. These methods are useful, but do not seem to be appropriate for comparison between two-dimensional and three-dimensional situations. Alternatively, is it worth considering movement as a Markov-type process (see Hartman et al. [22] and Curtis et al. [20]) or even as a fractal process (see [23])?

Activation

Macrophages and fibroblasts react to contact with grooves [24], even very shallow ones, by re-organization of the cytoskeleton and related tyrosine phosphorylation. This phosphorylation is apparent very soon after contact is made with the substratum (less than 5 min) and tends to mark the positions of the sharpest angles in the substratum. Actin and vinculin co-localization with the phosphorylation can be observed at about the same time. All of these reactions to topography are eliminated by the tyrosine-phosphorylation inhibitor herbimycin and by tyrphostins B56 and B50. Genistein was not inhibitory, suggesting that *rho* is not involved. This pattern of inhibition suggested that it would be rational to test whether *c-src* products are involved. Activation can also be seen in macrophages as an increase in their ability to phagocytose test particles (see Wojciak-Stothard et al. [25]).

Gene expression

We have begun to use reverse transcription–PCR (mRNA fingerprinting) methods to look for gene expression that appears or disappears on culturing cells on grooves. At least 30 changes in expression have been observed in epitenon cells. The expression of an elastin gene and of genes for various aminoglycosaminidases and for enzymes of phosphoinositide metabolism has so far been identified as being altered by topography.

Explanations: stretch reception

The earlier explanations of contact guidance were couched in terms of those events that can be visualized, namely actin cytoskeleton unbendability [26], focal contact alignment [27] and the reactions of cells to discontinuities [28]. These explanations do not accord well with other types of reaction to topography, such as the penetration of deep holes by cells, where the movement would be contrary to any of these three theories. The Dunn and Heath hypothesis [26] does not suggest that cells would react to nanometric topography, because angles of bending that the cells would need to make to accommodate themselves to the topography lie well within the known capabilities of cells. A further argument against these earlier theories comes from the fact that very early events, such as increased adhesion and tyrosine phosphorylation, occur before the alignment of the cells is detectable. These considerations suggest that other explanations of a more general type may be of greater value.

When cells react to topography, they spread preferentially in certain directions. This appears to place them under tension, resulting presumably from their own locomotory abilities. Harris [29] has shown how these tension-generating systems in cells may be visualized by growing the cells on very thin films of silicone rubber.

We have investigated the possibility that stretch receptors are involved in the reaction to substratum topography. A simple piece of evidence in favour of this hypothesis is that 50 μM gadolinium ions inhibits the topographic reactions without affecting cell adhesion. Several groups [30,31] claim that Gd ions are specific inhibitors of stretch reception. We suspect that these trivalent ions may also have other, less specific, actions. One possible candidate for the stretch receptor is a chloride channel. There are two reasons for this. First, there is high homology between the rat ClC5 chloride channel and the known stretch receptor mec-5 in Caenorhabditis elegans [D. Bassett (1996) at URL http://www.ncbi.nih.gov/XREFdb]. Secondly, it has recently been shown that the vertebrate stretch receptor is an asymmetrical pair-channel system that has three states [33]. The ability to compare x-direction movement with y-direction movement is an essential feature of a stretch sensor. When we grew cells on structures in low-chloride media (chloride replaced with nitrate), the cells showed much reduced stretching over the first few hours, and when the inhibitor tamoxifen was used, reaction to topography was almost completely abolished (Fig. 6). Obviously these experiments need to be taken further, but

(a)   (b)

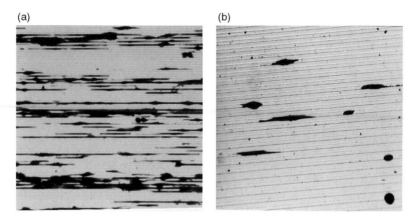

Fig. 6. Evidence for chloride-channel involvement in the topographic reaction of cells. (a) The controls are epitenon cells reacting after 24 h to 5 μm-wide grooves in normal chloride medium. (b) The same cell types in a medium where 85% of the chloride has been replaced with nitrate.

they suggest that a calcium-activated chloride channel may be involved, even at very early times, in stretch reception.

Stretch receptors are, of course, well known as volume regulators for cells, so cells clearly can react to internally generated forces [30]. Other reactions, possibly handled by the same receptors, are to externally generated forces (e.g. see Goldspink et al. [34]).

An explanation of the topographic reactions of cells in terms of the activity of their stretch receptors has advantages over the idea of reactions of cells to discontinuities [28], since cells can react to relatively smooth, curved surfaces, e.g. on fibres by extension along the fibres [3,4]. Such surfaces do not present discontinuities, but would be expected to cause severe distortion of the cell if it attempted to conform to the surface by extension along the circumference.

We can thus suggest that cells react to topography via the following sequence of events. First, increased adhesion occurs as the cell contacts the discontinuities. When two or more adhesions are formed, possibly within a small distance of each other, and cytoskeletal condensation has started in the cytoplasm around them, contractile forces develop asymmetrically and the cell starts elongation in an oriented way. This stretching may initiate the signal transduction that leads to tyrosine phosphorylation and related events. If this picture of events is correct, the changes in gene expression may take place further downstream, possibly as a consequence of nuclear distortion.

Conclusions

The reactions of cells to topography are important, not purely as evidence of the remarkable repertoire of things that cells can do, but primarily because cells encounter topography continuously in their life in the body.

Cell culture dishes are perhaps the only planar objects that cells encounter, and even these dishes may well have adventitious or unsuspected topography. Clearly, a great deal more remains to be done to obtain a full understanding of the reactions of cells to topography, but already we can appreciate that the following questions are in need of answers. (1) Do cells react by using their stretch receptors? (2) What is the sequence of events? This may require methods with time resolutions as short as 100 ms. (3) How are the signal transduction events, such as tyrosine phosphorylation, linked to the interaction with the edges of the structures? (4) Is there really a requirement for a discontinuity or a sharply bending topography for the surface to elicit topographic reactions? (5) Can the topographic reaction be divorced from possible chemical effects? Finally it may be useful to the reader to refer to other reviews; see [1,27,35].

References

1. Curtis, A.S.G. and Wilkinson, C.D.W. (1997) in Motion Analysis of Living Cells (Soll, D., ed.), pp. 141–156, Wiley-Liss, New York
2. Britland, S., Perridge, C., Denyer, M., Morgan, H., Curtis, A. and Wilkinson, C. (1996) Exp. Biol. Online 1, 2
3. Weiss, P. (1945) J. Exp. Zool. 100, 353–386
4. Curtis, A.G. and Varde, M. (1964) J. Natl. Cancer Inst. 33, 15–26
5. Clark, P., Connolly, P., Curtis, A.S.G., Dow, J.A.T. and Wilkinson, C.D.W. (1987) Development 99, 439–448
6. Clark, P., Connolly, P., Curtis, A.S.G., Dow, J.A.T. and Wilkinson, C.D.W. (1990) Development 108, 635–644
7. Clark, P., Connolly, P., Curtis, A.S.G., Dow, J.A.T. and Wilkinson, C.D.W. (1991) J. Cell Sci. 99, 73–77
8. Wojciak-Stothard, B., Curtis, A., Monaghan, W., Macdonald, K. and Wilkinson, C. (1996) Exp. Cell Res. 223, 426–435
9. Rosenberg, M.D. (1963) Science 139, 411–412
10. Tranquillo, R.T., Girton, T.S., Bromberek, B.A., Triebes, T.G. and Mooradian, D.L. (1996) Biomaterials 17, 349–357
11. Torbet, J. and Ronzere, M.C. (1984) Biochem. J. 219, 1057–1059
12. Wu, B.M., Borland, S.W., Giordano, R.A., Cima, L.G., Sachs, E.M. and Cima, M.J. (1996) J. Controlled Release 40, 77–87
13. Canham, L.T., Reeves, C.L., Wallis, D.J., Newey, J.P., Houlton, M.R., Sapsford, G.J., Godfrey, R.E., Loni, A., Simons, A.J., Cox, T.I. and Ward, M.C.L. (1997) Mater. Res. Soc. Symp. Proc. 452, 579–590
14. Rovensky, Y.A., Bershadsky, A.D., Givargizov, E.I., Obolenskaya, L.N. and Vasiliev, Y.M. (1991) Exp. Cell Res. 197, 107–112
15. Letourneau, P.C. and Cypher, C. (1991) Cell Motil. Cytoskeleton 20, 267–271
16. Thiery, J.P., Duband, J.L. and Tucker, G.C. (1985) Annu. Rev. Cell Biol. 1, 91–113
17. Britland, S., Morgan, H., Wojciak-Stothard, B., Riehle, M., Curtis, A. and Wilkinson, C. (1996) Exp. Cell Res. 228, 313–325
18. Rich, A.M. and Harris, A.K. (1981) J. Cell Sci. 50, 1–7
19. Nakata, I., Kawana, A. and Nakatsuji, N. (1993) Development 117, 401–408
20. Curtis, A., Wilkinson, C. and Wojciak-Stothard, B. (1995) J. Cell. Eng. 1, 35–38
21. Gail, M.H. and Boone, C.W. (1972) Exp. Cell Res. 70, 33–40
22. Hartman, R.S., Lau, K., Chou, W. and Coates, T.D. (1994) Biophys. J. 67, 2535–2545

23.   Hastings, H.M. and Sugihara, G. (1993) Fractals: A User's Guide for the Natural Sciences, Oxford Science Publications, Oxford
24.   Wojciak-Stothard, B., Curtis, A.S.G., Monaghan, W., McGrath, M., Sommer, I. and Wilkinson, C.D.W. (1995) Cell Motil. Cytoskeleton, **31**, 147–158
25.   Wojciak-Stothard, B., Madeja, Z., Korohoda, W., Curtis, A. and Wilkinson, C. (1995) Cell Biol. Int. **19**, 485–490
26.   Dunn, G.A. and Heath, J.P. (1976) Exp. Cell Res. **101**, 1–14
27.   Ohara, P.T. and Buck, R.C. (1979) Exp. Cell Res. **121**, 235–249
28.   Curtis, A.S.G. and Clark, P. (1990) Crit. Rev. Biocompat. **5**, 343–362
29.   Harris, A.K. (1982) in Cell Behaviour (Bellairs, R., Curtis, A. and Dunn, G., eds.), pp. 109–114, Cambridge University Press, Cambridge
30.   Robson, L. and Hunter, M. (1994) Pflugers Arch. **429**, 98–106
31.   Naruse, K., Asano, H. and Sokabe, M. (1993) Biophys. J. **64**, A93
32.   Reference deleted
33.   Ludewig, U., Pusch, M. and Jentsch, T.J. (1996) Nature (London) **383**, 340–343
34.   Goldspink, R., Goldspink, G., Scutt, A., Loughna, P.T., Wells, D.J., Jaenicke, T. and Gerlach, G.F. (1992) Am. J. Physiol. **262**, R356–R363
35.   Chehroudi, B. and Brunette, D.M. (1995) in Encyclopedic Handbook of Biomaterials and Bioengineering, Part S: Materials (Donald, D.J.T., Wise, L., Altobelli, D.E., Michael, J., Yaszenski, M.Y., Gresser, J.D. and Schwartz, E.R., eds.), pp. 813–842, Marcel Dekker, New York

Biochem. Soc. Symp. **65**, 27–42
Printed in Great Britain

3

Self-organization of tissue-equivalents: the nature and role of contact guidance

Robert T. Tranquillo

Department of Chemical Engineering and Materials Science,
University of Minnesota, Minneapolis, MN 55455, U.S.A.

Abstract

The morphology and behaviour of tissue cells when surrounded by a network of protein fibres, such as for a tissue-equivalent comprising cells entrapped in a type I collagen gel, is distinct from that when cells are cultured on a rigid surface, and physiologically relevant. The highly elongated and apparently bipolar morphology leads to a 'reversing' type of cell movement in gels, as opposed to a directionally persistent movement characteristic of highly spread, polar cells on surfaces. However, the hallmark of a tissue-equivalent is consolidation of the fibrillar network, or gel compaction, resulting from traction exerted by the cells. When the gel is mechanically constrained from compacting, alignment of the fibrils occurs, inducing cell alignment through a contact guidance response. In order to understand this 'self-organization' of tissue-equivalents, some relevant structural and mechanical properties of collagen gel are considered first, followed by a review of seminal studies of cell traction and tissue-equivalent compaction. Random cell migration in an isotropic gel is then discussed, including a modification of the persistent random walk model used to analyse cell migration on surfaces, followed by a review of contact guidance studies in gels with fibrils having defined alignment. With this background, observations of self-organization of mechanically constrained compacting tissue-equivalents are summarized and explained using a mechanical theory that relates traction-induced compaction to fibre alignment and consequent contact guidance, i.e. a strain-based rather than stress-based cell response to gel compaction. Data in support of this theory obtained from studies involving the controlled applied compression of tissue-equivalents are then presented. Finally, possible mechanisms of contact guidance are discussed.

Introduction

There has been growing interest in the study of tissue cell behaviour and metabolism in collagen gels, reflecting a growing appreciation of the use of 'tissue-equivalent' environments instead of tissue culture plastic and glass surfaces. The differences in behaviour can be significant. Although there has been a parallel increase in interest in the behaviour of leucocytes in three-dimensional environments, this chapter will consider only the behaviour of tissue cells, which, unlike leucocytes, exert sufficient traction to cause compaction of the gel and 'self-organization' of tissue-equivalents, the unifying subject of this review. For example, proliferation [1–3] and biosynthetic activity [3,4] are suppressed for both fibroblasts and smooth muscle cells cultured in floating collagen gels rather than on plastic.

Improved microscopy and imaging technology has facilitated studies using tissue-equivalents, and has revealed significant differences in morphology: fibroblasts cultured in collagen gels resemble the elongated bipolar or flattened stellate types observed *in vivo*, rather than the strongly polarized cells with a single broad lamella seen on glass or plastic [5,6] (Fig. 1). Although, strictly speaking, it may be more accurate to describe the elongated cell type as having a polarity that switches between the two ends of the cell, such cells will be referred to as bipolar. Such bipolar cells are the predominant type when the gel is mechanically constrained, as explained below. The development of tissue analogues, e.g. for skin [7–9] and blood vessels [10–12], based on tissue-equivalents, in order to circumvent biocompatibility problems associated with synthetic polymers, has been another incentive for studying tissue-equivalents.

A hallmark of a tissue-equivalent is the consolidation of the network of collagen fibrils resulting from traction exerted by the cells: a cell-induced syneresis, or compaction, of the gel. As will be discussed at length, an align-

(a)

(b)

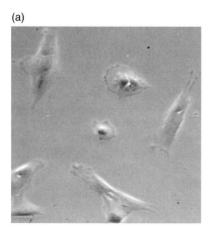 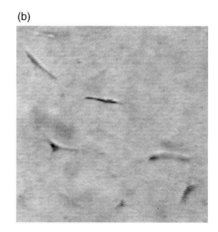

Fig. 1. Morphological differences of fibroblasts (a) on plastic compared with (b) in collagen gel. The elongated bipolar type of morphology exhibited in a collagen gel is shown. The degree of true bipolarity has not been well characterized, and the descriptor 'bipolar' is used herein with this qualification.

ment of the fibrils can also occur, inducing cell alignment through a contact guidance response, referred to as 'self-organization' of the tissue-equivalent. Indeed, this densification and alignment is essential for application as a tissue analogue. In order to understand compaction of tissue-equivalents, some relevant structural and mechanical properties of collagen gels are reviewed first. This is followed by sections addressing random cell migration in an isotropic gel, contact guidance in a gel with aligned fibrils, cell-traction-induced contact guidance and consequent self-organization of tissue-equivalents, organization of tissue-equivalents induced by applied compression, and possible mechanisms of contact guidance.

While the term 'tissue-equivalent' is only used in the literature to refer to tissue cells cultured within collagen gel, there are many similarities (whether or not tissue cells are present) between collagen gels and fibrin gels, the latter frequently being studied because of relevance to wound healing. Thus studies using fibrin gels are considered in this chapter as well, and usage of 'gel' or 'tissue-equivalent' without explicit reference to collagen or fibrin implies that the usage applies to either. Also, usage of 'tissue-equivalent' in the literature is generally reserved for the compacted state, but it will be used here to refer to any state.

Microstructural and rheological properties of collagen and fibrin gels

Collagen gel is typically reconstituted from a solution of pepsin-digested or acid-extracted type I collagen by restoring physiological pH and temperature, which induces self-assembly (end-to-end and lateral) of collagen molecules. Fibrin gel is typically prepared by combining fibrinogen and thrombin solutions containing calcium ions; thrombin-mediated cleavage of fibrinopeptides from fibrinogen leads to self-assembly in a similar manner. As seen for collagen gel in Fig. 2, a highly hydrated network [only 0.1–0.5% (w/w) protein] of long, highly entangled fibrils results, with the appearance of both gels being very similar in scanning and transmission electron microscopy [13,14]. The effective 'pore space' is about 1 μm which, along with the fibril diameter (0.05–0.5 μm), is small compared with the cell body dimension (50–100 μm) but comparable with that of a tip of a pseudopod. Tractional structuring causes consolidation of collagen fibrils around each cell (discussed further in the next section), so that, even before compaction leads to a significant reduction in pore size throughout the gel, pore size is rapidly reduced around the cells. In a fibrin gel this may not be true if fibrinolysis is not inhibited.

The similarity between collagen and fibrin gels extends to their mechanical behaviour, notably the way in which they respond to an applied force, which is determined by the interaction of their two component phases: the fibrillar network and the interstitial solution, typically tissue culture medium. The network effectively resists shear and extension, but has little compressive stiffness because fibrils buckle easily. Significant resistance to interstitial flow of the solution through the network (inversely related to the network permeability) can, however, lead to high solution pressures that will impart compressive stiffness to the gel. This is not of significance for tissue-equivalents, because of the

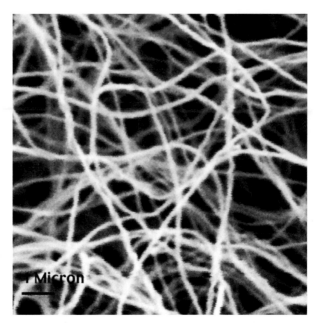

Fig. 2. Scanning electron micrograph of a collagen gel (scale bar = 1 μm).

very small force exerted by the cells on the network via cell traction, but is relevant when characterizing the rheology of the network, i.e. the way it deforms in response to an applied force, which is critical to interpreting the compaction of tissue-equivalents. Such a study has recently been accomplished for collagen gel in the case of an applied compression.

The results indicate that, qualitatively, the network flows like a viscoelastic fluid when compressed [15], as it does in response to an applied shear [16], although the values of the shear modulus and the viscosity depend on the type of deformation. (These results were obtained in an acellular collagen gel, so only apply to tissue-equivalents during their initial incubation before significant cell-induced compaction of the collagen and cell alteration of the collagen network by secretion or proteolysis.) An important consequence of fluid-like behaviour is that a stress in the network is completely relaxed with sufficient time. This stress relaxation property proves to be useful in elucidating the nature of the contact guidance signal in a gel with aligned fibrils (see later). The network of a fibrin gel similarly exhibits viscoelastic fluid behaviour in shear [17] and compression (D. Knapp and R. Tranquillo, unpublished work), which is not surprising given the similar protein concentration and fibrillar network microstructure.

Cell traction and compaction of tissue-equivalents

As noted above, traction exerted by tissue cells on the fibrillar network brings about macroscopic gel compaction. Pioneering work in the documentation of traction *in vitro*, and of its morphogenetic implications *in vivo*, was performed by Harris and co-workers. They first visualized cell traction by observing compression and tension wrinkles develop in a silicone rubber sheet on which various cells were cultured, with fibroblasts generating the greatest wrinkling among the various cell types examined [36]. (This experimental method has been refined subsequently [18,19] and a quantitative analysis developed [20].) They then documented the extensive consolidation of collagen fibrils that occurred over time around cells dispersed in collagen gel, termed tractional structuring [21,22]. The macroscopic manifestation of this phenomenon had been documented previously by Bell and co-workers in their seminal fibroblast-populated collagen lattice (FPCL) assay [23]. They reported that fibroblasts cultured in a small floating disc of collagen gel dramatically compacted the gel. (This will be termed free compaction, to distinguish it from cases where a mechanical constraint to compaction is imposed, such as the periphery of the disc being anchored; see section on constrained compaction below.) Notably, no observations were reported of any cell alignment in the FPCL. Our detailed analysis of freely compacting spheres of collagen and fibrin gels populated uniformly with fibroblasts (the spherical analogue of the FPCL assay) shows that cell orientation is random, i.e. there is no macroscopic alignment (B. Bromberek and R. Tranquillo, unpublished work).

The FPCL assay and its variants quickly became established as the primary means of investigating the ability of cells to interact mechanically with a tissue-like matrix (reviewed in [24]). Grinnell and co-workers conducted a series of investigations to elucidate the mechanism of fibril re-organization and collagen gel compaction [25–27]. Several important observations and conclusions were made. First, fibrils in the gel interior are rearranged even when the cells reside only at the gel surface, and disrupting the network connectivity inhibits compaction, both implying the transmission of traction force through a connected fibrillar network. Secondly, only 5% of the network is degraded, even though the gel volume may decrease by 85% or more, implying that compaction involves primarily a rearrangement of existing fibrils rather than their degradation and replacement. Furthermore, few covalent modifications of the collagen occur, and a partial re-expansion of compacted gel occurs after treatment of the cells with cytochalasin D or removal of the cells with detergent. Finally, cell-free gel compacted under centrifugal force exhibits a partial re-expansion, similar to a fibroblast-compacted gel. Based on these observations, a two-step mechanism was hypothesized for the mechanical stabilization of collagen fibrils during gel compaction: cells pull collagen fibrils into proximity via traction-exerting pseudopods, and, over a longer time scale, the fibrils become non-covalently cross-linked, independent of cell-secreted factors.

Cell migration in an isotropic gel

Noble [28] described tracks of fibroblasts in a gel using a computer-assisted optical sectioning system to study three-dimensional cell migration in time-lapse. In his approach, the co-ordinates of each cell were determined using a three-dimensional 'converging squares' algorithm in a cubic electronic tracking window around each cell. (Other tracking strategies have been described subsequently [33]; any three-dimensional cell tracking algorithm based on light-microscopic images is computationally intensive because of the difficulties presented by 'out-of-focus' cells.) It was shown that transformed 3T3 fibroblasts in collagen gel tended to move bidirectionally, periodically reversing their direction, although without preference for any particular direction, based on statistical analysis of displacement vectors [29]. This indicated that the regions of the collagen gels analysed were essentially isotropic (i.e. were far from surfaces, which always induce tangential alignment), and did not have alignment induced by free convection during fibrillogenesis, which can be significant. The reason for the pronounced reversing migration behaviour has not been elucidated definitively, but it appears to be correlated with the bipolar morphology and may reflect a local contact guidance phenomenon (see next section). Typical cell tracks in an isotropic collagen gel are presented in Fig. 3.

A more popular method of characterizing cell tracks in isotropic environments, whether a planar substratum or gel, is based on the correlated (or persistent) random walk model, which assumes that the cell tends to make small directional changes over short times. That is, the cells are biased towards continuing in their current direction of movement rather than behaving randomly, as assumed for Brownian motion. The most commonly used result from this model in the analysis of tracking data relates the time-dependence of the mean-squared total displacement, $<d^2(t)>$, in n_d dimensions ($n_d = 2$ for planar substrata and $n_d = 3$ for gel) to a diffusion (or random migration) coefficient, D, and a directional persistence time, P [30,31]:

$$<d^2(t)> = 2n_dD<t-P[1-\exp{(-t/P)]}>$$ (1)

where $D = S^2P/n_d$ (S is cell speed). However, fitting eqn. (1) to $<d^2(t)>$ data using conventional regression methods may be highly inaccurate, because inherent correlation between $<d^2(t)>$ for different values of t is neglected [32].

This model is certainly well founded for cells on a planar substratum having strong polarity. It would seem to be poorly founded for tissue cells in a gel, in light of their bipolar morphology and associated reversing migration behaviour. However, it provides excellent fits to data even in this latter case [33]. The resolution to this apparent paradox comes from a detailed mathematical analysis of a correlated random walk model upon which a process that generates a reversal in direction at random times is superimposed. In the case whereby this direction reversal is governed by the telegraph process (in which the probability for sustaining a correlated random walk before a direction reversal decreases exponentially with time), it can be shown that eqn. (1) is again derived for $<d^2(t)>$ (M. Wagle and R. Tranquillo, unpublished work). However, P is now an effective persistence time, which decreases monotonically with the frequency of reversal.

While there is no doubt that cell traction is necessary both for cell migration within a tissue-equivalent and for compaction of a tissue-equivalent, there is as yet no direct evidence demonstrating that cell migration is necessary for compaction, although this has been proposed based on circumstantial evidence

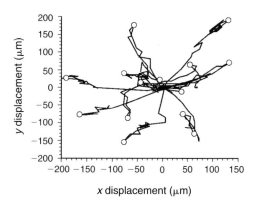

Fig. 3. Cell tracks in an isotropic collagen gel. Tracks were obtained using the 'high mag' algorithm described in Dickinson et al. [33] with a time lapse interval of 30 min, and were translated to a common origin. ○ denote reversals.

[34]. Our time-lapse recordings of freely compacting spheres of collagen and fibrin gels populated uniformly with fibroblasts reveal intensive cell motility (associated with cell spreading and continuous shape changes), but rarely any significant cell migration (in terms of displacement of the cell centroid by more than a cell diameter). Cell tracking in a compacting gel and the track analysis are obviously more complicated than on rigid substrata because of the convective movement of the cell associated with gel compaction being superimposed on any migration. Whether differences exist between cell migration in free compacting compared with mechanically constrained gels will be of interest, given the metabolic differences that have been reported [35]. Our preliminary data suggest that this is the case, with human foreskin fibroblasts in a mechanically constrained collagen gel (a disc adherent to a tissue culture well) migrating more quickly than cells in a free-floating collagen gel (an identical floating disc).

It will also be of interest to see whether a correlation analogous to that reported for cells on silicone rubber is found in gels (i.e. greater migration correlated with less traction, as Harris et al. [36] deduced by observing the extent of wrinkling of the rubber sheet). Migration might be modulated by the introduction of peptides that alter the adhesiveness of the cells to the network. There is considerable evidence, however, suggesting that adhesion-dependent traction is necessary for neutrophil migration on a surface, but that adhesion is not necessary for traction facilitating migration within a gel [37,38]. Traction in a gel might be attained by purely mechanical means, e.g. a (non-adhering) pseudopod that extends between and around fibrils providing numerous points of anchorage (i.e. traction) when the pseudopod attempts to retract [39]. It is not known whether tissue cell migration in a gel is adhesion-independent.

Finally, it should be mentioned that there are several studies involving migration following cell invasion of a gel, often aimed at assessing the effects on migration of other network-associating molecules present besides collagen or fibrinogen/thrombin during fibrillogenesis, either added as supplements [40,41] or present as contaminants [42]. However, the invasion step at the gel surface, not migration within the gel, may limit the gel invasion, which may render tenuous any conclusions about migration within the gel [43]. Further, elucidating the cause of any effects on migration among many possibilities (e.g. change in gel structure and/or mechanical properties, change in cell traction via change in cell adhesion or active signalling) is a difficult problem.

Cell contact guidance in an aligned gel

The movement of a cell exhibiting contact guidance is characterized as bidirectional, but, in contrast with an isotropic environment, the cell has maximum probabilities of migrating in opposite directions associated with some anisotropy of the substratum [44]. Since cells typically migrate in the direction of their orientation; contact guidance is also characterized by an equal and maximum probability of cell orientation in these two directions. The relevant anisotropy for a gel is associated with aligned fibrils, which can be generated by extrinsic methods (reviewed in [45]); a simple and reproducible one involves fibrillogenesis in a high-strength magnetic field. Typical cell tracks in a magnetically aligned collagen gel are presented in Fig. 4. An intrinsic method that exploits the ability of tissue cells to compact a gel and thereby align them, leading to contact guidance in a biomechanical feedback loop, is the subject of the next section. A similar process *in vivo* is believed to be operative in morphogenesis [22] and also several homoeostatic processes [46], such as wound contraction [39], emphasizing the fundamental importance of the study of contact guidance. The mechanism underlying contact guidance in an aligned gel is the subject of subsequent sections.

The method of magnetic-field-induced collagen alignment was used in the most quantitative characterization of contact guidance in an aligned gel to date [33]. Two population-averaged measures of contact guidance related to cell orientation were made as a function of the gel birefringence, Δn (a measure of the collagen alignment, taken to be in the x-direction). These measures,

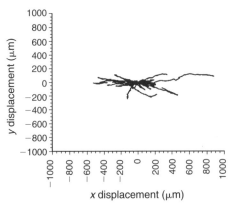

Fig. 4. Cell tracks in magnetically aligned collagen gel. For details, see the legend to Fig. 3.

based on q_c (the cell orientation with respect to the x-direction) are the cell orientation parameter O_C:

$$O_C = 2<\cos^2\theta_c>-1 = <\cos^2 2\theta_c> \tag{2}$$

and the anisotropic orientation parameter O_A:

$$O_A = <\cos^2\theta_c>/<\sin^2\theta_c> \tag{3}$$

O_C varies from 0 to 1 and O_A varies from 1 to ∞ as orientation varies from random to perfect alignment. Strictly speaking, orientation cannot be defined for a cell that appears bipolar, only alignment. We have used alignment to mean with respect to aligned fibrils. While O_C is more commonly used, O_A was defined so that it could be correlated with an anisotropic diffusion parameter D_A:

$$D_A = D_x/D_y \tag{4}$$

where D_x and D_y are diffusion coefficients in the x- and y-directions respectively, based on fitting eqn. (1) to the x- and y-components of the total displacement (i.e. $n_d = 1$). D_A varies from 1 to ∞ as migration varies from random to being strictly in the x-direction.

As expected, O_C increased with a saturating dependence on Δn, indicating that greater fibril alignment induces greater cell alignment. (Complicating form birefringence due to the difference in refractive index between the tissue culture medium and collagen did not allow Δn to be related to fibril alignment.) Also as expected, $D_A \approx O_A$, indicating that cells migrated in the direction in which they were oriented. More interestingly, D_A increased with Δn (fibril alignment), due mainly to a rapid increase in D_x at small values of Δn and a continued decrease in D_y at large values of Δn when D_x had attained a limiting value. Although low cell concentrations were used to minimize cell–cell interactions, such as the alignment of one cell induced by a neighbouring cell in a 'local guidance field' generated by tractional structuring, cell–cell interaction effects were not quantified. The existence of local guidance fields superimposed on the global magnetically induced guidance field was evident from the highly birefringent regions extending radially outwards from both ends of the bipolar cells superimposed on a uniformly birefringent background (Fig. 5). They might serve to amplify the response to a global guidance field.

Constrained compaction of tissue-equivalents: self-organization

There have been numerous observations of spontaneous cell alignment in tissue-equivalents that involves some mechanical constraint to the compaction (neither is true for the FPCL assay): (1) along the long axis of a rectangular slab adherent along its four sides [47]; (2) along the axis of a rectangular slab adherent at its two ends [48]; (3) in the plane of a disc adherent around its periphery [49]; (4) parallel to the base of a hemisphere adherent at its base [50], even if

Fig. 5. Local guidance fields associated with tractional structuring.
Polarized light reveals highly birefringent regions extending outwards from
both ends of cells, occasionally overlapping with those of neighbouring cells
that are co-aligned, superimposed on a birefringent background associated with
a magnetically aligned collagen gel.

cells were initially seeded only on the upper or lower surface of the hemisphere
[25]; (5) circumferentially in a tube compacting around a mandrel that constrained radial but not axial compaction, but axially in the case of an adherent
mandrel that constrained axial as well as radial compaction [11]; and (6) circumferentially in a sphere when cells were initially excluded from the core, but
radially when cells were initially only included in the core (B. Bromberek and
R. Tranquillo, unpublished work); in these cases the acellular regions of the gel
provide the mechanical constraint (Fig. 6).

Considered individually, each of these observations may appear con-
founding as to the origin of cell alignment. However, they can be explained by
a theory and thereby unified. This assumes that cells align, exert traction and
migrate preferentially in the direction in which surrounding fibrils are aligned
(i.e. they exhibit contact guidance in response to aligned fibrils), and that deformations of the fibrillar network (induced by cell traction in these observations)
that create an anisotropic strain cause fibril alignment. The interplay between
cell traction, fibrillar network deformation, fibril alignment and cell contact
guidance is depicted in Fig. 7. Simple examples of strain-based fibril re-orientation are depicted in Fig. 8. A continuum mechanical model that formalizes this
interplay is detailed in [51], along with predictions consistent with all the above
observations (except the first, which is neither spherically symmetrical nor axi-

(a)

(b)

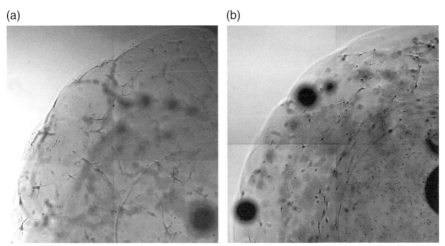

Fig. 6. Comparison of compaction and alignment in free and constrained compaction. Images of freely compacting spheres of fibrin gel after compaction showing one quadrant of the equatorial plane. (a) A sphere uniformly populated with fibroblasts shows that cell orientation remains isotropic. (b) A sphere where cells were initially excluded from the core, which is demarcated by small latex beads, shows that circumferential alignment develops.

symmetrical and therefore not amenable to the required numerical analysis of the model equations detailed in [52]).

Although this theory has similarities with the seminal traction-based continuum mechanical theory proposed by Oster and co-workers [53] as an alternative to morphogen-gradient-based theories for morphogenesis, this one,

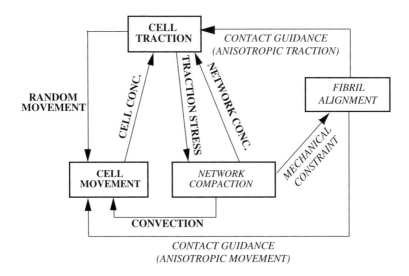

Fig. 7. Self-organizing biomechanical feedback mechanisms in tissue-equivalents.

in contrast, makes contact guidance central, consistent with observations in tissue-equivalents. Its predictions of pattern formation, relevant for morphogenesis in general and the observations of Harris and co-workers in particular [54], have not yet been examined, but similar models predict patterns for certain ranges of the various model parameters [55,56].

The success of the model in explaining observations of evolving alignment in tissue-equivalents, quantitatively as well as qualitatively in many cases [51,57] emphasizes the validity of its assumptions, including the strain-based origin of the contact guidance field and the nature of the contact guidance response described above. Independent and direct support for these key assumptions is provided by observation of the effects of an applied (rather than cell-induced) deformation of the network, as discussed in the next section. As with all models, this one is limited in its validity, and can only be applied to the relatively short time before the cells complicate the gel mechanics by extensive compaction and by secretion of significant amounts of extracellular matrix. In addition, the model does not allow for other potentially important modifiers of motility, such as chemotaxis (if the cells secrete chemotactic factors to which they respond) and contact inhibition of locomotion in its various forms (e.g. diminution of traction and/or reversal of migration direction upon contact). Experiments could be carried out at lower cell concentrations to make these complications of lesser significance, or the model could, in principle, be extended to include them.

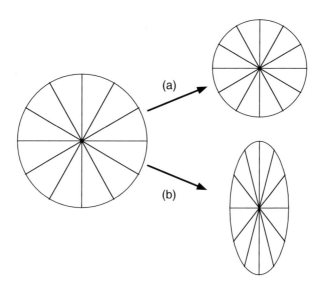

Fig. 8. Illustration of strain-based fibril re-orientation. Examples in two dimensions are given for simplicity. The consequences of compaction on an initially isotropic distribution of fibrils at a point are shown. (a) Homogeneous compaction, the case of a freely compacting tissue-equivalent, which maintains isotropy. (b) Confined compression, one example of an inhomogeneous compaction (or anisotropic strain) that can lead to alignment.

Confined compression of tissue-equivalents: direct evidence for strain-based contact guidance

That contact guidance in an aligned gel is correlated with fibril alignment is well established, although the mechanism by which cells sense and respond to aligned fibrils is still an open question, as discussed in the next section. However, this correlation does not necessarily imply that the cells exhibit contact guidance in direct response to aligned fibrils, since aligned fibrils associated with an anisotropic network strain (as illustrated in Fig. 8) will also generally be associated with an anisotropic network stress. Exactly how cells might sense an anisotropic stress directly is also an open question, and a moot point if the network is perfectly elastic, since stress is then proportional to strain and therefore mathematically equivalent, although it is far from irrelevant with respect to the cellular mechanism. However, the initial state of the gels for which all the observations are summarized in the previous section is as described in the second section, i.e. the network is viscoelastic, so that stress also depends on the strain rate, and the equivalence of stress and strain that applies in the case of an elastic gel does not apply. Moreover, the network exhibits fluid-like behaviour in which all stress is completely relaxed with sufficient time, unlike solid-like behaviour. This stress relaxation property allows an experiment to be conducted to directly verify the key assumptions about the nature of the contact guidance field and response, namely confined compression of a tissue-equivalent.

In this experiment (T. Girton and R. Tranquillo, unpublished work), a tissue-equivalent is formed within a rectangular chamber having an impermeable surface at one end and a flat porous piston at the other end that can be pushed a prescribed distance into the gel. The permeability of the porous piston allows the interstitial solution to flow out of the gel, but its pores are sufficiently small so that the network cannot. The fibril re-orientation assumed to occur in response to a confined compression is illustrated in Fig. 8(b). (If there were no chamber side walls at the time of compression, this would be termed an unconfined compression, which has a much more complicated deformation field associated with a bulging of the gel.) Two of the chamber walls are made of glass so that the network strain (based on displacement of marker beads), angle of extinction and birefringence (the direction and degree of fibril alignment respectively) and cell orientation can be measured using microscopy and image analysis. The key results are as follows: the direction of fibril alignment is perpendicular to the direction of compression (consistent with Fig. 8b), to a degree that increases with strain; cells align in the direction of fibril alignment, to a degree that increases with strain; and the degrees of both fibril alignment and cell alignment are approximately the same 24 h after the compression is applied as they are right after it is first applied. Taken together, these results verify the assumptions that cells align preferentially in the direction in which surrounding fibrils are aligned, and that deformations of the fibrillar network that create an anisotropic strain cause fibril alignment. The related assumption that cells migrate preferentially in the direction that surrounding fibrils are aligned is verified by the data presented in Dickinson et al.

[33]. The assumption that cells exert traction preferentially in the direction that surrounding fibrils are aligned is difficult to verify directly, but follows from the fact that cells align preferentially in that direction and from the pattern of birefringence around bipolar cells in gels, emanating radially outwards from both ends of the cells (Fig. 5).

Moreover, the observation that cell alignment is sustained at 24 h, at which time the stress should be virtually relaxed in the gel (since the longest measured relaxation time is 6 h), implies that a sufficient explanation is that the cells do in fact respond directly to the anisotropic strain (i.e. aligned fibrils) and not to any associated anisotropic stress. This is also supported by the observation of contact guidance in gels aligned with a magnetic field and using low cell concentrations [45], where little stress should exist. Of course, this does not preclude the possibility that cells can directly sense and respond to an anisotropic network stress in other situations, such as in an oscillatory strain field generated by external means, which has obvious cardiovascular relevance.

Insight into the mechanism of contact guidance

Beyond this discrimination between anisotropic strain (aligned fibrils) and anisotropic stress as the macroscopic signal, little progress has been made in distinguishing between the possibilities for the dominant cellular mechanism underlying contact guidance proposed by Dunn [44]. (1) Focal adhesions are confined to fibrils which, since aligned (chemical anisotropy), align the adhesions, hence locomotion. (2) The cell distorts differently when migrating in different directions; when aligned with fibrils it distorts less, due to the structural anisotropy, hence favouring locomotion oriented along the fibrils. (3) As elaborated by Haston et al. [58], since exertion of cell traction on the substratum is necessary for locomotion, pseudopods pulling in the direction of maximum elastic modulus (i.e. along fibrils) are more efficient, since the displacement of fibrils towards the cell is much less than for pseudopods pulling in the direction of the minimal modulus of elasticity (i.e. across fibrils). The result of this elastic anisotropy (more generally, a viscoelastic anisotropy) is again locomotion aligned with the fibrils.

The main reason for the lack of progress in distinguishing between these possibilities is the intrinsic difficulty in changing just one property of an aligned network (chemical, structural, mechanical anisotropy) without changing the others, although the lack of reproducibility of gel properties, let alone the alignment of components, has also been a problem.

Summary and conclusions

While contact guidance may not be well understood mechanistically nor therefore optimally realized, its consequences can be reasonably predicted in compacting tissue-equivalents. These predictions are based on a theory that assumes that cells align, exert traction and migrate preferentially in the direction in which surrounding fibrils are aligned (i.e. they exhibit contact guidance in response to aligned fibrils), and that deformations of the fibrillar network (induced by cell traction) that create an anisotropic strain cause fibril alignment. Most of these assumptions have been verified independently from characterizing the fibril and cell alignment that results from an applied compression, supporting the assertion that contact guidance in compacting tissue-equivalents is a response to fibril alignment associated with anisotropic network strain, not to an anisotropic network stress. There are limitations to the theory which create opportunities for investigation, such as accounting for contact inhibition of motility, change of cell phenotype and consequently cell behaviour, and evolution of the structural and mechanical properties of the gel, as compaction progresses. What are the molecular mechanisms by which cells exhibit contact guidance when surrounded by aligned fibrils? Reaching an answer to this question may entail an ultimate understanding of the mechanism and control of motility.

The research performed in my laboratory on these subjects over the years by Victor Barocas, Bruce Bromberek, Rich Dickinson, Tim Girton, Stefano Guido, Dave Knapp, Alice Moon and Mihir Wagle is very gratefully acknowledged, as are suggestions for this manuscript (T.G., D.K.) and funding provided by NSF (BES-9522758) and NIH (P01-GM50150-03S1).

References

1.   Sarber, R., Hull, B.C.M., Sorrano, T. and Bell, E. (1981) Mech. Aging Dev. **17**, 107–117
2.   Kono, T., Tanii, T., Furukawa, M., Mizuno, N., Kitajima, J., Ishii, M., Hamada, T. and Yoshizato, K. (1990) J. Dermatol. **17**, 2–10
3.   Thie, M., Schlumberger, W., Semich, R., Rauterberg, J. and Robenek, H. (1991) Eur. J. Cell Biol. **55**, 295–304
4.   Paye, M., Nusgens, B.V. and Lapiere, C.M. (1987) Eur. J. Cell Biol. **45**, 44–50
5.   Heath, J.P. and Hedlund, K.-O. (1984) Scanning Electron Microsc. **4**, 2031–2043
6.   Heath, J.P. and Peachey, L.D. (1989) Cell Motil. Cytoskeleton **14**, 382–392
7.   Bell, E., Ehrlich, H.P., Sher, S., Merrill, C., Sarber, R., Hull, B., Nakatsuji, T., Church, D. and Buttle, D.J. (1981) Plast. Reconstr. Surg. **67**, 386–392
8.   Michel, M., Germain, L. and Auger, F.A. (1993) In Vitro Cell. Dev. Biol. **29A**, 834–837
9.   Wilkins, L.M., Watson, S.R., Prosky, S.J., Meunier, S.F. and Parenteau, N.L. (1994) Biotechnol. Bioeng. **43**, 747–756
10.   Weinberg, C.B. and Bell, E. (1986) Science **231**, 397–400
11.   L'Heureux, N., Germain, L., Labbe, R. and Auger, F.A. (1993) J. Vasc. Surg. **17**, 499–509
12.   Tranquillo, R.T., Girton, T.S., Bromberek, B.A., Triebes, T.G. and Mooradian, D.L. (1996) Biomaterials **17**, 349–357
13.   Allen, T.D., Schor, S.L. and Schor, A.M. (1984) Scanning Electron Microsc. **1**, 375–390
14.   Muller, M.F., Ris, H. and Ferry, J.D. (1984) J. Mol. Biol. **174**, 369–384
15.   Knapp, D.M., Barocas, V.H., Moon, A.G., Yoo, K., Petzold, L.R. and Tranquillo, R.T. (1997) J. Rheol. **41**, 971–993
16.   Barocas, V.H., Moon, A.G. and Tranquillo, R.T. (1995) J. Biomech. Eng. **117**, 161–170
17.   Bale, M.D., Muller, M.F. and Ferry, J.D. (1985) Biopolymers **24**, 461–482
18.   Lee, J., Leonard, M., Oliver, T., Ishihara, A. and Jacobson, K. (1994) J. Cell Biol. **127**, 1957–1964
19.   Oliver, T., Dembo, M. and Jacobson, K. (1995) Cell Motil. Cytoskeleton **31**, 225–240
20.   Dembo, M., Oliver, T., Ishihara, A. and Jacobson, K. (1996) Biophys. J. **70**, 2008–2022

21. Harris, A.K., Stopak, D. and Wild, P. (1981) Nature (London) **290**, 249–251
22. Stopak, D. and Harris, A.K. (1982) Dev. Biol. **90**, 383–398
23. Bell, E., Ivarsson, B. and Merrill, C. (1979) Proc. Natl. Acad. Sci. U.S.A. **76**, 1274–1278
24. Tranquillo, R.T., Durrani, M.A. and Moon, A.G. (1992) Cytotechnology **10**, 225–250
25. Grinnell, F. and Lamke, C.R. (1984) J. Cell Sci. **66**, 51–63
26. Guidry, C. and Grinnell, F. (1985) J. Cell Sci. **79**, 67–81
27. Guidry, C. and Grinnell, F. (1986) Collagen Relat. Res. **6**, 515–529
28. Noble, P.B. (1987) J. Cell Sci. **87**, 241–248
29. Noble, P.B. and Boyarsky, A. (1988) Exp. Cell Biol. **56**, 289–296
30. Dunn, G.A. (1983) Agents Actions Suppl. **12**, 14–33
31. Alt, W. (1990) in Biological Motion, vol. 89 (Alt, W. and Hoffmann, G., eds.), pp. 254–268, Springer-Verlag, Berlin
32. Dickinson, R.B. and Tranquillo, R.T. (1993) AIChE J. **39**, 1995–2010
33. Dickinson, R.B., Guido, S. and Tranquillo, R.T. (1994) Ann. Biomed. Eng. **22**, 342–356
34. Ehrlich, H.P. and Rajaratnam, J.B. (1990) Tissue Cell **22**, 407–417
35. Mochitate, K., Pawelek, P. and Grinnell, F. (1991) Exp. Cell Res. **193**, 198–207
36. Harris, A.K., Wild, P. and Stopak, D. (1980) Science **208**, 177–179
37. Brown, A.F. (1982) J. Cell Sci. **58**, 455–467
38. Schmalstieg, F.C., Rudloff, H.E., Hillman, G.R. and Anderson, D.C. (1986) J. Leukocyte Biol. **40**, 677–691
39. Lackie, J.M. (1986) Cell Movement and Cell Behaviour, Allen & Unwin, London
40. Schor, S.L., Schor, A.M. and Bazill, G.W. (1981) J. Cell Sci. **48**, 301–314
41. Docherty, R., Forrester, J.V., Lackie, J.M. and Gregory, D.W. (1989) J. Cell Sci. **92**, 263–270
42. Brown, L.F., Lanir, N., McDonagh, J., Tognazzi, K., Dvorak, A.M. and Dvorak, H.F. (1993) Am. J. Pathol. **142**, 273–283
43. Dickinson, R.B., McCarthy, J.B. and Tranquillo, R.T. (1993) Ann. Biomed. Eng. **21**, 679–697
44. Dunn, G.A. (1982) in Cell Behaviour (Bellairs, R., Curtis, A. and Dunn, G., eds.), pp. 247–280, Cambridge University Press, Cambridge
45. Guido, S. and Tranquillo, R.T. (1993) J. Cell Sci. **105**, 317–331
46. Katz, M.J. and Lasek, R.J. (1980) Cell Motil. **1**, 141–157
47. Klebe, R.J., Caldwell, H. and Milam, S. (1989) Matrix **9**, 451–458
48. Kolodney, M.S. and Elson, E.L. (1993) J. Biol. Chem. **268**, 23850–23855
49. Lopez Valle, C.A., Auger, F.A., Rompre, R., Bouvard, V. and Germain, L. (1992) Br. J. Dermatol. **127**, 365–371
50. Tuan, T.L., Song, A., Chang, S., Younai, S. and Nimni, M.E. (1996) Exp. Cell Res. **223**, 127–134
51. Barocas, V.H. and Tranquillo, R.T. (1997) J. Biomech. Eng. **119**, 137–145
52. Barocas, V.H. and Tranquillo, R.T. (1997) J. Biomech. Eng. **119**, 261–269
53. Oster, G.F., Murray, J.D. and Harris, A.K. (1983) J. Embryol. Exp. Res. **78**, 83–125
54. Harris, A.K., Stopak, D. and Warner, P. (1984) J. Embryol. Exp. Morphol. **80**, 1–20
55. Murray, J.D. and Oster, G.F. (1984) J. Math. Biol. **19**, 265–279
56. Ngwa, G.A. and Maini, P.K. (1995) J. Math. Biol. **33**, 489–520
57. Barocas, V.H., Girton, T.S. and Tranquillo, R.T. (1998) J. Biomech. Eng., in the press
58. Haston, W.S., Shields, J.M. and Wilkinson, P.C. (1983) Exp. Cell Res. **146**, 117–126

Biochem. Soc. Symp. **65**, 43–62
Printed in Great Britain

4

Extracellular regulation of cancer invasion: the E-cadherin–catenin and other pathways

V. Noë*, E. Chastre†, E. Bruyneel*, C. Gespach† and M. Mareel*[1]

*Laboratory of Experimental Cancerology, University Hospital, De Pintelaan 185, B-9000 Gent, Belgium, and †INSERM U55, Hôpital Saint-Antoine, Paris, France

Abstract

The E-cadherin–catenin complex is pivotal for the regulation of cancer invasion. It not only serves cell–cell adhesion but also transduces signals from the micro-environment to other molecular complexes possibly implicated in invasion. Both functions are disturbed when the extracellular part of E-cadherin is cleaved off. Moreover, upon release into the environment, the E-cadherin fragments may interfere with intact complexes, as indicated by experiments with His-Ala-Val (HAV)-containing peptides that are homologous to parts of the first extracellular domain of E-cadherin. Scatter factor/hepatocyte growth factor (SF/HGF), on binding to its c-met tyrosine kinase receptor, can induce invasion through tyrosine phosphorylation of β-catenin. SF/HGF-induced invasion is also associated with phosphorylation of pp125^{FAK}, and both invasion and phosphorylation are inhibited by platelet-activating factor (PAF). Activation of the membrane-bound non-receptor tyrosine kinase pp60^{src} can also induce invasion. Signal transduction pathways starting from pp60^{src} include E-cadherin-associated β-catenin as well as the focal adhesion kinase pp125^{FAK}. Whereas all invasion-inducing pathways implicate phosphoinositide 3-kinase, the PAF pathway seems to be E-cadherin–catenin-independent. We conclude that cancer cell invasion is regulated by paracrine and autocrine factors that are released upon cross-talk with the host cells.

---

[1]To whom correspondence should be addressed.

Introduction

Cancer invasion competes with epithelial organization. Normal epithelial cells are polarized and closely apposed to each other on top of a basement membrane. Invasive epithelial cells are irregularly shaped, lose contact with their neighbours and break through the basement membrane. The architecture and barrier function of epithelia are maintained by cell–cell and cell–substratum junctional complexes that not only allow structural adhesion but also have crucial functions in signal transduction. Indeed, qualitative and quantitative alterations in molecules such as cadherins and integrins, which participate in the structure and function of such junctional complexes, can suppress or promote invasion, as will activators and inhibitors of these molecules [1]. Such molecules are implicated in the regulation of cellular activities that control invasion, namely cell–cell adhesion, cell–substratum adhesion, breakdown of the extracellular matrix and chemotaxis.

The present chapter considers the E-cadherin–catenin complex (where E-denotes epithelial) as pivotal for the regulation of epithelial invasion. Although we have concentrated on E-cadherin, it is only one of a family of cell–cell adhesion proteins; many other members of the family have not been investigated with regard to their participation at invasion to the same extent as E-cadherin. One reason for focusing on the E-cadherin complex is that it is disturbed in most, if not all, human cancers. In addition, invasion can be induced or suppressed through manipulation of the complex in experimental systems [2]. Another reason is that cell–cell adhesion mechanically counteracts invasion, although release of invasive cells from the epithelial structure is less obvious in some types of invasive cancers than in others [3]. A third and possibly more important reason is that the E-cadherin–catenin complex signals to other complexes that are implicated in other functions of invasive cells, such as cell–substratum adhesion, motility and proteolysis. We shall illustrate these points by results from ongoing, as yet unpublished, studies.

During the transition from the non-invasive to the invasive phenotype and vice versa, changes in the E-cadherin–catenin complex go hand in hand with changes in other complexes. For example, during invasion of the cytotrophoblast into the uterine wall, a well-known example of non-cancerous epithelial invasion, cells of the anchoring villi lose E-cadherin and express a novel cadherin, vascular VE-cadherin. At the same time these cells lose $\alpha_6\beta_4$ integrin and start to express $\alpha_5\beta_1$ integrin [4,5]. The latter integrin is also associated with cancer invasion [6]. Syndecan-1 is a cell surface proteoglycan involved in cell–substratum adhesion. Genetic manipulation of E-cadherin expression via transfection with sense or antisense E-cadherin cDNA, leading to changes in invasion, was accompanied by translational up- or down-regulation of syndecan-1, while β_1 integrin was not affected [7]. Conversely, suppression of syndecan-1 by antisense RNA led to down-regulation of E-cadherin [8]. Under these circumstances, invasion may be regulated synergistically by binding of syndecan-1 to extracellular matrix molecules and by homophilic homotypic intercellular adhesion via E-cadherin. In Caco-2 cells, Ha-*ras*- or polyoma middle T-induced progression towards a more tumorigenic phenotype was associated with decreased syndecan-1 expression at the level of the protein, but not the mRNA [9]. This progression did not, however, lead to acquisition of the invasive phenotype. Restriction of motility by the E-cadherin–catenin complex is suggested by the inverse correlation between the expression of autocrine motility factor and E-cadherin in bladder cancer [10]. Lack of E-cadherin or addition of anti-E-cadherin antibodies leads to positional instability of cells in confluent cultures, and this was accompanied by succinate dehydrogenase-mediated mitochondrial conversion of 3-(4,5-dimethylthiazol-2-yl-2,5-diphenyl)tetrazolium bromide [11]. Similarly, down-regulation of E-cadherin by a decompacting monoclonal antibody leads to increased secretion of urokinase, a hydrolase acting as an invasion promoter in both experimental and clinical cancer [12]. These examples strongly indicate that manipulation of invasion-promoter or invasion-suppressor complexes not only causes mechanistic changes to the complex itself but also signals to other synergistic or antagonistic complexes.

To this complexity of molecular pathways interacting with each other in the same cell we need to add another level of complexity. Invasion occurs within the context of microecosystems in which infiltrated host cells, such as endothelial cells, leucocytes and myofibroblasts, participate together with cancer cells (Fig. 1). This concept has been discussed previously [13]. Of particular interest in the microecosystem is the extracellular matrix, which can be considered as a molecular 'bulletin board', allowing indirect communication between the cells [14]. Soluble molecules are posted in a more or less stable matrix and cells express receptors for reading these messages.

Genomic changes responsible for cancer progression towards an invasive and metastatic disease take place in the cancer cell itself [15]. Such genomic changes may bring about an immediate phenotypic shift through signalling to the host cells and receiving invasion-promoter responses. The momentum for phenotypic changes, given the appropriate genomic status of the cancer cell, may also depend upon intercurrent phenomena, e.g. inflammation, bringing the necessary helper cells into the microecosystem of invasion [16]. To our knowledge, genomic changes in the infiltrated host cells responsible for cancer progression have not been demonstrated. Nevertheless, inflammation may cause (e.g. through production of oxygen radicals) genetic alterations in the cancer cell and thus may be partially responsible for tumour progression [17].

Molecular aspects of the E-cadherin–catenin invasion-suppressor complex

E-cadherin is a 120 kDa transmembrane glycoprotein that was originally described as a homophilic homotypic calcium-dependent cell–cell adhesion molecule, linked to the actin cytoskeleton via the catenins (Fig. 2). It is expressed early in development and can be found in most embryonic and adult epithelia. More recently, homophilic heterotypic interactions, e.g. between E-cadherin on keratinocytes and E-cadherin on melanocytes, and heterophilic heterotypic interactions, e.g. between E-cadherin on enterocytes and $\alpha_E\beta_7$ integrin on T-lymphocytes or internalin on invasive bacteria, have been

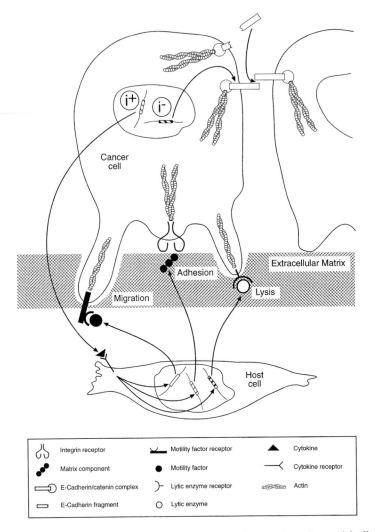

Fig. 1. Microecosystem of invasion at the primary site of an epithelial cancer. Arrows indicate the production and transport of molecules that participate in the cross-talk between cancer cells and host cells. i⁺, invasion-promoter genes; i⁻, invasion-suppressor genes.

demonstrated [18,19]. Furthermore, E-cadherin was found, quite unexpectedly, on erythroblasts and during formation of osteoclasts [20,21].

It is still not fully understood how the homophilic interactions between the extracellular domains of E-cadherins on neighbouring cells occur. The most recent concept attributes a crucial role to dimerization of E-cadherin molecules at the cell surface via the EC1 and EC2 domains (extracellular domains 1 and 2 respectively) and adhesion between dimers on neighbouring cells via the EC1 domain [22–24]. At the cytoplasmic side, the number of molecules that have been demonstrated to associate directly with or signal through the E-cad-

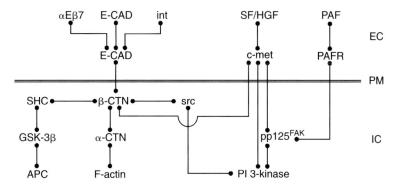

●———●, binds to (shown by co-immunoprecipitation or two-hybrid system) and/or signals to

Fig. 2. E-cadherin–catenin–actin signal transduction pathway. Schematic representation of signalling through the E-cadherin–catenin–actin complex. EC, extracellular; PM, plasma membrane; IC, intracellular; α-CTN, α-catenin; $α_Eβ_7$, integrin expressed by intraepithelial T-lymphocytes; APC, adenomatous polyposis coli protein; β-CTN, β-catenin; c-met, HGF/SF receptor; E-CAD, E-cadherin; GSK-3β, glycogen synthase kinase-3β; Int, internalin; PAF, platelet-activating factor; PAFR, PAF receptor; pp125^{FAK}, focal adhesion kinase; PI 3-kinase, phosphoinositide 3-kinase; SF/HGF, scatter factor/hepatocyte growth factor; SHC, protein with homology to src and collagen; src, phosphoprotein encoded by the src (proto)oncogene. Adapted from [69].

herin–catenin complex is steadily growing. Fig. 2 shows how changes in the E-cadherin–catenin complex go hand in hand with changes in other complexes, and indicates putative signalling pathways.

Invasion in experimental and clinical cancers may result from changes in different elements of the E-cadherin–catenin complex, and these changes may occur at different levels [2]. The gene encoding E-cadherin (*CDH1*) is an invasion-suppressor gene in accordance with the Knudson hypothesis, which states that both alleles should be lost to induce the tumour phenotype. For example, in lobular carcinoma of the breast, mutations of *CDH1* are accompanied by loss of the other allele, as shown by sequencing and by loss of heterozygosity [25]. Hypermethylation of the 5′ CpG island of E-cadherin, silencing the E-cadherin gene, is held responsible for the invasiveness of breast and prostate cancers [26].

Whereas genomic changes may inactivate E-cadherin irreversibly, post-translational modifications do so in a reversible manner. For example, tyrosine phosphorylation of β-catenin, mediated by the oncogenic *src* tyrosine kinase, destabilizes the E-cadherin–catenin complex and induces invasion [27]. Glycosylation of E-cadherin causes its accumulation at the cell surface due to a reduced turnover rate and inhibition of its release from the cell surface [28,29].

Experimental strategies

For the analysis of the E-cadherin–catenin complex and related pathways, we have used mainly tissue culture cell lines, and we have tested these in *in vitro* assays for invasion or for other functions of the complex.

The MDCKts.*src* cell line is of particular interest from the experimental point of view, because it permits manipulation of the invasive phenotype and of the E-cadherin–catenin complex in an almost physiological manner, namely by changing the temperature [27]. This cell line originated from the Madin–Darby canine kidney cell line MDCK after transformation with a temperature-sensitive mutant of an avian sarcoma virus.

The PC/AA human enteric cell line is of interest because it is one of the rare cell lines derived from an adenoma in a patient with familial adenomatous polyposis coli [30]. Such adenomas frequently progress to carcinomas. PC/AA cells carry one activated Ki-*ras* allele and a mutation in the APC (adenomatous polyposis coli) gene, leading to expression of a truncated form of the APC protein. The PCm*src* cell line was derived from the PC/AA cell line after transfection with a mutated (Tyr-527 → Phe) chicken *src* gene [31].

In the present study, two assays for invasion were used: confronting organ cultures of cancer cells with embryonic chick heart fragments and evaluation on histological sections [32]; and seeding of cancer cells on top of gels made from collagen type I or Matrigel and counting of cells inside the gel in living cultures [33]. The importance of the microecosystem, comprising different elements in different assays, is underscored by the fact that some cells are invasive in one assay and not in another [34]. For example, most human colon cancer cell lines do not invade collagen type I, despite their origin from invasive cancers and their clear-cut invasiveness in the chick heart assay. One possibility is that they need helper cells, as suggested by experiments with a chemically induced rat colon cancer [35]. Cloned epithelial PROb cancer cells failed to invade in either the chick heart or the collagen type I or Matrigel invasion assay. They did, however, form invasive cancers when injected into syngeneic rats. Cells freshly isolated from such tumours were invasive in all three *in vitro* assays, and the same was true if PROb cells were mixed with myofibroblasts.

There are a number of assays that explore the function of the E-cadherin–catenin complex and are more rapid or simpler than the invasion assays mentioned above (Table 1). With the majority of cell lines tested so far, results from these simpler assays are in agreement with those from invasion assays. However, even in these simpler assays, variation in the elements of the microecosystem may lead to different results with some cell types. For example, some human breast cancer cells of the MCF-7 family display a functional E-cadherin–catenin complex when they are on solid tissue culture substrata, but not when they are kept in suspension, in contrast with other members of the MCF-7 family in which the complex functions under both circumstances [36].

Many of the molecular changes in the E-cadherin–catenin complex have been detected by morphotypic changes in cultures on solid substrata. These morphotypic transitions have been coined E (epithelioid) to F (fibroblastoid), E to M (mesenchymal), E to R (round) or S (forming smooth edged colonies)

Table 1. Assays of the function of the E-cadherin–catenin complex.

Function	Assay	Reference
Cell–cell adhesion and compaction	Fast aggregation	[70]
	Slow aggregation	[70]
	Spheroid formation	[71]
	Colony formation in matrix	[41]
	Dispersion by pipetting	[71]
Segregation	Sorting out	[72]
Epithelial organization	Morphotype on solid substrata	[38]
	Transepithelial resistance	[73]
	Morphotype in confronting organ culture	[38]
	Integration in epithelial monolayer	[74]
Anti-invasion	Confronting organ culture	[50]
	Collagen type I Matrigel®	[38]
Vectorial transport	Dome formation	[38]
	Cyst formation in confronting organ culture	[38]

to R [37–39]. Scattering can be considered as an E to F or E to M transition. All these transitions cannot simply be interpreted as a loss of cell–cell adhesion; they suggest also changes in cell motility and in cell–substratum adhesion.

Each assay emphasizes particular aspects of the function of the complex and of its functional relationships with other complexes [40–42]. The fast (30 min) aggregation assay, as evaluated by counting the number of particles or by measuring the particle size distribution after incubation in suspension under gyratory shaking, explores the ability of cells to remain attached to each other under the shear-force conditions of the assay. Slow (1–3 days) aggregation assays reveal the capacity of cells not only to attach to one another but also to compact and to form organoid structures. In order to examine branching and tubule formation, the cells have to be included in a gel of collagen or extracellular matrix. Aggregation and formation of multicellular spheroids does not always necessitate the expression of E-cadherin. For the analysis of the effects of chemotherapeutic agents on growth and invasion, we have routinely used multicellular spheroids of mouse fibrosarcoma MO_4 cells that are E-cadherin-negative [43]. The E-cadherin specificity of aggregation or of other activities is indicated by Ca^{2+}-dependency; it is formally proven by inhibition with E-cadherin-specific antibody. It is, however, more difficult to obtain direct evidence for E-cadherin specificity in the case of loss of function.

Natural and synthetic agents that specifically activate or inhibit certain molecules are used to elucidate the pathway that leads from the point of activation, e.g. the extracellular domain of E-cadherin or the scatter factor/hepatocyte growth factor (SF/HGF) receptor kinase c-met or the non-receptor membrane-associated kinase Src, to cellular functions such as motility and cell–cell adhesion that are crucial for the invasive or the non-invasive phenotype. This strategy has two major drawbacks. First, most inhibitors are specific not for a single molecule but, at best, for a family of molecules. Notable examples are kinase and phosphatase inhibitors. Secondly, a single molecule may be implicated in many signalling pathways. A striking example is the phosphoinositide 3-kinase (PI 3-kinase) family of dual lipid and Ser/Thr kinases that regulate many functions in pathways related to trafficking, adhesion, actin rearrangement, cell growth and cell survival [44].

Extracellular part of E-cadherin

There are strong indications that release of the extracellular part of E-cadherin may down-regulate the function of the truncated and also of the still intact E-cadherin–catenin complex, and in this way may induce or enhance invasion. Theoretically, down-regulation of the complex may occur in several ways (Fig. 3): (1) cleavage of the extracellular part of E-cadherin not only inactivates this molecule but also leads to conformational changes in the complex followed by signalling to invasion promoters; (2) the extracellular fragment may interact with the extracellular domain of intact E-cadherin molecules, causing functional disturbance of the complex. The latter interaction may also cause conformational changes followed by signalling to invasion-promoter molecules such as hydrolases, which in turn contribute to cleavage and creation of more soluble fragments.

Table 2 summarizes data from the literature on the inhibition of E-cadherin–catenin functions by peptides and antibodies. The existence of E-cadherin was first suspected due to the decompacting effect of antibodies or their Fab fragments against gp80 (glycoprotein of 80 kDa) on embryonal carcinoma cells and eight-cell mouse embryos [45]. On mouse mammary epithelial cells, decompaction could be induced by gp80 purified from the serum-free conditioned medium of cultures of human MCF-7 mammary cancer cells [46,47]. To obtain this effect on epithelioid cells, highly purified undenatured gp80 was required at concentrations between 0.005 and 0.01 nM. This concentration-dependence explains why gp80 has no autocrine effect on the MCF-7 cells from which it was isolated. Decompaction was reversible and could be neutralized by antibody against gp80 at equimolar concentrations. By contrast with these observations, gp84 trypsinized from mouse PCC4 embryonal carcinoma cells failed to inhibit compaction of mouse pre-implantation embryos [45]. In line with the latter result, a recombinant 84 kDa protein (r-gp84) secreted by baculovirus-infected insect cells had no decompacting activity on preformed aggregates of PCC4 cells and did not inhibit aggregation in cell adhesion assays. r-gp84 exhibited no effect on the compaction of the morula or on the development of blastocysts [48]. r-gp84 harbours the epitope of the anti-

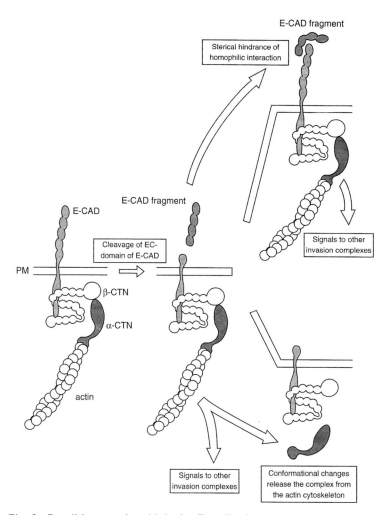

Fig. 3. Possible ways in which the E-cadherin–catenin complex may be down-regulated as a consequence of the cleavage of the extracellular domain of E-cadherin. Possible mechanisms are indicated by open arrows. E-CAD, E-cadherin; β-CTN, β-catenin; α-CTN, α-catenin; EC, extracellular; PM, plasma membrane.

E-cadherin antibody DECMA-1, is resistant to proteolytic degradation in the presence of Ca^{2+} and has an identical N-terminal amino acid sequence. There are several explanations for the lack of activity of r-gp84. The authors think that structural alterations during biosynthesis and purification of r-gp84 are responsible for its lack of biological activity. The lower degree of glycosylation as compared with the native peptide may render r-gp84 more sensitive to proteolytic degradation, as mentioned above for the complete E-cadherin molecule.

Inhibition of cadherin functions by synthetic peptides has been described by others (see Table 2). It is interesting that 17-mer peptides were more

Table 2. Inhibition of E-cadherin–catenin functions by (poly)peptides or antibodies specific for the extracellular part of cadherin. PCC4 cells, mouse embryonal carcinoma cells; MMTE, mouse mammary tumour epithelium; ARM cells, mouse fibrosarcoma cells transfected with chicken N-cadherin; n.m., not mentioned; r-, recombinant; EC SOD B, extracellular superoxide dismutase B.

(Poly)peptide/antibody	Active concentration	Target	Effect	Reference
Anti-gp84 Fab	0.25 ng/ml	Mouse egg PCC4 cells	Decompaction	[45]
PCC4 gp84	n.m.	Mouse egg	None	[45]
Anti-gp80	n.m.	Mouse egg MMTE	Decompaction	[46]
MCF-7 gp80	5 pM–10 pM	MMTE	Scattering	[47]
LRAHAVDVNG (avian N-cadherin)	8.35 mM	Mouse egg	Decompaction	[75]
LRAHAVDING (mouse N-cadherin)	8.35 mM	Myoblasts	No fusion	[76]
r-gp84	25 pM–2.5 mM	Mouse egg	None	[48]
LYSHAVSSNG (mouse E-cadherin)	550 mM	Osteoclast precursors	No fusion	[21]
REMHAVSRVQ (rat EC SOD B)	170 μM–1.7 mM	ARM cells	Inhibition of aggregation	[49]

Table 3. Species-specificity of the inhibition of E-cadherin-dependent cellular aggregation by HAV-containing peptides.
Values are percentages of those in untreated controls; each experiment with two measurements was repeated at least once; *significantly different from untreated control ($P < 0.005$) (Student's t-test).

Treatment		Aggregation (%)		
	Cell line...	MDCKts.srcCl2 cells	MCF-7/AZ cells	2B2 cells
	E-cadherin species....	Canine	Human	Chicken
Peptide sequence Based on	Amino acid sequence....	LY**SHAV**SSNG	LF**SHAV**SSNG	LL**SHAV**SASG
LF**SHAV**SSNG Human E-cadherin		31*	47*	82
LRA**HAV**DING Human N-cadherin		99	84	91

Table 4. Peptide sequence and induction of invasion into collagen of mouse mammary epithelial cells that do or do not express E-cadherin. Values are percentages of those of untreated controls; each experiments was repeated at least twice; *significantly different from untreated control ($P < 0.005$) (Student's t-test). The NM-f-*ras*TD1 cells are already invasive, and their invasion is not enhanced by the decapeptides. EC SOD, extracellular superoxide dismutase.

Treatment		Invasion (%)	
		NM-e-*ras*	NM-f-*ras*TD1
Cell line...		Mouse	None
E-cadherin species...			
Peptide sequence	Based on	Amino acid sequence...	
		LY**SHAV**SSNG	
LF**SHAV**SSNG	Human E-cadherin	247*	126
LFG**HAV**SENG	Human P-cadherin	143	121
REM**HAV**SRQV	Rat EC SOD	85	84
GTL**HAA**SQVQ	Human EC SOD	169	102
GTL**HAV**SQVQ	Human EC SOD (mutated)	256*	100

efficient than 10-mer peptides in preventing fusion of monocytic osteoclast precursors, indicating a role for the His-Ala-Val (HAV) flanking sequences [21].

In a first series of experiments, we demonstrated that the decapeptide REMHAVSRVQ, which is identical to part of the rat extracellular superoxide dismutase B, inhibited cadherin-dependent aggregation of cells expressing chicken N-cadherin (amino acid sequence in the region of interest: LRA-HAVDING), but not of cells expressing E-cadherin (LLSHAVSASG) [49]. Furthermore, we expanded these observations by exploring the species-specific matching of the peptide with the E-cadherin expressed on the target cell (Table 3). We concluded that HAV is essential for the inhibition of cadherin functions, and that homology between flanking amino acids in the peptide and those flanking the HAV motif in the cadherin expressed on the cell surface defines the specificity of the peptide. The matching of peptide sequences and their effect on the E-cadherin–catenin complex was tested further in the collagen type I assay for invasion (Table 4). Mutation of the human extracellular super-oxide dismutase decapeptide to contain an HAV instead of an HAA sequence activated the decapeptide, confirming the essential role of the HAV sequence. Although the flanking amino acid sequence of the mutated human extracellular superoxide dismutase is not specific, it might have the right configuration to inactivate the complex, indicating that, besides cadherin fragments, HAV-con-taining fragments of other molecules might also interfere with cadherin functions.

Possible scenarios for the inactivation of the E-cadherin–catenin complex by extracellular fragments comprise the following steps. (1) The peptide binds to the extracellular domain of E-cadherin. Flow cytometry has shown binding of E-cadherin decapeptides, but little or no binding of N-cadherin decapep-tides, to the surface of E-cadherin-expressing cells. Direct evidence that E-cadherin is the binding site has, however, not yet been obtained. We have mentioned previously that the the natural binding site of the HAV domain is not yet known. (2) The putative interaction between the peptide and the extra-cellular domain hampers the homophilic interaction. Alternatively or additionally, the interaction may lead to conformational changes to the intra-cellular part of E-cadherin, with loss of proper linkage with the actin cytoskeleton via the catenins. (3) We cannot exclude negative or positive sig-nalling to other invasion-promoter or invasion-suppressor complexes respectively. Circumstantial evidence in favour of signalling comes from the fact that the effect of the peptides has been demonstrated in various assays. Some of the phenotypes (e.g. segregation and invasion into collagen type I or into chick heart in organ culture) can hardly be explained on the basis of release of cell–cell adhesion only. We have only just started to investigate possible sig-nal transduction pathways. Molecules of interest are β-catenin, glycogen synthase kinase-3β and SHC (*src* homology/collagen) (see Fig. 2). (4) Regardless of the molecular mechanisms hampering the function of the com-plex, several explanations remain open as to the cellular activities that lead to invasion. One obvious explanation is that weakening or loss of E-cadherin-dependent cell–cell adhesion permits the cells to move more freely and in any direction. Hampering of cell–cell adhesion by HAV peptides was demonstrated in the fast aggregation assay. That neutralization of E-cadherin may not be sufficient to induce invasion is demonstrated by the behaviour of variants from the mouse NMuMG (normal murine mammary gland) cell family [50]. Clones with a fibroblastic morphotype (NM-f) that lacked E-cadherin were non-invasive, like their E-cadherin-positive epithelioid congeners (NM-e). After transfection with an activated *ras* oncogene, NM-f-*ras* cells were invasive, in contrast with NM-e-*ras* cells, demonstrating that invasion necessitates both the activation of an invasion-promoter pathway and the inactivation of E-cadherin (see also Fig. 1). Antibody-mediated neutralization of E-cadherin leads to loss of contact inhibition of motility [11] by as yet unknown pathways. One possibility is that the complex loses its control over the actin cytoskeleton. In the collagen type I invasion assay, peptide-induced invasion may involve single cells as well as smaller islands. In this case it is hard to imagine how loss of cell–cell adhesion may acount for invasion. (5) The relevance to invasion in experimental or human cancers of our *in vitro* observations with the synthetic HAV peptides and the recombinant EC1$^{1/2}$ (recombinantly made E-cadherin fragment containing extracellular domain 1 and half of extracellular domain 2) extracellular proteins requires answers to the following questions. First, do cells release into their micromilieu fragments of E-cadherin from their surface, for example as a result of natural turnover, proteolytic cleavage or secretion of soluble forms of the molecule? Such forms may result from mutations in the transmembrane domain, as suggested by Berx et al. [51]. Secondly, do such fragments disturb the function of the complex? Thirdly, is this autocrine loop different in invasive as compared with non-invasive cell populations? Differences may be sought in the factors affecting the natural turnover of E-cadherin and in proteolytic degradation by invasion-related hydrolases, as well as in the signalling as a result of the manipulation of the extracellular part of the E-cadherin–catenin complex.

Invasion-promoter function of SF/HGF

Paracrine factors are major players in the microecosystem of invasion as depicted in Fig. 1. Here, SF/HGF acts as an invasion promoter, both in experimental and in clinical cancers [52].

SF (a 92 kDa glycoprotein) was originally purified from conditioned media on the basis of its capacity to dissociate epithelial cells [53]. It was later shown to be identical to HGF [54]. During embryonic development SF/HGF acts as a regulator of epithelial–mesenchymal interactions involved in migration, growth and organogenesis. This is in line with its pleiotropic activity, as demonstrated also *in vitro*. SF/HGF binds to the heterodimeric receptor tyrosine kinase c-met, a 190 kDa transmembrane glycoprotein [55]. One of the major changes in colonic cancer is overexpression of c-met secondary to activation of other oncogenes [56].

The invasion-promoter function of SF/HGF is mostly paracrine [31,57]. Interestingly, the stimulation of cancer cell invasion by SF/HGF results from a cross-talk between the cancer cells, secreting interleukin-1, basic fibroblast growth factor and platelet-derived growth factor or combinations of these, and stromal fibroblasts responding to these cytokines by production of SF/HGF [58]. This is an illustration of the microecosystem (see Fig. 1), with the cancer cells as the creators and the host cells as helpers. Progression to a more malignant phenotype renders the cancer cells more sensitive to the invasion-inducing effects of SF/HGF. In our experiments with the FAP colon PC/AA cell family, only cells expressing a mutated *src* oncogene responded to SF/HGF. Autocrine stimulation of invasion has been described as well. After transfection with simian virus 40 large T, but not with the *ras* oncogene, MDCK cells became invasive into collagen type I and into chick heart; their conditioned medium contained SF/HGF [59]. SF/HGF is secreted as an inactive zymogen that is converted into the active form by limited proteolysis [60].

Possible scenarios of SF/HGF-induced invasion are as follows (see also Fig. 2). (1) Little doubt exists that the initial step of paracrine stimulation of the cancer cell is the binding of SF/HGF to its receptor c-met, followed by activation of the receptor tyrosine kinase domain. (2) The signal transduction pathway leading to the invasion-promoter complex is less clear. Molecules of interest are the small GTPase Rac [61], the focal-adhesion-associated kinase pp125^{FAK} [62], PI 3-kinase, the platelet-activating factor (PAF) receptor [63] and pp60^{src} (see below). In PCmsrc cells, SF-induced invasion was inhibited by PAF; this effect was abolished by the PAF antagonists WEB2086 and SR27417, which block access of PAF to its receptor. The invasion-inducing activity of SF/HGF was also counteracted by the PI 3-kinase inhibitors wortmannin and LY294002. (3) On following the behaviour of cells after addition of SF/HGF, one observes first cell spreading with centrifugal extension of cell colonies, and secondly disruption of cell–cell contacts with cell scattering. This sequence and the well-documented re-organization of the actin cytoskeleton points to motility as a crucial activity in SF/HGF-induced invasion. The SF/HGF-induced phosphorylation of pp125^{FAK}, which is sensitive to inhibition by PAF, points to a role for changes in cell–substratum adhesion as well. The pleiotropic effects of the activation of c-met again illustrate the crucial role of the microecosystem in the behaviour of the stimulated cell population. As a result of this activation, MDCK cells on solid substrata scatter with loss of epithelioid characteristics, whereas in collagen type I branching tubules are formed with a gain in epithelioid characteristics [42,59]. When cells are lying on top of the collagen, the changes in focal adhesion contacts mentioned above, together with increased membrane ruffling, may advance the cell's leading edge and permit a new anchorage deeper in the collagen. The ruffle moves into the gel rather than gliding over its surface because of a gradient built up by the unequal distribution of collagen over the cell surface. When a cell equalizes this distribution by moving into the collagen, further invasion should stop. Indeed, the SF/HGF-induced invasion mentioned above, and its inhibition, are marked by changes in the number of cells in the upper layers of the collagen and not in the depth profiles.

It is not yet clear whether or not the signalling pathway of SF/HGF-induced invasion is independent from the E-cadherin–catenin pathway. In the *in vitro* systems, invasion is conceivable without changes in the E-cadherin–

catenin complex. In the case of collagen invasion by solitary cells or smaller groups of cells, there are free cell borders that can move into the matrix, eventually pulling in the other cells. In the case of an intact polarized epithelium, even invagination of intact epithelial buds is hard to conceive without alterations of junctional complexes. PAF, at concentrations that inhibit invasion, permits HGF-induced phosphorylation of β-catenin. Of course, this observation does not invalidate the possibility that phosphorylation of β-catenin is involved in the SF/HGF invasion pathway, another step of which may be blocked by PAF. In human cancer cells of gastrointestinal origin, SF/HGF-induced scattering was accompanied by phosphorylation of catenins on tyrosine residues [64]. In MDCK(LT) cells with an autocrine SF/HGF loop, E-cadherin is down-regulated [59]. This is not, however, the only molecular alteration that might contribute to the acquisition of the invasive phenotype. MDCK(LT) cells also show down-regulation of desmosomal proteins and production of a 62 kDa gelatinase. In other systems, SF/HGF induces the expression of urokinase plasminogen activator and urokinase plasminogen activator receptor [65].

Role of *src* in cancer cell progression towards invasiveness

The need for progression of cancer cells towards a state of responsiveness to paracrine or autocrine induction of invasion appears from experiments with fragments of the extracellular part of E-cadherin as well as with SF/HGF. The kinds of cells used in this work suggest a role for activation of the *src* oncogene in this progression.

src is a well-known oncogene (growth promoter gene). In colon carcinogenesis, activation of membrane-bound pp60^{c-src} is considered to be an early event [66]. Normally, pp60^{c-src} acts as a non-receptor tyrosine kinase; it is widely expressed and is particularly abundant in platelets, neural tissue and osteoclasts. In osteoclasts, c-*src* associates with microtubules that traffic proteins to the cell surface [67].

There is experimental evidence that *src* also acts as an invasion-promoter gene. MDCKts.*src* cells are non-invasive, but the invasive phenotype can be induced by shifting the temperature from 40 to 35°C [27]. Morphotypically, this shift is translated into scattering of the cells in culture on solid substrata within 30 min. MDCKts.*src* cells fail to form aggregates in suspension culture at 35°C, but not at 40°C.

Multiple signal transduction pathways start from pp60^{src}, and some interact with invasion-promoter or invasion-suppressor complexes, such as the integrin extracellular matrix complex.

As described also for SF/HGF, invasion may result from stimulated motility due to increased membrane ruffling and from loss of cell–cell adhesion. The latter has been explained through inactivation of the E-cadherin–catenin complex, although other mechanisms are certainly not excluded. Activation of *src* causes tyrosine phosphorylation of β-catenin [27,41], leading to release of α-catenin from the complex and, as a consequence, to severing of the actin cytoskeleton [68]. This is in line with the finding that, in immunoprecipitates from *src*-activated (35°C) MDCKts.*src* cells with antibody against β-catenin, the amount of E-cadherin is not lower when there is increased tyrosine phosphorylation of β-catenin. However, we did not see differences in the α-catenin content of such immunoprecipitates.

The relationship between *src* and SF/HGF probably differs in different cancer cells due to the nature of the upstream signals, the integration of downstream signals and the type of effectors. Transfection of PC/AA cells with the activated (Tyr-527 → Phe) form of pp60^{src} did not make the transfectants PCm*src* invasive. Invasion of PCm*src* cells can, however, be induced by addition of SF/HGF [31]. Activation of resident *ras* or introduction of *PY middle T* did not make PC/AA cells invasive, either spontaneously or after addition of SF/HGF. PCm*src* cells are marked by increased tyrosine phosphorylation of E-cadherin, but not of the catenins. In MDCKts.*src* cells, activation of the *src* oncogene is sufficient to induce invasion. However, at the non-permissive temperature of activity of the oncogenic *src*, the MDCKts.*src* cells become invasive after addition of SF/HGF. In both situations there is increased tyrosine phosphorylation on β-catenin. It thus appears that in PC/AA cells activation of both *src* and c-*met* is required to induce invasion, whereas MDCK cells can be rendered invasive by activation of either pathway alone. Both kinds of invasion were sensitive to PAF and to inhibitors of PI 3-kinase, suggesting common or partly overlapping signalling pathways. There are two reasons to think that the PAF pathway does not involve E-cadherin: (1) PAF has no effect on scattering of MDCKts.*src*Cl2 cells through *src* activation (35°C) or on addition of HGF/SF; and (2) PAF inhibits the invasion of DHD-FIB, a rat myofibroblast cell line that does not express E-cadherin.

Conclusions

Manipulation of the E-cadherin–catenin complex by extracellular elements may induce invasion of cancer cells. HAV-containing peptides homologous to parts of the first extracellular domain of E-cadherin inactivate the E-cadherin–catenin complex, leading to invasion. We have inferred from this observation the possibility that release of fragments from the extracellular part of E-cadherin may inactivate intact complexes. Such inactivation may be realized through steric hindrance of homophilic extracellular interactions, which might initiate a signalling cascade, or through intracellular signalling by the truncated complex to other invasion-related complexes.

SF/HGF induces invasion of cancer cells provided that they have reached a certain degree of progression. Activation of the *src* oncogene plays a role in such progression towards sensitivity to paracrine or autocrine invasion promoters.

The present as well as previous observations support our concept of invasion occurring within microecosystems in which there is a continuous molecular cross-talk between cancer cells and host cells.

We thank Lieve Baeke and Jean Roels van Kerckvoorde for technical assistance. This work was supported by the ASLK/VIVA-verzekeringen (Brussels, Belgium), the FWO-Vlaanderen (Brussels, Belgium), l'Association de la Recherche contre le Cancer (ARC; France), and la Ligue Nationale contre le Cancer (France).

References

1. Mareel, M.M., Van Roy, F.M. and Bracke, M.E. (1993) Crit. Rev. Oncol. **4**, 559–594
2. Bracke, M.E., Van Roy, F.M. and Mareel, M.M. (1996) in Attempts to Understand Metastasis Formation I (Günthert, U. and Birchmeier, W., eds.), pp. 123–161, Springer, Berlin
3. Gabbert, H. (1985) Cancer Metastasis Rev. **4**, 293–309
4. Redman, C.W.G. (1997) Nature Med. **3**, 610–611
5. Zhou, Y., Fisher, S.J., Janatpour, M., Genbacev, O., Dejana, E., Wheelock, M. and Damsky, C.H. (1997) J. Clin. Invest. **99**, 2139–2151
6. Albelda, S.M. (1993) Lab. Invest. **68**, 4–17
7. Leppä, S., Vleminckx, K., Van Roy, F. and Jalkanen, M. (1996) J. Cell Sci. **109**, 1393–1403
8. Kato, M., Saunders, S., Nguyen, H. and Bernfield, M. (1995) Mol. Biol. Cell **6**, 559–576
9. Levy, P., Munier, A., Baron-Delage, S., Di Gioia, Y., Gespach, C., Capeau, J. and Cherqui, G. (1996) Br. J. Cancer **74**, 423–431
10. Otto, T., Birchmeier, W., Schmidt, U., Hinke, A., Schipper, J., Rübben, H. and Raz, A. (1994) Cancer Res. **54**, 3120–3123
11. Bracke, M.E., Depypere, H., Labit, C., Van Marck, V., Vennekens, K., Vermeulen, S.J., Maelfait, I., Philippé, J., Serreyn, R. and Mareel, M.M. (1997) Eur. J. Cell Biol. **74**, 342–349
12. Frixen, U.H. and Nagamine, Y. (1993) Cancer Res. **53**, 3618–3623
13. Van Roy, F. and Mareel, M. (1992) Trends Cell Biol. **2**, 163–169
14. Hedgecock, E.M. and Norris, C.R. (1997) Trends Genet. **13**, 251–253
15. Hedrick, L., Cho, K.R. and Vogelstein, B. (1993) Trends Cell Biol. **3**, 36–39
16. Opdenakker, G. and Van Damme, J. (1992) Immunol. Today **13**, 463–464
17. Atfi, A., Chastre, E. and Gespach, C. (1996) Lettre du cancérologue Supplément, Décembre, 6–19
18. Cepek, K.L., Shaw, S.K., Parker, C.M., Russell, G.J., Morrow, J.S., Rimm, D.L. and Brenner, M.B. (1994) Nature (London) **372**, 190–193
19. Mengaud, J., Ohayon, H., Gounon, P., Mège, R.-M. and Cossart, P. (1996) Cell **84**, 923–932
20. Armeanu, S., Bühring, H.-J., Reuss-Borst, M., Müller, C.A. and Klein, G. (1995) J. Cell Biol. **131**, 243–249
21. Mbalaviele, G., Chen, H., Boyce, B.F., Mundy, G.R. and Yoneda, T. (1995) J. Clin. Invest. **95**, 2757–2765
22. Shapiro, L., Fannon, A.M., Kwong, P.D., Thompson, A., Lehmann, M.S., Grübel, G., Legrand, J.F., Als-Nielsen, J., Colman, D.R. and Hendrickson, W.A. (1995) Nature (London) **374**, 327–337
23. Nagar, B., Overduin, M., Ikura, M. and Rini, J.M. (1996) Nature (London) **380**, 360–364
24. Koch, A.W., Pokutta, S., Lustig, A. and Engel, J. (1997) Biochemistry **36**, 7697–7705
25. Berx, G., Cleton-Jansen, A.-M., Nollet, F., de Leeuw, W.J.F., van de Vijver, M.J., Cornelisse, C. and Van Roy, F. (1995) EMBO J. **14**, 6107–6115
26. Graff, J.R., Herman, J.G., Lapidus, R.G., Chopra, H., Xu, R., Jarrard, D.F., Isaacs, W.B., Pitha, P.M., Davidson, N.E. and Baylin, S.B. (1995) Cancer Res. **55**, 5195–5199
27. Behrens, J., Vakaet, L., Friis, R., Winterhager, E., Van Roy, F., Mareel, M.M. and Birchmeier, W. (1993) J. Cell Biol. **120**, 757–766
28. Yoshimura, M., Ihara, Y., Matsuzawa, Y. and Taniguchi, N. (1997) J. Biol. Chem. **271**, 13811–13815

29. Leroy, A., Noë, V., Mareel, M. and Nelis, H. (1997) Biochem. Soc. Trans. **25**, 228–234
30. Paraskeva, C., Buckle, B.G., Sheer, D. and Wigley, C.B. (1984) Int. J. Cancer **34**, 49–56
31. Empereur, S., Djelloul, S., Di Gioia, Y., Bruyneel, E., Mareel, M., Van Hengel, J., Van Roy, F., Comoglio, P., Courtneidge, S., Paraskeva, C., et al. (1997) Br. J. Cancer **75**, 241–250
32. Bracke, M.E. and Mareel, M.M. (1994) in Cell and Tissue Culture: Laboratory Procedures (Doyle, A., Griffiths, J.B. and Newell, D.G., eds.), vol. 5A:4, pp. 1–16, John Wiley & Sons, Chichester
33. Vakaet, Jr., L., Vleminckx, K., Van Roy, F. and Mareel, M. (1991) Invasion Metastasis **11**, 249–260
34. Mareel, M., Bracke, M., Gao, Y. and Van Roy, F. (1991) Eur. Arch. Biol. **102**, 185–188
35. Dimanche-Boitrel, M.T., Vakaet, Jr., L., Pujuguet, P., Chauffert, B., Martin, M.S., Hammann, A., Van Roy, F., Mareel, M. and Martin, F. (1994) Int. J. Cancer **56**, 512–521
36. Bracke, M.E., Van Larebeke, N.A., Vyncke, B.M. and Mareel, M.M. (1991) Br. J. Cancer **63**, 867–872
37. Jouanneau, J., Gavrilovic, J., Caruelle, D., Jaye, M., Moens, G., Caruelle, J.-P. and Thiery, J.-P. (1991) Proc. Natl. Acad. Sci. U.S.A. **88**, 2893–2897
38. Vermeulen, S.J., Bruyneel, E.A., Bracke, M.E., De Bruyne, G.K., Vennekens, K.M., Vleminckx, K.L., Berx, G.J., Van Roy, F.M. and Mareel, M.M. (1995) Cancer Res. **55**, 4722–4728
39. van Hengel, J., Gohon, L., Bruyneel, E., Vermeulen, S., Cornelissen, M., Mareel, M. and Van Roy, F. (1997) J. Cell Biol. **137**, 1103–1116
40. Takeichi, M. (1977) J. Cell Biol. **75**, 464–474
41. Matsuyoshi, N., Hamaguchi, M., Tanigushi, S., Nagafuchi, A., Tsukita, S. and Takeichi, M. (1992) J. Cell Biol. **118**, 703–714
42. Montesano, R., Schaller, G. and Orci, L. (1991) Cell **66**, 697–711
43. Mareel, M.M. and De Mets, M. (1984) Int. Rev. Cytol. **90**, 125–168
44. Carpenter, C.L. and Cantley, L.C. (1996) Curr. Opin. Cell Biol. **8**, 153–158
45. Hyafil, F., Morello, D., Babinet, C. and Jacob, F. (1980) Cell **21**, 927–934
46. Damsky, C.H., Richa, J., Solter, D., Knudsen, K. and Buck, C.A. (1983) Cell **34**, 455–466
47. Wheelock, M.J., Buck, C.A., Bechtol, K.B. and Damsky, C.H. (1987) J. Cell. Biochem. **34**, 187–202
48. Herrenknecht, K. and Kemler, R. (1993) J. Cell Sci. Suppl. **17**, 147–154
49. Willems, J., Bruyneel, E., Noë, V., Slegers, H., Zwijsen, A., Mège, R.-M. and Mareel, M. (1995) FEBS Lett. **363**, 289–292
50. Vleminckx, K., Vakaet, Jr., L., Mareel, M., Fiers, W. and Van Roy, F. (1991) Cell **66**, 107–119
51. Berx, G., Cleton-Jansen, A.-M., Strumane, K., de Leeuw, W.J.F., Nollet, F., Van Roy, F. and Cornelisse, C. (1996) Oncogene **13**, 1919–1925
52. Humphrey, P.A., Zhu, X., Zarnegar, R., Swanson, P.E., Ratliff, T.L., Vollmer, R.T. and Day, M.L. (1995) Am. J. Pathol. **147**, 386–396
53. Stoker, M., Gherardi, E., Perryman, M. and Gray, J. (1987) Nature (London) **327**, 239–242
54. Weidner, K.M., Arakaki, N., Hartmann, G., Vandekerckhove, J., Weingart, S., Rieder, H., Fonatsch, C., Tsubouchi, H., Hishida, T., Daikuhara, Y. and Birchmeier, W. (1991) Proc. Natl. Acad. Sci. U.S.A. **88**, 7001–7005
55. Naldini, L., Weidner, K.M., Vigna, E., Gaudino, G., Bardelli, A., Ponzetta, C., Narsimhan, R.P., Hartmann, G., Zarnegar, R., Michalopoulos, G.K., et al. (1991) EMBO J. **10**, 2867–2878
56. Di Renzo, M.F., Olivero, M., Giacomini, A., Porte, H., Chastre, E., Mirossay, L., Nordlinger, B., Bretti, S., Bottardi, S., Giordano, S., et al. (1995) Clin. Cancer Res. **1**, 147–154
57. Weidner, K.M., Behrens, J., Vandekerckhove, J. and Birchmeier, W. (1990) J. Cell Biol. **111**, 2097–2108

58. Nakamura, T., Matsumoto, K., Kiritoshi, A., Tano, Y. and Nakamura, T. (1997) Cancer Res. **57**, 3305–3313

59. Martel, C., Harper, F., Cereghini, S., Noë, V., Mareel, M. and Crémisi, C. (1997) Cell Growth Differ. **8**, 165–178

60. Shimomura, T., Denda, K., Kitamura, A., Kawaguchi, T., Kito, M., Kondo, J., Kagaya, S., Qin, L., Takata, H., Miyazawa, K. and Kitamura, N. (1997) J. Biol. Chem. **272**, 6370–6376

61. Ridley, A.J., Comoglio, P.M. and Hall, A. (1995) Mol. Cell. Biol. **15**, 1110–1122

62. Matsumoto, K., Matsumoto, K., Nakamura, T. and Kramer, R.H. (1994) J. Biol. Chem. **269**, 31807–31813

63. Kotelevets, L., Noë, V., Bruyneel, E., Myssiakine, E., Chastre, E., Mareel, M. and Gespach, C. (1998) J. Biol. Chem. **273**, 14138–14145

64. Shibamoto, S., Hayakawa, M., Takeuchi, K., Hori, T., Oku, N., Miyazawa, K., Kitamura, N., Takeichi, M. and Ito, F. (1994) Cell Adhes. Commun. **1**, 295–305

65. Pepper, M.S., Matsumoto, K., Nakamura, T., Orci, L. and Montesano, R. (1992) J. Biol. Chem. **267**, 20493–20496

66. Cartwright, C.A., Coad, C.A. and Egbert, B.M. (1994) J. Clin. Invest. **93**, 509–515

67. Abu-Amer, Y., Ross, F.P., Schlesinger, P., Tondravi, M.M. and Teitelbaum, S.L. (1997) J. Cell Biol. **137**, 247–258

68. Takahashi, K., Suzuki, K. and Tsukatani, Y. (1997) Oncogene **15**, 71–78

69. Vermeulen, S., Van Marck, V., Van Hoorde, L., Van Roy, F., Bracke, M. and Mareel, M. (1996) Pathol. Res. Pract. **192**, 694–707

70. Bracke, M.E., Vyncke, B.M., Bruyneel, E.A., Vermeulen, S.J., De Bruyne, G.K., Van Larebeke, N.A., Vleminckx, K., Van Roy, F.M. and Mareel, M.M. (1993) Br. J. Cancer **68**, 282–289

71. Takeda, H., Nagafuchi, A., Yonemura, S., Tsukita, S., Behrens, J., Birchmeier, W. and Tsukita, S. (1995) J. Cell Biol. **131**, 1839–1847

72. Friedlander, D.R., Mège, R.M., Cunningham, B.A. and Edelman, G.M. (1989) Proc. Natl. Acad. Sci. U.S.A. **86**, 7043–7047

73. Gumbiner, B. and Simons, K. (1986) J. Cell Biol. **102**, 457–468

74. Rebel, J.M.J., Thijssen, C.D.E.M., Vermey, M., Delouvée, A., Zwarthoff, E.C. and Van der Kwast, T.H. (1994) Cancer Res. **54**, 5488–5492

75. Blaschuk, O.W., Sullivan, R., David, S. and Pouliot, Y. (1990) Dev. Biol. **139**, 227–229

76. Mège, R.M., Goudou, D., Diaz, C., Nicolet, M., Garcia, L., Geraud, G. and Rieger, F. (1992) J. Cell Sci. **103**, 897–906

Biochem. Soc. Symp. **65**, 63–78
Printed in Great Britain

5

Towards a structural model of an integrin

Martin J. Humphries

Wellcome Trust Centre for Cell-Matrix Research, School of Biological Sciences, University of Manchester, 2.205 Stopford Building, Oxford Road, Manchester M13 9PT, U.K.

Abstract

Integrins are currently viewed as the principal family of extracellular matrix receptors. The interactions mediated by integrins are responsible for certain typical properties of adhesive cells, such as attachment and migration, but these molecules are also recognized to contribute to intracellular signalling processes, either by transducing signals themselves or by enabling and/or co-ordinating signalling via other receptor systems. As yet, the structural basis of integrin function is unknown, although detailed computer-based predictions have suggested working models for integrin tertiary structure. In this chapter, I will review this information and discuss recent studies examining the molecular basis of integrin regulation using stimulatory and inhibitory monoclonal antibodies (mAbs). Through the use of sensitive isolated integrin-binding assays, stimulatory mAbs have been found to function either by inducing shape changes in integrins or by selectively recognizing and stabilizing active and ligand-occupied conformations of integrins, while blocking mAbs were found to be allosteric inhibitors of ligand binding that report specific ligand engagement events. This information has improved our understanding of the composition of the integrin ligand-binding pocket and the structural basis of integrin activation.

Introduction

Cells undergo a complex, bidirectional interaction with their adhesive environment. On the one hand, they sense and respond to the different matrices that they encounter, using this information to inform their decisions about whether to move or stay still; on the other, cells are able to regulate the adhesiveness of their own receptor complement, such that the cells acquire adhesive activity when it is needed. The ability to react in this way necessitates the existence of a dynamic adhesion receptor system that can be regulated in two directions: (1) as an environmental sensor, interacting in a graded manner with the extracellular matrix and converting this information into cytoplasmic signals, and (2) as a guidance system permitting cells to anchor themselves or to migrate when physiology demands it (Fig. 1)

Integrins endow cells with these properties. They are one of the most, if not the most, complex family of adhesion molecules (reviewed in [1–6]). All integrins are dimeric proteins, containing an α-subunit and a β-subunit. Currently, 17 α-subunits and eight β-subunits have been identified in mammals, and these combine into 23 different dimers. Integrins recognize a wide variety of extracellular matrix and cell surface proteins, and each integrin dimer displays a specific ligand-binding profile. In most cases, integrins bind multiple ligands with a consistent hierarchy of affinities.

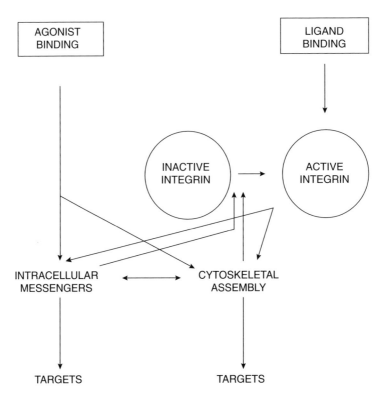

Fig. 1. Schematic representation of bidirectional signalling through integrins. As with all receptors, ligand occupancy of the active form of an integrin is hypothesized to transduce a signal or signals that results in alterations in second messenger pathways or in cytoskeletal architecture. In turn, these changes regulate downstream targets of adhesion, including cell migration and gene expression. Conversely, the binding of certain agonists to their cellular receptors can elicit signals that lead to the activation of integrins from within. As yet, the signalling molecules that interact proximally with integrins, and that are responsible for integrin activation and signalling, have not been identified.

As with most receptors, the ligand-binding activity of integrins can be regulated, but this is by an unusual combination of two different phenomena. First, integrins can cluster together in the plane of the plasma membrane; secondly, they can undergo conformational changes that correlate with changes in ligand-binding capacity (Fig. 1). Clustering can generate a higher avidity interaction with the ligand, but it can also have the same consequences for molecular interactions within the cell and consequently affect signalling. The combination of extracellular and intracellular regulation of integrin clustering and conformation not only provides a means of rapidly and dynamically regulating cell adhesion, but also makes the quest for a molecular understanding of integrin function an extremely difficult and complex one.

Arguably, the single most important advance that could be made in the adhesion field would be the solution of the three-dimensional structure of an

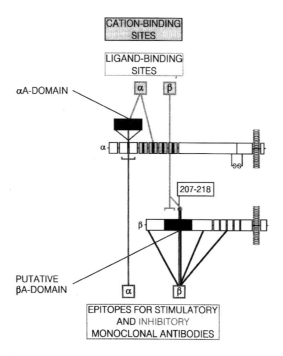

Fig. 2. Cartoon of integrin domain structure. The extracellular domains of the α- and β-subunits are shown to the left. The α-subunit contains a seven-fold repeated structure at its N-terminus, with the final three or four repeats containing a potential bivalent-cation-binding site. One group of α-subunits contains an inserted αA-domain; a different group contains a post-translational cleavage site near to the membrane. The β-subunit contains an A-domain-like region and a series of epidermal growth factor-like repeats. Both A-domains probably contain bound bivalent cations. Ligand-binding sites are found in both subunits (see the text for details). The epitopes for inhibitory and stimulatory mAbs are indicated by grey and black lines respectively. One region of the β_1 subunit (residues 207–218) contains epitopes for both classes of antibody.

integrin dimer, since this would shed light on the processes of integrin–ligand binding and activation. However, to date this goal has proved intractable. Both integrin subunits are large, varying in molecular mass from 90 to 180 kDa, they are glycosylated, and they are both intercalated in the plasma membrane. Each of these properties, coupled with the inherent conformational flexibility and the difficulty of isolating large amounts of integrins, has retarded progress. Although soluble, recombinant versions of integrins have been generated [7,8], as yet none of these have crystallized.

In the absence of a three-dimensional structure, a great deal of research effort has been focused on a biochemical dissection of integrin function. A number of complementary approaches have been used to identify the sites involved in ligand binding and receptor activation (Fig. 2). Data from covalent cross-linking of RGD (Arg-Gly-Asp) peptides, the effects of point mutations and the expression of recombinant fragments are all consistent with both integrin subunits making a contribution to the ligand-binding pocket (reviewed in [2,3]). Where present, the α-subunit A-domain appears to play a key role in ligand binding: to date, it is the only region of an integrin that has been expressed in isolation and shown to retain ligand-binding activity, and it also contains the epitopes for anti-functional mAbs [9–14]. Where there is no A-domain in the α-subunit, the N-terminal half of the subunit still appears to be important, since ligands cross-link to this region and anti-functional antibodies map there. For all integrins, the putative β-subunit A-domain plays a major role; ligands cross-link to this region and it contains epitopes for both inhibitory and stimulatory mAbs (see [2,3,15]). Finally, the likely presence of cation-binding sites within both the seven-fold repeat that makes up the N-terminal half of the α-subunit and the α- and β-subunit A-domains is consistent with the bivalent-cation dependence of integrin–ligand binding [16].

Mechanism of action of stimulatory mAbs

The three-way link between ligand-binding capacity, bivalent cation occupancy and conformation has been reinforced by other studies which have investigated the mechanism of action of stimulatory anti-integrin mAbs. This subset of mAbs has the unusual property of stimulating receptor function. Thus addition of these mAbs to cells results in higher levels of attachment to immobilized adhesion proteins, and addition to purified integrins in solid-phase assays elicits higher levels of ligand binding. Previous studies had reported cation-regulated binding of stimulatory anti-integrin mAbs for β_3 integrins [17–19], and therefore we performed similar studies for β_1 integrins in order to investigate the relationship between bivalent-cation-binding sites and integrin conformation. One of the most intriguing findings that we obtained was that the different stimulatory mAbs responded differently to cations [19a].

Stimulatory anti-β_1 mAbs fell into two classes. Some mAbs, such as the frequently used TS2/16, were virtually unaffected by cations, while others (such as 9EG7, 15/7 and 12G10) recognized epitopes that were altered by cation binding. Most strikingly, the binding of one anti-β_1 mAb, 12G10, that was generated in this laboratory was exquisitely regulated by bivalent cations. Mn^{2+} was a potent promoter of epitope expression and Mg^{2+} also stimulated binding, but Ca^{2+} suppressed the formation of the site. Importantly, this over-

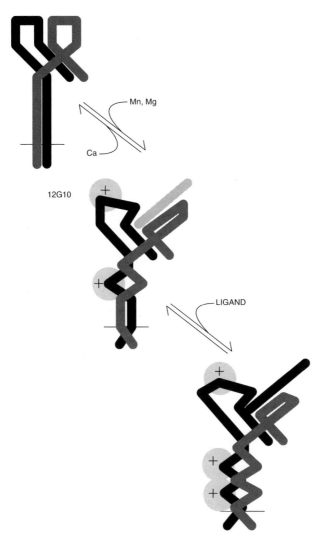

Fig. 3. Schematic model of the major conformational states of integrins, showing the effects of bivalent cation occupancy and ligand binding. The α-subunit is shown in grey and the β-subunit in black. Ca^{2+}-occupied integrin does not express epitopes for stimulatory mAbs (e.g. 12G10, 9EG7; shown by +), but Mg^{2+} or Mn^{2+} binding elicits a major conformational change and generates an active conformer with the potential to bind ligand. This change also leads to the expression of epitopes for some stimulatory mAbs. The activation step may also cause a re-orientation of integrin cytoplasmic domains and a concomitant change in basal signalling. Ligand engagement may induce further shape changes, also reported by stimulatory mAbs.

all pattern was remarkably similar to that found in parallel studies of the cation dependence of ligand binding. This finding indicates not only that cation occupancy alters the shape of integrins, and is required for ligand binding, but also that cations cause very specific changes in shape that result in the expression of epitopes that report the potential of the integrin to bind ligand. These findings suggest that it is possible to define the existence of three conformational states of an integrin: inactive, active and ligand-bound (Fig. 3). In turn, this suggests that some stimulatory mAbs, typified by 12G10, enhance ligand binding by selectively recognizing and stabilizing active and ligand-bound forms of the integrin. The other epitopes for stimulatory mAbs that are not altered by bivalent cations probably function by forcing integrins to change shape into an active conformation.

Mechanism of action of inhibitory mAbs

mAbs with the ability to perturb integrin function have been extremely useful as probes for studying the contribution of integrins to biological function. In addition, since it has been assumed, almost always implicitly, that the antibodies work by sterically blocking ligand binding, they have also been used as a means of pinpointing active sites within integrins. The argument is that, by identifying the epitopes recognized by anti-functional mAbs, it can be predicted that the ligand contact sites will lie close by. The key assumption that mAbs act sterically is reasonable, but it is possible that they might have other modes of inhibition. Furthermore, several paradoxes exist which are difficult to reconcile with the concept of steric inhibition. Most obviously, an explanation is needed for the observation that a short 12-amino-acid stretch of the β_1 subunit (residues 207–218) contributes to the epitopes of both inhibitory and stimulatory mAbs [20]. Since it is assumed that stimulatory mAbs bind simultaneously with ligand, either inducing or stabilizing a conformational change that correlates with ligand-binding potential, it is difficult to imagine how an epitope for an inhibitory antibody might overlap. Nonetheless, a large number of studies have reported the mapping of mAb epitopes using a combination of inter-species chimaeras and point mutants [21–26]. Taking β_1 integrins as an example, the results point to the centre of the putative β-subunit A-domain and an extended region, including sites in both the second and third blades of the putative α-subunit propeller, as containing these epitopes.

Initial studies with the anti-β_1 mAb 13 showed that its inhibitory effect on ligand binding in an isolated receptor–ligand binding assay could not be outcompeted by raising the ligand concentration [27]. Specifically, in experiments measuring the ability of RGD peptide to block the binding of mAb 13 to immobilized $\alpha_5\beta_1$ integrin, for a 10-fold increase in antibody concentration there was a <2-fold shift in the concentration of RGD peptide required for half-maximal inhibition. Importantly, these findings indicated that mAb 13 was acting as an allosteric, rather than a competitive, inhibitor. Double-reciprocal plots demonstrated that ligand binding reduced the affinity of antibody binding by approx. 50-fold, suggesting that ligand binding led to the destruction of the mAb 13 epitope, and that antibody and ligand binding might be

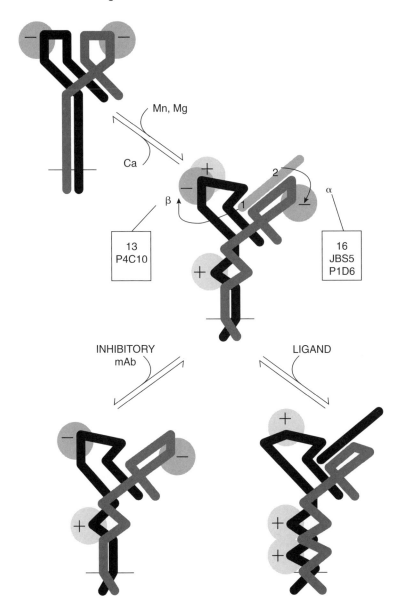

Fig. 4. Both the inactive and active conformers of the $\alpha_5\beta_1$ integrin contain epitopes for inhibitory mAbs. The inhibitory mAbs ($-$) shown are mAb 13 and P4C10 on β_1; and mAb16, JBS5 and P1D6 on α_5. Ligand binding to active integrin via sites (1 and 2) on both subunits destroys the epitopes for inhibitory mAbs on both subunits and induces further stimulatory/ligand-induced binding site epitopes (bottom right). Ultimately, these changes are propagated across the plasma membrane to re-orient cytoplasmic domains into a full signalling conformation. Binding of inhibitory mAbs to active integrin induces a shape change which destroys the ligand-binding pocket and may also change the cytoplasmic domain conformation (bottom left). Ligand binding and inhibitory mAb binding are mutually exclusive.

mutually exclusive. This was proven to be the case when the binding of both molecules was performed in parallel. mAb 13 can therefore be considered to recognize a ligand-attenuated binding site. The mechanism by which mAb 13 blocks ligand binding appears to be an allosteric disruption of the ligand-binding pocket. Thus, although the epitope itself might not overlay the ligand-binding site, it is an important part of the ligand-binding process.

These studies were then extended to other anti-functional anti-integrins, directed against both the α- and β-subunits. Similar results were obtained for P4C10 (anti-β_1) and JBS5, P1D6 and mAb 16 (all anti-α_5) [28]. Thus both integrin subunits appear to contain key sites that alter their shape in response to ligand binding. The fact that all of the anti-integrin mAbs tested to date work in this way suggests that their epitopes are located in highly antigenic regions of the molecule, and that the active-site pocket is either non-immunogenic or inaccessible to antibodies. Importantly, it can be predicted that anti-functional anti-integrin antibodies work by stabilizing regions of the receptor that need to change shape in order to accommodate ligand, and therefore preclude this from taking place (Fig. 4). The discovery that these mAbs are allosteric inhibitors explains the paradox that epitopes for stimulatory and inhibitory anti-β_1 mAbs map to the same place; they partially recognize different conformations of the same peptide.

The fact that ligand binding results in the structural perturbation of sites in both integrin subunits suggests that specific receptor-binding sites within the ligand are responsible for these changes. The active sites within the ligand that interacts with $\alpha_5\beta_1$ integrin, i.e. fibronectin, are well characterized; the major site is the RGD tripeptide located in the tenth type III repeat of the molecule [29], but this works in concert with a so-called synergy site, represented by the sequence Pro-His-Ser-Arg-Asn (PHSRN) in type III repeat 9 [30]. To examine which regions of the fibronectin molecule are responsible for shape changes in the α_5 and β_1 subunits as a consequence of ligand engagement, we expressed a series of recombinant fragments containing various combinations of active sites. These fragments were then tested for their ability to perturb the binding of mAbs directed against either integrin subunit. The results showed that fragments containing the RGD sequence alone, the RGD peptide itself and the snake venom disintegrin kistrin were only able to block anti-β_1 mAb binding, whereas wild-type fragments mutated specifically in the synergy region lost the ability to block anti-α_5 mAb binding [28]. These results imply that binding of RGD to β_1 destroys inhibitory mAb epitopes on that subunit, while synergy engagement blocks binding of anti-α mAbs. These results provided the first direct evidence that the topology of binding of fibronectin active sites is RGD to β_1 and synergy to α_5.

Role of bivalent cations

Since it is clear that ligand binding and bivalent cation occupancy are intimately linked, it is of interest to determine their precise mode of action. Several models have been presented in which bivalent cations are involved in a direct bridging between receptor and ligand [31,32]. In this scenario, the aspartate or glutamate residues that are a critical feature of the active sites of many integrin ligands are viewed as providing a key co-ordination group for an integrin-bound cation. Although this may indeed take place, the models do not take into account the contribution of multiple cation-binding sites to ligand binding. An equally likely possibility is that cations regulate integrin function allosterically.

To investigate this idea, we have examined the cation dependence of the binding of function-blocking mAbs to integrin $\alpha_5\beta_1$. The binding of some, but not all, mAbs was cation-sensitive, but for those antibodies that were affected, the concentration dependence of cation-regulated mAb binding correlated closely with the concentration dependence of ligand binding. This finding suggested that the same sites were involved in both processes. Interestingly, opposing effects of cations were seen on anti-functional anti-α and anti-β mAbs. For example, the binding of the anti-functional anti-β_1 mAb 13 was inhibited by Mn^{2+} and Ca^{2+}, and slightly by Mg^{2+}, while binding of anti-functional anti-α_5 mAbs, such as JBS5 and mAb 16, was stimulated by the cations in the same relative order. This suggested that the cations had opposing effects on the two subunits, and to test this idea the effect of cations on cross-blocking between anti-α and anti-β mAbs was examined. The feasibility of these studies was increased by the finding that the RGD and PHSRN sites in fibronectin specifically perturbed the binding of anti-β and anti-α mAbs respectively, because the distance between these two sites is known from the crystal structure of the region, and this constrains the α–β separation to 3–4 nm, a distance likely to result in antibody cross-blocking.

As predicted, anti-functional anti-α_5 and anti-β_1 mAbs inhibited each other's binding to $\alpha_5\beta_1$. The binding of anti-α_5 mAbs was almost totally inhibited by other anti-α_5 mAbs, but was only partially inhibited by anti-β_1 mAbs. The converse was also true. When the effect of cations on cross-blocking was tested, the extent of inhibition was altered; in the presence of EDTA, substantial cross-inhibition was observed, while in the presence of any cation, and particularly Mn^{2+}, the extent of cross-inhibition was lower. These findings were interpreted to suggest a role for cations in increasing the distance between the α- and β-subunits, thereby opening up the integrin dimer and exposing ligand-binding sites.

Taken together, these results also suggest that cations in general, and Ca^{2+} ions in particular, have at least two distinct effects on integrin conformation that affect ligand-binding potential. First, Ca^{2+} stimulates opening of the integrin dimer, since it enhances expression of the epitopes for anti-functional anti-α_5 mAbs and reduces expression of those for anti-functional anti-β_1 mAbs, and because it reduces the ability of anti-α and anti-β mAbs to cross-block each other's binding. Secondly, Ca^{2+} suppresses ligand binding, since it fails to induce expression of the 12G10 epitope. As described above, this epitope is a reporter for the ability of an integrin to bind ligand, and the data suggest that Ca^{2+} has a local effect on the conformation of the region containing the 12G10 epitope that results in its disruption. Thus, although Ca^{2+} may elicit integrin α-subunit activation, it fails to induce a conformation in the β-subunit that is compatible with ligand binding.

Integrin structure

Much is now known about the sites involved in integrin activation and ligand binding, but a major inhibition to further progress is the lack of a structural perception of these sites. In the absence of authentic structural information, several laboratories have resorted to structure prediction. Many algorithms are available for these studies, some more accurate than others, and data obtained with any of them must be treated with some caution. Nonetheless, the availability of a model allows predictions to be tested and can be of enormous value in guiding structure–function studies.

Several years ago, we adopted an approach based on multiple alignments to predict the structure of the N-terminal half of integrin α-subunits [33]. As described above, this region is implicated in ligand binding, and is composed of seven homologous polypeptide modules, each of approx. 60 amino acids. The last three or four of these repeats, depending on the α-subunit, contain linear, acidic sequences that resemble the EF-hand cation-binding motif characterized in calmodulin and parvalbumin. In addition, in a subset of integrins there is a polypeptide module of approx. 200 amino acids inserted between the second and third modules that is homologous to the A-domain found in von Willebrand factor, other extracellular matrix proteins and several complement factors. Any model of this region of the integrin needs to account both for the seven-fold repeat and for the presence or absence of an α-subunit A-domain.

At the time of our original studies, 16 integrin α-subunit sequences were available, making a total of 112 repeat modules. An optimized multiple alignment of these sequences, which included several gaps, was generated by a combination of computer-based fitting and manual adjustment. By then calculating the level of conservation and noting the types of amino acid tolerated at different positions, it was possible to apply several criteria to the alignment which would lead to a secondary structure prediction. These criteria were as follows: (i) regions where there are insertions or deletions mark the borders of elements of secondary structure; (ii) within stretches of sequence where there is high identity, substitutions of charged residues can only be tolerated at the surface of the protein; and (iii) within stretches of sequence of low identity, conserved hydrophobic residues are likely to be present in the interior of the protein. This prediction was compared with results obtained from other secondary structure algorithms to produce the consensus model shown in Fig. 5. This model consists of three (possibly four) anti-parallel β-strands and possibly one short α-helix. It was also predicted that the seven modules would align in parallel, either as a rod or, more likely, as a single structure requiring contributions from all repeats to fold. The latter idea is consistent with previous studies that analysed the protease susceptibility of integrins and their low-resolution structure by rotary-shadowed electron microscopy.

Recently, Springer built on this work and extended it to predict the structure for the entire seven-fold repeat region (Fig. 5) [15]. The structure that appears most likely is a β-propeller, in which each of the seven repeats forms a blade of a cyclical propeller. In this structure, the regions previously implicated in ligand binding through analyses of mAb epitopes locate to surface loops

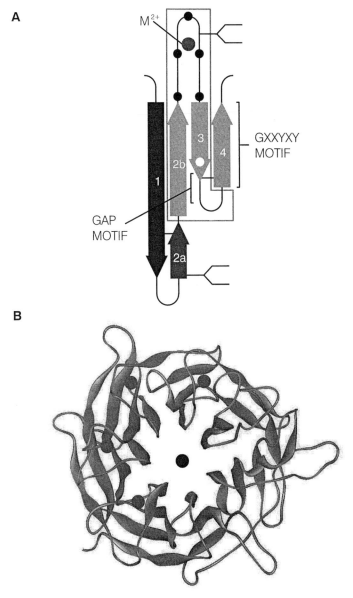

Fig. 5. Structure prediction for the N-terminal seven-fold repeat in integrin α-subunits. (A) This structure is based on multiple alignments and incorporates constraints from biochemical data (based on Tuckwell et al. [33]). Elements 1, 2b, 3 and 4 represent β-strands. Element 2a represents an α-helix. The cation-binding loop is shown co-ordinated to a bivalent ion, M^{2+}. (B) Based on the model of Springer [15], this region is predicted to fold into a β-propeller domain containing seven four-stranded β-sheet blades arranged in a doughnut-like ring. Where present, an αA-domain would be inserted between blades 2 and 3. A putative bivalent cation is shown as a small sphere at the centre of the ring, and cations co-ordinated by three EF-hand-like sequences are shown as large spheres on the left.

A

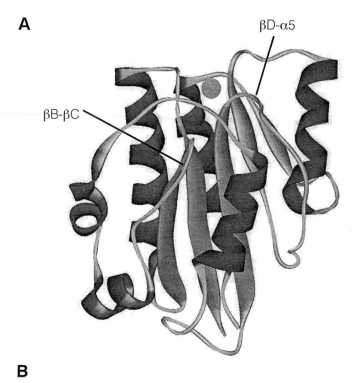

βD-α5

βB-βC

B

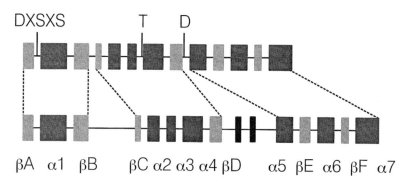

DXSXS   T   D

βA α1 βB   βC α2 α3 α4 βD   α5 βE α6 βF α7

Fig. 6. Structures of α- and β-subunit A-domains. (A) Ribbon diagram showing the tertiary structure of the A-domain from the α1 subunit as determined by X-ray crystallography. The module adopts a Rossmann (or dinucleotide-binding) fold, with a central sheet of β-strands decorated peripherally by six α-helices. At one end of the module, a series of loops co-ordinate to a bivalent cation. Since both integrin–ligand and A-domain–ligand binding are dependent on cations, this site appears to be of major importance. (B) Schematic representation of the predicted structure of an A-domain-like region in integrin β-subunits. The region is predicted to contain the same secondary structure elements (boxes) as an A-domain, and to have these arranged in the same order, but to contain two insertions between the βB and βC strands and the βD strand and α5 helix. These loops are shown in (A).

(joining β-strands 2 and 3, and 4 and 1) on the top of the propeller. Intriguingly, where present, the α-subunit A-domain would be inserted between the second and third blades, and would lie on top of the propeller.

Several crystal structures of A-domains from integrin α-subunits are now available [11–13,34]. In each case, the module is homologous to a Rossmann or dinucleotide-binding fold. The core of the module is made up of six β-strands, and these are decorated peripherally by a series of seven α-helices. At one end of the module there is a bivalent cation co-ordinated directly or indirectly by residues from three loops. Mutations in each of these co-ordinating residues have been shown to abolish ligand binding, suggesting either an important contribution of this site to direct ligand engagement or a role for it in maintaining the conformation of the ligand-binding site.

Binding of ligand and of function-regulating anti-$β_1$ mAbs appears to involve a highly conserved region close to the N-terminal end of the subunit. When the structure of an α-subunit A-domain was first published, Liddington and co-workers noted a similarity, at least in hydrophilicity, of the conserved region in the β-subunit to an A-domain [12]. Subsequent structure predictions support this conclusion, but point to a slight modification of the overall module, possibly including two peptide insertions (Fig. 6). Intriguingly, the use of a β-propeller and a Rossmann fold to form the ligand-binding pocket of a multimeric protein also occurs in heterotrimeric G-proteins. In this case, the Rossmann fold in the α-subunit forms a direct contact via its α2–α3 helix region with the top of the β-subunit propeller, maintaining the complex in an inactive state. Activation of the G-protein via a change in nucleotide binding causes an alteration in the conformation of the α2–α3 helix (the so-called switch region), and a separation of the α- and β-subunits. It is tempting to speculate that an integrin may share the domain organization of such a G-protein, and it is particularly notable that the region of the integrin β-subunit that alters its shape in response to the binding of regulatory mAbs locates to the putative α2–α3 helix region (A.P. Mould and M.J. Humphries, unpublished work; [19a]).

In the current structural model of the integrin dimer presented in Fig. 7, the activation process is pictured as a separation of the two subunits, and an exposure of the ligand-binding sites in the α-subunit propeller. This conformational change can be induced by bivalent cations, suggesting that there is a conformational link between cation-binding sites at the base and putatively in the centre of the propeller and in the β-subunit A-domain, and the ligand-binding pocket. It is currently unclear whether cation regulation is a physiological process, or whether it is accomplished by the binding of other molecules. The similarity of the integrin to a G-protein makes this latter possibility both feasible and an exciting prospect.

Future prospects

In the immediate future, the availability of new structural models for the integrin dimer will aid further structure–function studies. Key tasks that remain include localizing the cation-binding sites that influence ligand binding and activation, the sites that interact directly with ligand and the sites that

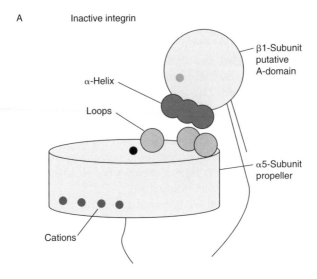

A   Inactive integrin

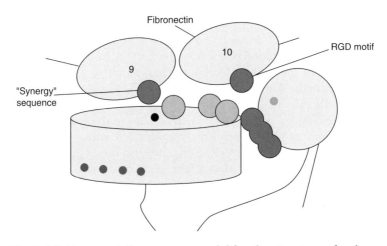

B   Ligand-occupied integrin

Fig. 7. Highly speculative cartoon model for the structure of an inactive (A) and ligand (fibronectin)-occupied (B) form of the $\alpha_5\beta_1$ integrin based on similarity to heterotrimeric G-proteins. In the inactive form of the integrin, the $\alpha 2$ helix of the β_1 subunit putative A-domain interacts with loops projecting from the top surface of the α_5 subunit β-propeller. Conformational changes in the $\alpha 2$ helix, induced by cation binding or addition of stimulatory mAbs that recognize this region, elicit a major repositioning of the integrin subunits, an opening of the integrin dimer and a concomitant exposure of ligand-binding sites in the α_5 subunit. The RGD motif in type III repeat 10 of fibronectin then engages the bound cation in the β-subunit A-domain, and the 'synergy' sequence in type III repeat 9 engages a site in the α_5 subunit β-propeller.

undergo conformational changes in response to activation and ligand binding. In addition, the model may even expedite the production of integrin crystals for X-ray studies by informing the construction of recombinant receptor domains. Recent data indicate that neither integrin subunit is able to fold in the absence of its partner, and therefore it may be necessary to co-express truncated versions of each. Finally, a major challenge will be to determine whether and/or how the conformational changes induced by integrin occupancy are translated into cytoplasmic signals.

References

1. Giancotti, F.G. (1997) Curr. Opin. Cell Biol. **9**, 691–700
2. Humphries, M.J. (1996) Curr. Opin. Cell Biol. **8**, 632–640
3. Loftus, J.C., Smith, J.W. and Ginsberg, M.H. (1994) J. Biol. Chem. **269**, 25235–25238
4. Schwartz, M.A., Schaller, M.D. and Ginsberg, M.H. (1995) Annu. Rev. Cell Dev. Biol. **11**, 549–599
5. Stewart, M., Thiel, M. and Hogg, N. (1995) Curr. Opin. Cell Biol. **7**, 690–696
6. Yamada, K.M. and Miyamoto, S. (1995) Curr. Opin. Cell Biol. **7**, 681–689
7. Briesewitz, R., Epstein, M.R. and Marcantonio, E.E. (1993) J. Biol. Chem. **268**, 2989–2996
8. Wippler, J., Kouns, W.C., Schlaeger, E.J., Kuhn, H., Hadvary, P. and Steiner, B. (1994) J. Biol. Chem. **269**, 8754–8761
9. Calderwood, D.A., Tuckwell, D.S., Eble, J., Kuhn, K. and Humphries, M.J. (1997) J. Biol. Chem. **272**, 12311–12317
10. Kamata, T. and Takada, Y. (1994) J. Biol. Chem. **269**, 26006–26010
11. Lee, J.O., Bankston, L.A., Arnaout, M.A. and Liddington, R.C. (1995) Structure **3**, 1333–1340
12. Lee, J.O., Rieu, P., Arnaout, M.A. and Liddington, R. (1995) Cell **80**, 631–638
13. Qu, A. and Leahy, D.J. (1995) Proc. Natl. Acad. Sci. U.S.A. **92**, 10277–10281
14. Randi, A.M. and Hogg, N. (1994) J. Biol. Chem. **269**, 12395–12398
15. Springer, T.A. (1997) Proc. Natl. Acad. Sci. U.S.A. **94**, 65–72
16. Gailit, J. and Ruoslahti, E. (1988) J. Biol. Chem. **263**, 12927–12932
17. Frelinger, III, A.L., Du, X., Plow, E.F. and Ginsberg, M.H. (1991) J. Biol. Chem. **266**, 17106–17111
18. Frelinger, III, A.L., Lam, S.C.T., Plow, E.F., Smith, M.A., Loftus, J.C. and Ginsberg, M.H. (1988) J. Biol. Chem. **263**, 12397–12402
19. Honda, S., Tomiyama, Y., Pelletier, A.J., Annis, D., Honda, Y., Orchekowski, R., Ruggeri, Z. and Kunicki, T.J. (1995) J. Biol. Chem. **270**, 11947–11954
19a. Mould, A.P., Garrett, A.N., Puzon-McLaughlin, W., Takada, Y. and Humphries, M.J. (1998) Biochem. J. **331**, 821–828
20. Takada, Y. and Puzon, W. (1993) J. Biol. Chem. **268**, 17597–17601
21. Irie, A., Kamata, T., Puzon-McLaughlin, W. and Takada, Y. (1995) EMBO J. **14**, 5550–5556
22. Irie, A., Kamata, T. and Takada, Y. (1997) Proc. Natl. Acad. Sci. U.S.A. **94**, 7198–7203
23. Kamata, T., Irie, A., Tokuhira, M. and Takada, Y. (1996) J. Biol. Chem. **271**, 18610–18615
24. Kamata, T., Puzon, W. and Takada, Y. (1994) J. Biol. Chem. **269**, 9659–9663
25. Kamata, T., Puzon, W. and Takada, Y. (1995) Biochem. J. **305**, 945–951
26. Kamata, T., Puzon, W. and Takada, Y. (1996) Biochem. J. **317**, 959 (erratum to the above)
27. Mould, A.P., Akiyama, S.K. and Humphries, M.J. (1996) J. Biol. Chem. **271**, 20365–20374
28. Mould, A.P., Askari, J.A., Aota, S., Yamada, K.M., Irie, A., Takada, Y., Mardon, H.J. and Humphries, M.J. (1997) J. Biol. Chem. **272**, 17283–17292
29. Pierschbacher, M.D., Hayman, E.G. and Ruoslahti, E. (1985) J. Cell. Biochem. **28**, 115–126
30. Obara, M., Kang, M.S. and Yamada, K.M. (1988) Cell **53**, 649–657

31.   Corbi, A.L., Miller, L.J., O'Connor, K., Larson, R.S. and Springer, T.A. (1987) EMBO J. **6**, 4023–4028

32.   Humphries, M.J. (1990) J. Cell Sci. **97**, 585–592

33.   Tuckwell, D.S., Humphries, M.J. and Brass, A. (1994) Cell Adhes. Commun. **2**, 385–402

34.   Emsley, J., King, S.L., Bergelson, J.M. and Liddington, R.C. (1997) J. Biol. Chem. **272**, 28512–28517

Biochem. Soc. Symp. **65**, 79–99
Printed in Great Britain

6

Integrin-mediated cell adhesion: the cytoskeletal connection

David R. Critchley[1], Mark R. Holt[2], Simon T. Barry[3], Helen Priddle[4], Lance Hemmings[5] and Jim Norman

Department of Biochemistry, University of Leicester, University Road, Leicester LE1 7RH, U.K.

Abstract

Members of the integrin family of cell adhesion molecules play a pivotal role in the interaction between animal cells and the extracellular matrix. This article reviews the evidence (i) that the integrin β-subunit cytoplasmic domain is important in the localization of integrins to focal adhesions, and for integrin-mediated cell adhesion/spreading; and (ii) that the integrin β-subunit can be linked to F-actin via the actin-binding proteins talin, α-actinin and filamin. Talin has two or more actin-binding sites, and three binding sites for the cytoskeletal protein vinculin. Because vinculin can also bind F-actin, it may cross-link talin and actin, thereby stabilizing the interaction. In addition, vinculin contains a binding site for VASP (vasodilator-stimulated phosphoprotein), a protein which may serve to recruit a profilin/G-actin complex to talin, which has actin-nucleating activity. Evidence that talin, vinculin and α-actinin are important in the assembly of focal adhesions, obtained using anti-sense technology and protein microinjection, is reviewed. To analyse the role of talin in focal adhesions, we have disrupted both copies of the talin gene in mouse embryonic stem (ES) cells. Undifferentiated talin $(-/-)$ ES cell mutants are unable to assemble focal adhesions when plated on fibronectin, whereas vinculin $(-/-)$ ES cells are able to do so. Finally, the role of small GTP-binding proteins in the assembly of focal adhesions is discussed, along with our

[1]To whom correspondence should be addressed.
[2]Present address: Department of Physiology, UCL, London WC1E 6JJ, U.K.
[3]Present address: Receptor Systems Unit, Glaxo Wellcome Medicines Research Centre, Gunnels Wood Road, Stevenage, Herts. SG1 2NY, U.K.
[4]Present address: Centre for Genome Research, University of Edinburgh, The King's Buildings, West Mains Road, Edinburgh EH9 3JQ, U.K.
[5]Present address: Perkin Elmer, Applied Biosystems Division, 7 Kingsland Grange, Woolston, Warrington, Cheshire WA1 4SR, U.K.

recent studies using streptolysin-O-permeabilized Swiss 3T3 cells which suggest that the GTP-binding protein ADP-ribosylation factor-1 (ARF-1) is important in targeting the protein paxillin to focal adhesions.

Introduction

The interaction between animal cells and the extracellular matrix (ECM) is important in a wide range of biological phenomena, including cell proliferation, suppression of apoptosis, cell migration, regulation of gene expression, differentiation, blood clotting and wound healing. As a consequence, there has been a major effort to define the mechanisms involved. Much of the initial progress in understanding the molecular basis of cell adhesion to the ECM has come from studying structures called focal adhesions that are formed when fibroblasts are cultured on rigid supports (Fig. 1). Cell–ECM interactions are frequently mediated by members of the integrin family of cell adhesion molecules ($\alpha\beta$ heterodimers), the cytoplasmic domains of which are thought to be linked to the actin cytoskeleton via a series of interacting proteins, including talin, vinculin and α-actinin. In this chapter, we will review the data that have contributed to the development of this model, and outline studies that we have undertaken to elucidate the structure and function of cytoskeletal proteins in integrin-mediated cell adhesion.

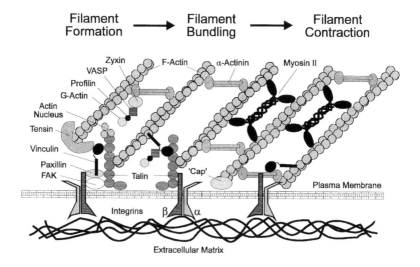

Fig. 1. The focal adhesion. The model depicts some of the interactions shown to occur *in vitro* between proteins that co-localize with integrins in focal adhesions. VASP, vasodilator-stimulated phosphoprotein; FAK, focal adhesion kinase.

Integrin β-subunit cytoplasmic domains are required for localization of integrins to focal adhesions, cell adhesion and cell spreading

Much of the work assessing the role of integrin cytoplasmic domains in mediating the link to cytoskeletal actin has been carried out on β_1 and β_3 integrins. A chimaeric protein containing the cytoplasmic domain of the β_1 or β_3 integrin subunit (but not the α_5 subunit) fused to the transmembrane and extracellular domains of the interleukin-2 receptor localized to focal adhesions when expressed in human fibroblasts [1]. A similar protein containing the β_5 integrin subunit cytoplasmic domain also localized to focal adhesions, but less efficiently [2]. When overexpressed, these chimaeric proteins behaved as dominant-negative mutants, blocking both the adhesive function of integrins and signalling induced by ligand binding [2,3]. Cross-linking the integrin β_1, β_3 and β_5 chimaeric proteins with an antibody to the interleukin-2 receptor stimulated phosphorylation of the focal adhesion tyrosine kinase (pp125^{FAK}) [4], confirming that the cytoplasmic domains of these integrins interact with the intracellular signalling machinery. Dominant-negative effects were also seen when the β_1 integrin cytoplasmic domain was expressed as a fusion protein with the transmembrane and extracellular domain of CD4 in cultured human embryonic kidney cells [5]. The CD4/β_1 integrin chimaera blocked adhesion mediated by both integrins β_1 and β_5 [5], indicating that the cytoplasmic domains of the two integrins interact with common cytoplasmic factors, even though the β_1 and β_5 heterodimers appear to have different biological roles. While β_1 integrins are associated with focal adhesions, β_5 integrins appear diffuse and are thought to be involved in cell migration [6].

Mutagenesis studies on the chicken β_1 integrin subunit expressed in mouse NIH-3T3 cells revealed that three regions of the cytoplasmic domain were required for localization to focal adhesions; these were named cyto 1 (amino acids 764–774), cyto 2 (785–788) and cyto 3 (797–800). Although the cyto 1 sequence is only found in β_1 integrins, the cyto 2 and cyto 3 sequences are found in other β-subunits, and contain two NPXY (Asn-Pro-Xaa-Tyr) motifs, which are predicted to form two tight turns. The first NPXY motif is conserved between all the β integrin subunits (with the exception of β_4), and the second is conserved between β_1, β_2 and β_7 [7], but it is somewhat divergent in the β_3, β_5 and β_6 subunits (Fig. 2). Disruption of either NPXY motif is sufficient to inhibit recruitment of β_1 integrins to focal adhesions [8]. Mutating the β_3 subunit NPXY motifs inhibits integrin function, blocking spreading and migration but without affecting ligand binding [9,10]. However, disruption of these sequences in $\alpha_L\beta_2$ [11,12] and β_6 [13] integrins does inhibit ligand binding. Mutation or deletion of other residues (723–733) within the cytoplasmic domain of the β_3 integrin subunit also has inhibitory effects on cell spreading and focal adhesion recruitment, suggesting that the NPXY motifs are not the only sequences required for targeting of integrins to focal adhesions [14]. Cell-permeable peptides have also been used to study the role of NPXY motifs in integrin function. Thus peptides corresponding to the second NPXY motifs of the β_1 and β_3 integrin subunits inhibited cell spreading in a cell-type- and

```
              α–actinin        talin
              _____        _____

β1Ac  KLLMIIHDRREFAKFEKEKMNAKWDTGENPIYKSAVTTVVNPKYEGK  (757-803)

β1Ah  KLLMIIHDRREFAKFEKEKMNAKWDTGENPIYKSAVTTVVNPKYEGK  (752-798)
β2    KALIHLSDLREYRRFEKEKLKSQWNND-NPLFKSATTTVMNPKFAES  (724-769)
β3A   KLLITIHDRKEFAKFEEERARAKWDTANNPLYKEATSTFTNITYRGT  (742-788)
β5    KLLVTIHDRREFAKFQSERSRARYEMASNPLYRKPISTHTVDFTFNKFNKSYNGTVD  (743-799)
β6    KLLVSFHDRKEVAKFEAERSKAKWQTGTNPLYRGSTSTFKNVTYKHREKQKVDLSTDC  (731-788)
β7    RLSVEIYDRREYSRFEKEQQQLNWKQDSNPLYKSAITTTINPRFQEADSPTL  (747-798)
```

Fig. 2. Alignment of the cytoplasmic domains of β integrins. Sequence alignment of the integrin β-subunit cytoplasmic domains from the membrane proximal lysine. $β_{1A}$ is one of four cytoplasmic domain splice variants; $β_{1A}$c represents the chicken cytoplasmic domain sequence, and $β_{1A}$h is the equivalent human sequence. All other sequences are from the human. The residue numbers are in parentheses. The major binding sites for α-actinin (FAKFEKEKMN) and talin (WDTGENPIYK) in the chicken $β_1$ integrin cytoplasmic domain, as determined using in vitro binding assays, are indicated.

matrix-dependent manner. In the endothelial cell line ECV304, a $β_3$ peptide (but not a $β_1$ peptide) inhibited adhesion to vitronectin (mediated by $αvβ_3$), whereas in fibroblasts a $β_1$ peptide (but not a $β_3$ peptide) inhibited adhesion to fibronectin (which is mediated by $β_1$ integrins) [15].

A number of isoforms of $β_1$ and $β_3$ integrins have been identified which arise through alternative splicing of exons encoding cytoplasmic domain sequences. The A, B, C and D variants of $β_1$ integrin differ only in the C-terminal half of the cytoplasmic domain. Integrin subunits $β_{1A}$ and $β_{1D}$, which contain the two conserved NPXY motifs, localize to focal adhesions and support cell spreading [16], while $β_{1B}$, which has neither NPXY motif, acts in a dominant-negative manner, blocking adhesion and migration [17,18]. Taken together, these results provide clear evidence that the cytoplasmic domains of integrins are important for integrin function and localization.

Evidence that integrin cytoplasmic domains associate with the actin cytoskeleton

The view that activated integrins do associate with the actin cytoskeleton is supported by the following observations. A subpopulation of the integrin $α_{IIb}β_3$ on activated platelets is associated with the Triton X-100-insoluble cytoskeletal fraction [19]. Ligand binding resulted in redistribution of the integrin from a detergent-insoluble fraction containing talin and other cytoskeletal proteins to a fraction which also contained actin filaments. This transition is suggested to stabilize the high-affinity ligand-bound state of this integrin, a process that can be inhibited by disrupting the actin cytoskeleton by the addition of cytochalasin D [20]. Disruption of the actin cytoskeleton affects the function of a number of integrins. In activated lymphocytes, strong binding of the $β_2$ integrin LFA-1 (lymphocyte function-associated antigen-1) to ICAM-1 (where ICAM is intercellular adhesion molecule) is blocked by cytochalasin D [21], as well as by deletion of the β integrin cytoplasmic domain [11,12,22].

Interestingly, LFA-1 appears to be negatively regulated by the actin cytoskeleton in resting lymphocytes, as treatment with cytochalasin D stimulates binding to ICAM-1 [21]. The ability of β_1 integrins in fibroblasts to support assembly of a fibronectin matrix can also be inhibited by cytochalasin D, as well as by deletion of the cytoplasmic domain of the β_1 subunit [23]. Similarly, deletion of the cytoplasmic domain of the β_3 subunit of integrin $\alpha_{IIb}\beta_3$ inhibits fibrin clot retraction, without blocking fibrinogen binding [24], presumably by preventing coupling to the actomyosin contractile machinery.

The dynamic properties of integrins at the leading edge of migrating mouse fibroblasts expressing chicken β_1 integrin has been investigated by engaging the integrins with either non-blocking antibodies or the β_1 integrin-binding domain of fibronectin (residues 7–10) coupled with various particles [25]. In the absence of ligand, integrins diffused freely in the membrane around the leading edge. However, upon binding of fibronectin-coated particles, the integrins ceased diffusing and began to move rearwards, away from the leading edge, demonstrating that ligand-mediated activation of the integrin initiates coupling to the cytoskeleton. Particles bound via the non-blocking antibody also showed cytoskeletal engagement if the fibronectin-related peptide GRGDS was added to mimic ligand occupancy. Moreover, these effects were inhibited by deleting the cytoplasmic domain of the β_1 integrin. These observations have been extended using laser trap technology to position and restrain antibody- or matrix-coated beads on the dorsal surface of cells [26]. The level of integrin occupancy/activation was found to determine both the stability of the linkage to the cytoskeleton and the strength of this association [27].

Beads coated with anti-integrin antibodies or GRGDS peptides have been used to investigate the effects of integrin clustering and ligand binding on the localization of various cytoplasmic proteins in human foreskin fibroblasts [28,29]. Clustering of integrins on the apical surface of these cells (in the presence of a tyrosine kinase inhibitor) was sufficient to induce co-clustering of pp125[FAK] and tensin, suggesting that these two proteins are constitutively associated with β_1 integrin. Removal of the kinase inhibitor resulted in the further co-localization of a variety of signalling molecules with the clustered integrin, including members of the Src family of protein tyrosine kinases, Rho family GTPases, members of the lipid signalling and mitogen-activated protein kinase pathways, and cortactin. Ligand occupancy was required for co-localization of proteins such as talin and vinculin, which are thought to link integrins to actin filaments. Integrin clustering and addition of a soluble peptide ligand (in the presence of a tyrosine kinase inhibitor) induced the co-localization of talin, vinculin and α-actinin to the cell–bead interface, along with tensin and pp125[FAK]. However, the further recruitment of paxillin, filamin and actin was only achieved when the integrins were clustered and occupied in the absence of the tyrosine kinase inhibitor. These observations suggest that, to recruit cytoskeletal proteins, the integrin must be in the ligand-bound state, perhaps in the open-hinge conformation ([30]; reviewed in [31]), revealing binding sites for cytoskeletal proteins in the integrin cytoplasmic domain. However, the integrin must be in its highest activation state to recruit cytoskeletal actin. These results appear to be consistent with the observations of Sheetz and colleagues (see Chapter 14 in the present volume), which also suggest that integrin clustering and ligand occupancy, rather than clustering alone, is required for association of integrins with the actin cytoskeleton.

Integrin α-subunits appear to play a negative role in regulating integrin association with the actin cytoskeleton, and prevent the integrin heterodimer defaulting to focal adhesions [24]. The cytoplasmic domain of the α-subunit is thought to mask the β integrin cytoplasmic domain unless the heterodimer is in the ligand-bound state, when the two domains move apart to reveal binding sites for cytoplasmic proteins [30,32,33]. Although little attention has been given to the α-subunits, they are not generally regarded as contributing directly to the link between integrins and the cytoskeleton. However, the α_2 subunit has been shown to bind actin *in vitro* [34], and α_{IIb} also interacts with talin *in vitro* [35].

Cytoskeletal proteins that bind to the cytoplasmic face of integrins

Talin was the first cytoplasmic protein found to interact *in vitro* with β_1 integrins [36], via residues 780–789 (chicken protein) [37]. Talin also binds directly to the cytoplasmic domain of β_2 [38] and β_3 integrins, as well as the α_{IIb} subunit [35]. The interaction between talin and integrin within the cell requires both NPXY motifs of the chicken β_1 subunit (residues 780–789 and 791–799) [7,39]. Talin appears to bind as a homodimer, providing a potential mechanism by which talin might cross-link integrins [8]. It has proved difficult to demonstrate the association between integrins and cytoskeletal proteins by co-immunoprecipitation, but talin has recently been shown to co-immunoprecipitate with β_1 integrin, although only if the integrin ligand GRGDS was included in the lysis buffer. Although disruption of either NPXY motif in the β_1 integrin subunit is sufficient to inhibit localization to focal adhesions, talin binding is only abolished if both NPXY motifs are mutated [8]. This suggests that the interaction of integrins with talin is by itself sufficient to localize β_1 integrins to focal adhesions.

The actin cross-linking protein α-actinin provides another potential link between integrins and the actin cytoskeleton. *In vitro*, α-actinin binds to β_1 integrin cytoplasmic domain peptides [40], and co-immunoprecipitates with β_2 integrins [38,41]. The binding site within the chicken β_1 subunit has been mapped to two regions (residues 768–778 and 785–794) [42], and in the β_2 subunit to residues 733–742 [38]. These data are in close agreement with the α-actinin binding sites identified from *in vivo* studies, where residues 776–790 in chicken β_1 integrin are critical [39].

Filamin [also known as ABP-280 (actin-binding protein of 280 kDa)] is a large actin-cross-linking protein that binds to β_2 integrin cytoplasmic domain peptides and co-immunoprecipitates with the β_2 integrin Mac-1, and it may directly anchor Mac-1 to the cytoskeleton [38]. The binding site for filamin (residues 723–727) appears to be N-terminal to, but overlaps with, the α-actinin-binding site (residues 733–742). Whether filamin interacts with other than β_2 integrins remains to be established. Under conditions where talin co-immunoprecipitates with β_1 integrin, α-actinin was absent [8]. Conversely, where filamin and α-actinin co-immunoprecipitated with β_2 integrin, no talin was detectable [38]. These data suggest that the cytoskeletal proteins that link integrins to the actin cytoskeleton may vary depending on the integrin subclass involved.

Structure of integrin-associated cytoskeletal proteins

Talin

The cytoskeletal protein talin was originally discovered as a component of focal adhesions in cultured cells, although it is also found in membrane ruffles in certain cell types [43]. Unlike vinculin, it is not found in adherens-type cell–cell junctions [44], although it is present in the intercellular junctions formed between T-cells and antigen-presenting cells [45]. Nucleotide sequences for mouse, chicken, *Caenorhabditis elegans* and *Dictyostelium discoideum* talins have been published (see [46]). Chicken and mouse talin cDNAs both encode polypeptides containing 2541 amino acid residues, with a predicted molecular mass of 270 kDa. The chicken and mouse proteins show 89% identity, and comparisons with *D. discoideum* talin show that the N- and C-terminal regions are the most conserved [46].

Analysis of talin by electron microscopy reveals a rod-shaped molecule (56 ± 7 nm) composed of a series of globular domains, and chemical cross-linking data suggest that it is a dimer [47], although whether the subunits are parallel or anti-parallel is a matter of controversy. It can be separated into an N-terminal 47 kDa fragment and a 190 kDa C-terminal fragment containing multiple alanine-rich repeats [48] by the calcium-dependent protease calpain II, which cleaves between residues 433 and 434 [49] (Fig. 3). Such cleavage occurs upon platelet activation [50], but the significance of this is unclear. Intact talin, and both the 47 kDa and 190 kDa fragments, localize to cell–matrix junctions when microinjected into cells [51], whereas the 190 kDa fragment also targets cell–cell contacts. This suggests that the 47 kDa N-terminal region of the protein is an important determinant of the intracellular localization of talin. It is known to interact with charged lipids and can bind to membranes [52]. Residues 165–373 within the N-terminal region of talin share identity with the band 4.1, ezrin, radixin, moesin (ERM) family of actin-binding proteins [53]. ERM proteins are also found in focal adhesions and may provide an additional link between the actin cytoskeleton and the membrane. The N-terminal region of band 4.1 protein, which acts as a linker between the membrane and the spectrin/actin-based cytoskeleton in erythrocytes, contains binding sites for the cytoplasmic domain of the integral membrane protein glycophorin C as well as for protein p55, a peripheral membrane protein of the MAGUK family [54]. This raises the possibility that the N-terminal region of talin also contains binding sites for both an integral and a peripheral membrane protein. Indeed, this region of talin will bind to pp125[FAK] [55].

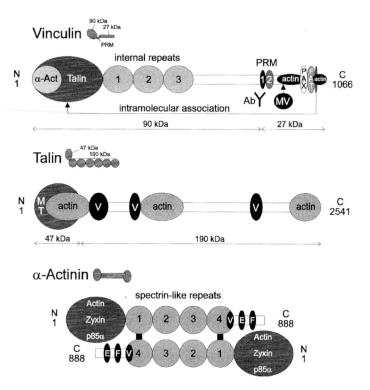

Fig. 3. Structural features of the cytoskeletal proteins vinculin, talin and α-actinin. Ellipsoids represent the protein-binding sites within the molecule, and the name(s) of the binding partner(s) is/are indicated. The co-ordinates of the protein-binding sites are given the the text. Abbreviations: α-Act, α-actinin; MV, metavinculin splice variant; PRM1 and PRM2, proline-rich sequences in vinculin; PAX, paxillin; FAT, focal adhesion targeting sequence; MT, membrane targeting sequence; V, vinculin; p85α, p85 subunit of phospho-inositide 3-kinase; EF, EF-hand motif; Ab, antibody. N and C indicate N- and C-termini respectively, with the residue numbers indicated. Double-headed arrows indicate proteolytic fragments of the protein.

The 190 kDa fragment of talin binds to the cytoplasmic domain of β_1 and β_3 integrins [35–37], but this site has not been defined further. Evidence for the interaction of talin and the 190 kDa fragment with vinculin and F-actin is more extensive, and the binding sites for vinculin [56] and F-actin [46,57] have been defined in significant detail. Initial experiments identified three binding sites for vinculin within talin residues 489–656, 852–950 and 1929–2029 [56]. These have been further defined using the yeast two-hybrid system to residues 607–656, 852–889 and 1950–1969 (M.D. Bass and D.R. Critchley, unpublished work). The N-terminal sites show limited sequence identity, whereas the more C-terminal site shows identity with a vinculin-binding peptide identified using phage display [58]. At least three actin-binding sites have been identified in talin using glutathione S-transferase fusion proteins and actin co-sedimentation assays [46,57]. The smallest talin fusion proteins found to bind actin were those spanning residues 102–497, 952–1328 and 2304–2463. The latter site shares similarity with the yeast actin-binding protein Sla2P [59]. Talin polypeptides from this region will quantitatively inhibit binding of intact talin to actin [60], suggesting that the C-terminal actin-binding site plays the key role in the interaction between these two proteins. Talin and the 190 kDa fragment of talin have also been reported to possess actin nucleating activity [61]. The fact that talin is recruited to newly forming focal adhesions before vinculin [62], and is localized in membrane ruffles, suggests that it might play a role in both the recruitment of pre-existing actin filaments to the cytoplasmic domain of integrins and the *de novo* assembly of actin filaments.

Ezrin and moesin contain cryptic F-actin-binding sites which are hidden by an intramolecular association between the N-terminal region of the protein, which contains a membrane-binding site, and the C-terminal region, which contains the actin-binding site [63]. Moreover, it appears that ERM protein activity may be controlled by phosphatidylinositol 4,5-bisphosphate [PtdIns(4,5)P_2], as well as by the Rho family of small GTP-binding proteins [64–66]. It will be important to establish whether there is an intramolecular association between the N- and C-terminal regions of talin which obscures the binding sites for integrins, vinculin and F-actin. Although electron microscopy studies indicate that talin is generally rod-shaped, various other forms have been observed, depending on salt concentration and pH, including a globular species [67]. In contrast, the 190 kDa domain is always seen as an elongated rod, and it has been reported to bind vinculin with higher affinity than intact talin [68].

Vinculin

Vinculin (see Fig. 3) is a well-characterized cytoskeletal protein found in adherens-type cell–matrix and cell–cell contacts [44]. Human, mouse, chicken, *C. elegans* and *Drosophila melanogaster* sequences are available, and these show extensive similarity to one another. In vertebrates, vinculin contains 1066 amino acid residues, with an approximate molecular mass of 117 kDa [69]. Rotary shadowing and electron microscopy suggest that vinculin can be divided into an N-terminal globular head and a C-terminal rod-like domain or tail [70]. The globular head (90 kDa) can be liberated from the extended tail domain (27/29 kDa) by cleavage with the *Staphylococcus aureus* protease V8 [71], which cleaves at two points within the proline-rich domain spanning residues 837–878. *In vitro* studies have identified numerous binding sites within vinculin for other focal adhesion proteins, including talin, which binds to residues 1–258 [72], α-actinin [73], which may bind to residues 1–107 [74], paxillin [75], which binds to residues 978–1000 [76], F-actin, which binds to two separate sites (residues 893–985 and 1016–1066) [77], and vasodilator-stimulated phosphoprotein (VASP), which binds to the FPPPP motif present between residues 839 and 850 in the proline-rich domain [78,79]. Wood et al. [76] have also presented evidence for a focal adhesion targeting sequence (residues 1000–1028) that is separate from, but contiguous with, the paxillin-binding site. It is not known what binds to this sequence, although other molecules reported to bind vinculin include tensin [80], protein kinase C-α [81] and the polyphosphates $InsP_4$ and $InsP_6$ [82]. A muscle-specific isoform, metavinculin, arises from the inclusion of an alternatively spliced 204-nucleotide exon, which results in the insertion of an additional 68 amino acids between residues 915 and 916 of the human sequence [83]. It is unclear what additional properties are conferred on vinculin by this sequence.

Interestingly, residues within the C-terminal region of vinculin can bind to the N-terminal region of the same polypeptide [84]. This intramolecular interaction is mediated by residues 1–258 and 1029–1036 [85]. The intramolecular interaction inhibits the binding of vinculin to several of its ligands, including talin [84], α-actinin [74] and F-actin [85,86], but apparently not paxillin [87]. Acidic phospholipids, such as $PtdIns(4,5)P_2$, inhibit this intramolecular interaction *in vitro* [85,87], exposing binding sites for talin and F-actin. Predicted $PtdIns(4,5)P_2$ binding sites exist within residues 946–974 and 1020–1040 [88]. However, evidence that $PtdIns(4,5)P_2$ regulates the activity of vinculin within the cell is limited. *In vivo*, vinculin is thought to contain bound acidic phospholipids [89] and is purified with bound $PtdIns(4,5)P_2$ [90]. The level of $PtdIns(4,5)P_2$ bound to vinculin decreases upon stimulation of cells with platelet-derived growth factor (PDGF) [89], which activates phospholipase C-γ. It is possible that hydrolysis of $PtdIns(4,5)P_2$ leads to inactivation of vinculin and contributes to the dissolution of focal adhesions seen in response to PDGF [91]. Moreover, microinjection of an anti-$PtdIns(4,5)P_2$ antibody into serum-starved cells blocks formation of focal adhesions and stress fibres in response to serum [87].

α-Actinin

α-Actinin (see Fig. 3) is a rod-shaped, anti-parallel homodimer with a subunit molecular mass of approx. 100 kDa [92]. It is a member of the spectrin family of F-actin cross-linking proteins, which also includes dystrophin and atrophin [93]. The N-terminal globular head contains the actin-binding site, which can be subdivided into three regions [94]. The central region of the protein contains four spectrin-like repeats which are responsible for formation of the α-actinin homodimer. Dimerization is thought to be mediated by an interaction between the first spectrin-like repeat from one subunit and the fourth repeat from the other subunit [95]. The C-terminal region of the protein contains two EF-hand-like calcium-binding motifs. There are muscle and non-muscle isoforms of α-actinin, and in humans one smooth muscle and two skeletal muscle genes have been identified [96]. In chickens, non-muscle isoforms are expressed from the smooth muscle and skeletal muscle α-actinin genes, and reflect alternative splicing of an exon encoding the second part of the first EF-hand calcium-binding motif [97,98]. This may account for the different calcium sensitivities of the isoforms [99,100].

α-Actinin binds to vinculin via residues 713–749 at the end of the fourth repeat [73], and to peptides derived from integrin β_1 cytoplasmic domains via undefined sequences within the spectrin-like repeats [42]. α-Actinin also binds to the cytoplasmic regions of other cell surface receptors, including selectins [101], ICAM-1 [102] and ICAM-2 [103]. Zyxin, another focal adhesion component, binds to the N-terminal globular head of α-actinin [104], while components of signalling pathways also bind α-actinin, including the p85α subunit of phosphoinositide 3-kinase [105] and protein kinase N [106], a downstream effector of the Rho GTPase [107]. In addition, α-actinin can bind the acidic phospholipid PtdIns(4,5)P_2 [108] and may be present in an α-actinin–PtdIns(4,5)P_2–vinculin ternary complex [89]. The F-actin-binding activity of α-actinin is regulated by PtdIns(4,5)P_2, and talin augments actin cross-linking by α-actinin [108]. Stimulation of cells with PDGF results in a decrease in the amount of PtdIns(4,5)P_2 bound to α-actinin [89], and this is associated with cytoskeletal re-organization.

VASP

VASP is a vinculin-binding protein [78] which is localized in the focal adhesions of well-spread cultured cells and periodically along stress fibres in a manner reminiscent of α-actinin and zyxin [109,110]. In spreading cells, VASP is found in microspikes, and this localization is dependent on Cdc42 [111], a GTPase of the Rho family. VASP is also found in the ruffling membranes of migrating fibroblasts and the cell–cell contacts of epithelial cells. It is possible that it localizes to these regions via its ability to bind vinculin [78,79] and zyxin [109], which are also found in these regions. It migrates on SDS/PAGE as a 46/50 kDa doublet [112]. The low-mobility band results from phosphorylation. Three sites of phosphorylation have been identified, i.e. Ser-157, Ser-239 and Thr-278 [110,112], and Ser-157 phosphorylation correlates with inhibition of binding of platelet integrin $\alpha_{IIb}\beta_3$ to fibrinogen [113].

Human and canine cDNAs have been cloned [110]. The derived primary amino acid sequences are highly similar and encode polypeptides with predicted molecular masses of about 40 kDa. The native protein exists as a homotetramer. Recently, a VASP-related family has been identified [114]. This family consists of VASP, the *Drosophila* protein enabled (Ena), the mouse homologue of Ena (Mena), and an Ena/VASP-like cDNA from mouse (Evl). There are three regions of identity: an N-terminal EVH1 domain (EVH = Ena/VASP homology), a central proline-rich region and a C-terminal EVH2 domain. A possible inclusion in this family is WASP (Wiskott–Aldrich syndrome protein), which possesses limited similarities with the EVH1 domain and has an adjacent proline-rich region [115,116]. WASP has been implicated as a downstream effector of Cdc42, and also in actin polymerization [116]. Wiskott–Aldrich syndrome is characterized by thrombocytopoenia (an inability to produce platelets of the correct size, possibly due to a cytoskeletal defect), T- and B-cell signalling defects, and also eczema.

Deletion of the C-terminal region encompassing the EVH2 domain from VASP abolishes focal adhesion localization [110] and vinculin binding [78], although this may be due to disruption of tertiary/quaternary structures rather than to loss of binding motifs. In fact, recent data demonstrate that vinculin (and zyxin) binding is actually via the N-terminal EVH1 domain of VASP family members [114]. The EVH1 domain binds to FPPPP (FP$_4$)-containing motifs present in vinculin and zyxin [117]. Approximately half of the proline residues in the central proline-rich region of VASP are arranged into four GPPPP (GP$_5$) motifs [110]. Profilin, a G-actin-binding protein, can bind to GP$_5$-motif-containing peptides [118], and also to VASP in a manner that is inhibited by GP_5 peptides [119]. These observations suggest that VASP could be important in the delivery of G-actin to points of focal adhesion assembly.

Evidence that talin, vinculin and α-actinin are involved in the assembly of focal adhesions

The hypothesis that talin is an important structural component of focal adhesions is based on the following observations. Microinjection of a poly-clonal antibody raised against chicken talin inhibits the ability of chicken embryo fibroblasts to spread and migrate on fibronectin [120]. Similar studies with anti-talin monoclonal antibodies recognizing epitopes in the N- and C-terminal regions of talin have indicated that talin is an essential component of focal adhesions. Microinjection of these antibodies into human MRC5 lung fibroblasts disrupted focal adhesions and stress fibres and also inhibited cell motility [57]. Talin is one of the first components to be recruited to nascent focal adhesions; this occurs before the incorporation of vinculin [62]. Interestingly, HeLa cells expressing antisense talin mRNA show reduced sur-face expression of β_1 integrins [121]. Such cells spread more slowly on fibronectin, while the size and distribution of focal adhesions, and the number of stress fibres, are reduced.

Microinjection into fibroblasts of monoclonal antibodies to vinculin also causes dissolution of stress fibres and focal adhesion breakdown [122]. Reductions in the levels of expression of vinculin, by antisense down-regula-tion [123], viral transformation [124] or random mutagenesis of cells [125], is correlated with a reduction in the number and size of focal adhesions and stress fibres. Overexpression of vinculin results in cells with more, and larger, focal adhesions [126] and reduced cell motility, suggesting that vinculin stabilizes focal adhesions and their link with actin stress fibres. Surprisingly, vinculin gene disruption in *Drosophila* does not create an observable phenotypic change [127].

Several lines of evidence suggest that α-actinin is important for focal adhesion and stress fibre integrity. Microinjection of proteolytic fragments of α-actinin into fibroblasts disrupts focal adhesions and actin stress fibres [128]. This is effected by fragments containing the actin-binding domain and the spectrin-like repeats/EF-hands. The α-actinin gene has been disrupted by homologous recombination in *D. discoideum* [129,130]. Surprisingly, no detectable change in phenotype was apparent. However, additional deletion of another actin-cross-linking protein, filamin, produced slime moulds defective in developmental processes. These were able to aggregate and differentiate into various cell types, but were unable to undergo further morphogenesis, i.e. for-mation of migrating slugs and fruiting bodies did not take place. This suggests functional redundancy between actin-cross-linking proteins. Mutation of α-actinin in *D. melanogaster* only results in minor muscle abnormalities, sugges-tive of a role in the formation and stabilization of cytoskeletal architecture [131]. However, the lack of severity is again suggestive of redundancy of func-tion. A role for α-actinin in cytoskeletal architecture has been demonstrated in studies aimed at modulating α-actinin expression levels [132]. Down-regulation of α-actinin levels by expression of an α-actinin antisense mRNA in BALB/c 3T3 fibroblasts resulted in cells with fewer, smaller, focal adhesions and stress fibres. Such cells are more motile than control cells. Conversely, overexpression of a chick α-actinin cDNA resulted in cells with larger adhesions which were less motile. These results are similar to observations from identical studies on vinculin [123,126].

Studies of the roles of talin and vinculin in the assembly of focal adhesions using gene disruption

While the above experiments provide compelling evidence that talin, vinculin and α-actinin play a key role in cell adhesion and the architecture of the actin cytoskeleton, it is difficult to extend these studies to address questions about the importance of the proposed interactions between the various focal adhesion proteins. We have therefore used homologous recombination to target both alleles of the talin and vinculin genes in mouse embryonic stem (ES) cells (H. Priddle, L. Hemmings and D.R. Critchley, unpublished work). The ES cell talin (−/−) mutants express normal levels of vinculin, but do not express intact talin, although they do express low levels of a truncated talin polypeptide. The phenotypic characteristics of these cells when plated on gelatin or fibronectin are consistent with the conclusion that talin is essential to the adhesion of cells to the ECM. Thus, unlike wild-type ES cells, the talin (−/−) mutants are unable to spread on either substrate and instead form large cellular aggregates. Immunofluorescence studies show that the talin (−/−) mutants are unable to form focal adhesions, and vinculin and F-actin remain diffusely distributed within the cell. The phenotype of the vinculin (−/−) mutants is somewhat different. Although the cells show a tendency to grow as small aggregates, they do spread on gelatin and fibronectin and are able to form talin-containing focal adhesions and associated actin stress fibres. The phenotype of the vinculin mutants is similar to that described for mouse F9 teratocarcinoma cells [133], in which both copies of the vinculin gene were targeted as above. The results clearly establish that vinculin is not an essential component of the focal adhesion, at least in ES cells.

This information can be incorporated into a model for focal adhesion assembly in which talin is envisaged to act as a key component, on the one hand binding to integrins and on the other hand acting as a template for the assembly of actin filaments (Fig. 4). The fact that the talin dimer has the potential to cross-link integrins and that it has actin-nucleating and cross-linking activity makes it ideally suited for such a role. Talin has three vinculin-binding sites, suggesting that the interaction between these proteins is likely to be important to focal adhesion assembly and/or dynamics. Cells containing little or no vinculin are more motile than cells overexpressing the protein [123,124,126], suggesting that vinculin might stabilize focal adhesions. Such a stabilization function might be attributed to the fact that vinculin can bind both talin and F-actin. However, vinculin can also bind to VASP [78], a protein which binds profilin, itself a G-actin-binding protein. This raises the possibility that vin-

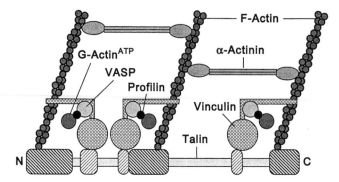

Fig. 4. Role of talin in the assembly of focal adhesions. Talin is represented as a long rod-shaped molecule containing three actin-binding sites adjacent to three vinculin-binding sites. Vinculin binds to both talin and actin as well as to VASP, a protein that can bind profilin, which is in turn a G-actin-binding protein. As such, vinculin might stabilize the interaction between talin and actin and provide G-actin monomers for nucleation of actin filament assembly by talin. The diagram implies that all three binding sites for actin and vinculin may serve equivalent functions, but this has not been investigated.

culin serves to recruit monomeric G-actin to talin, facilitating actin filament assembly.

The talin $(-/-)$ and vinculin $(-/-)$ cell lines offer the opportunity to test this model. It would be predicted that expression of the full-length talin cDNA in the mutant ES cells should rescue the defect in cell–ECM interactions. If so, we can test the importance of the interaction of talin with vinculin by introducing talin cDNAs containing mutations that inactivate one or more of the vinculin-binding sites. Based on the phenotype of the vinculin null mutants, it would be predicted that such mutations in talin would not compromise the ability to form focal adhesions, but that the assembly process might be significantly slower due to the lack of an efficient mechanism for delivering G-actin monomers to talin. Similar experiments can be envisaged with the vinculin $(-/-)$ cell line.

Role of small GTP-binding proteins in the assembly of focal adhesions

Studies by Hall and his colleagues have provided new insights into the mechanisms regulating the assembly of focal adhesions [134–136]. Thus the small GTP-binding proteins Cdc42, Rac and RhoA regulate the assembly of filopodia, membrane ruffling and actin stress fibres respectively. Subsequent studies have led to the identification of the downstream effector molecules that mediate these responses. For example, the assembly of actin stress fibres by RhoA is mediated, at least in part, by p160^{rho} kinase, which phosphorylates and inhibits the activity of a myosin light chain phosphatase [137]. As a consequence, the steady-state levels of myosin light chain phosphorylation increase, and this favours the formation of an actomyosin complex. Contraction of acto-myosin is then thought to lead to the alignment of actin filaments into bundles. Moreover, evidence has been presented that this results in the clustering of integrins and associated cytoskeletal proteins in the plane of the plasma membrane to form focal adhesions [138].

Much of the progress in understanding the role of small GTP-binding proteins in the assembly of focal adhesions has involved the microinjection of proteins into serum-starved Swiss 3T3 cells, which lack focal adhesions and actin stress fibres [134–136]. To facilitate analysis of the biochemical events associated with focal adhesion assembly, we have introduced proteins into serum-starved Swiss 3T3 cells permeabilized with streptolysin-O. We have demonstrated that these cells retain the intracellular pathways required for the Rho-dependent assembly of actin stress fibres (J. Norman, S.T. Barry, D. Jones, S. Cockcroft and D.R. Critchley, unpublished work). Thus addition of guanosine 5′-[γ-thio]triphosphate (GTP[S]) drives stress fibre assembly via a Rho-dependent pathway, as the response is blocked by C3 transferase, which ADP-ribosylates and inactivates Rho. In serum-starved cells, much of the cellular paxillin is localized in the perinuclear region and redistributes to the ends of actin stress fibres, but this response is rapidly lost over time. In contrast,

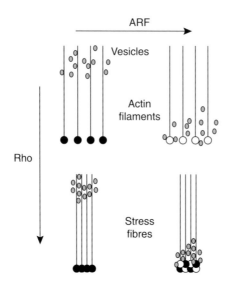

Fig. 5. Roles of the small GTP-binding proteins Rho and ARF-1 in the assembly of focal adhesions. Rho is thought to lead to contraction of the actomyosin cytoskeleton, the alignment of actin filaments into large bundles (stress fibres), and the clustering of integrins into focal adhesions. However, these adhesions lack paxillin. ARF-1 drives the recruitment of paxillin to the ends of sub-membranous actin filaments independently of Rho, and in the absence of actin stress fibres. Activation of both Rho and ARF-1 leads to the formation of actin stress fibres and focal adhesions containing paxillin. Because the best characterized function of ARF-1 is in vesicular transport in the Golgi apparatus, we speculate that paxillin is localized in a vesicular compartment.

paxillin, which is also perinuclear in serum-starved cells, still redistributes to the ends of actin filaments when the GTP[S] is added up to 10 min after permeabilization. However, paxillin redistribution is only partially inhibited by C3 transferase (which completely blocks stress fibre assembly) and is not stimulated by RhoA. Thus paxillin can be recruited to focal-adhesion-like structures in the absence of the formation of stress fibres. These results show that it is possible to dissociate responses that appear to be co-ordinately regulated in intact cells.

The finding that GTP[S]-induced recruitment of paxillin to focal adhesions is largely Rho-independent suggests a role for another GTP-binding protein. Interestingly, we have observed that Q71L-ARF-1, the constitutively active mutant of ADP-ribosylation factor-1 (ARF-1), will drive the redistribution of paxillin to focal adhesions. ARF proteins leak out of streptolysin-O-permeabilized cells such that little remains after 15 min, although paxillin is retained by the cell. Addition of ARF-1 (but not the inactive Δ17ARF-1 mutant) to these cells re-sensitizes the response to GTP[S]. Furthermore, the Δ17ARF-1 mutant acts as a dominant-negative mutant when added to cells permeabilized with streptolysin-O for 8 min, blocking the GTP[S]-induced recruitment of paxillin to focal-adhesion-like structures. The results suggest that Rho proteins regulate the assembly of actin stress fibres, but that ARF-1 regulates the distribution of paxillin (but not vinculin) to focal-adhesion-like structures.

A role for the ARF family of proteins in regulating cell adhesion is not wholly unexpected. Recent studies have shown that the activation state of β_2 integrins is controlled by a protein, cytohesin, which also activates ARF [139,140]. The small GTP-binding protein Rac, which controls membrane ruffling, binds to a protein POR1 (a Rac-1-interacting protein) which also binds ARF [141]. Interestingly, ARF-1 and Cdc42, which drive the formation of filopodia and the assembly of small focal-adhesion-like structures, are both localized in the Golgi compartment [142,143], raising the possibility that vesicular transport plays a significant role in the assembly of filopodia and focal adhesions (Fig. 5).

The work on the role of ARF proteins was carried out in collaboration with D. Jones and S. Cockcroft (Department of Physiology, UCL, London, U.K.). The work was funded by the Medical Research Council and the Wellcome Trust.

References

1. LaFlamme, S.E., Akiyama, S.K. and Yamada, K.M. (1992) J. Cell Biol. **117**, 437–447
2. LaFlamme, S.E., Thomas, L.A., Yamada, S.S. and Yamada, K.M. (1994) J. Cell Biol. **126**, 1287–1298
3. Smilenov, L., Briesewitz, R. and Marcantonio, E.E. (1994) Mol. Biol. Cell **11**, 1215–1223
4. Akiyama, S.K., Yamada, S.S. and Yamada, K.M. (1994) J. Biol. Chem. **269**, 15961–15964
5. Lukashev, M.E., Sheppard, S. and Pytela, S. (1994) J. Biol. Chem. **269**, 18311–18314
6. Pasqualini, R. and Hemler, M.E. (1994) J. Cell Biol. **125**, 447–460
7. Reszka, A.A., Hayashi, Y. and Horwitz, A.F. (1992) J. Cell Biol. **117**, 1321–1330

8. Vignoud, L., Albiges-Rizo, C., Frachet, P. and Block, M.R. (1997) J. Cell Sci. 110, 1421–1430

9. Filardo, E.J., Brooks, P.C., Deming, S.L., Damsky, C. and Cheresh, D.A. (1995) J. Cell Biol. 130, 441–450

10. O'Toole, T.E., Ylanne, J. and Culley, B.M. (1995) J. Biol. Chem. 270, 8553–8558

11. Hibbs, M.L., Jakes, S., Stacker, S.A., Wallace, R.W. and Springer, T.A. (1991) J. Exp. Med. 174, 1227–1238

12. Hibbs, M.L., Xu, H., Stacker, S.A. and Springer, T.A. (1991) Science 251, 1611–1613

13. Cone, R.I., Weinacker, A., Chen, A. and Sheppard, D. (1994) Cell Adhes. Commun. 2, 101–113

14. Ylanne, J., Huuskonen, J., O'Toole, T.E., Ginsberg, M.H., Virtanen, I. and Gahmberg, C.G. (1995) J. Biol. Chem. 270, 9550–9557

15. Liu, X.-Y., Timmons, S., Lin, Y.-Z. and Hawiger, J. (1996) Proc. Natl. Acad. Sci. U.S.A. 93, 11819–11824

16. Belkin, A.M., Zhidkova, N.I., Balzac, F., Altruda, F., Tomatis, D., Maier, A., Tarone, G., Koteliansky, V.E. and Burridge, K. (1996) J. Cell Biol. 132, 211–226

17. Balzac, F., Belkin, A.M., Koteliansky, V.E., Balabanov, Y.V., Altruda, F., Silengo, L. and Tarone, G. (1993) J. Cell Biol. 121, 171–178

18. Balzac, F., Retta, S.F., Albini, A., Melchiorri, A., Koteliansky, V.E., Geuna, M., Silengo, L. and Tarone, G. (1994) J. Cell Biol. 127, 557–565

19. Fox, J.E., Taylor, R.G., Taffarel, M., Boyles, J.K. and Goll, D.E. (1993) J. Cell Biol. 120, 1501–1507

20. Fox, J.E., Shattil, S.J., Kinlough-Rathbone, R.L., Richardson, M., Packham, M.A. and Sanan, D.A. (1996) J. Biol. Chem. 271, 7004–7011

21. Lub, M., van Kooyk, Y., van Vliet, S.J. and Figdor, C.G. (1997) Mol. Biol. Cell 8, 341–351

22. Peter, K. and O'Toole, T.E. (1995) J. Exp. Med. 181, 315–326

23. Wu, C., Keivens, V.M., O'Toole, T.E., McDonald, J.A. and Ginsberg, M.H. (1995) Cell 83, 715–724

24. Ylanne, J., Chen, Y., O'Toole, T.E., Loftus, J.C., Takada, Y. and Ginsberg, M.H. (1993) J. Cell Biol. 122, 223–233

25. Felsenfeld, D.P., Choquet, D. and Sheetz, M.P. (1996) Nature (London) 383, 438–440

26. Schmidt, C.E., Horwitz, A.F., Lauffenburger, D.A. and Sheetz, M.P. (1993) J. Cell Biol. 123, 977–991

27. Choquet, D., Felsenfeld, D.P. and Sheetz, M.P. (1997) Cell 88, 39–48

28. Miyamoto, S., Akiyama, S.K. and Yamada, K.M. (1995) Science 266, 1719–1723

29. Miyamoto, S., Teramoto, H., Coso, O.A., Gutkind, J.S., Burbelo, P.D., Akiyama, S.K. and Yamada, K.M. (1995) J. Cell Biol. 131, 791–805

30. O'Toole, T.E., Katagiri, Y., Faull, R.J., Peter, K., Tamura, R., Quaranta, V., Loftus, J.C., Shattil, S.J. and Ginsberg, M.H. (1994) J. Cell Biol. 124, 1047–1059

31. Schwartz, M.A., Schaller, M.D. and Ginsberg, M.H. (1995) Annu. Rev. Cell Dev. Biol. 11, 549–599

32. Ginsberg, M.H. (1995) Biochem. Soc. Trans. 23, 439–446

33. Hughes, P.E., Diaz-Gonzalez, F., Leong, L., Wu, C., McDonald, J.A., Shattil, S.J. and Ginsberg, M.H. (1996) J. Biol. Chem. 271, 6571–6574

34. Kieffer, J.D., Plopper, G., Ingber, D.E., Hartwig, J.H. and Kupper, T.S. (1995) Biochem. Biophys. Res. Commun. 217, 466–474

35. Knezevic, I., Leisner, T.M. and Lam, S.C.T. (1996) J. Biol. Chem. 271, 16416–16421

36. Horwitz, A., Duggan, K., Buck, C., Beckerle, M.C. and Burridge, K. (1986) Nature (London) 320, 531–533

37. Tapley, P., Horwitz, A., Buck, C., Duggan, K. and Rohrschneider, L. (1989) Oncogene 4, 325–333

38. Sharma, C.P., Ezzell, R.M. and Arnaout, M.A. (1995) J. Immunol. 154, 3461–3470

39. Lewis, J.M. and Schwartz, M.A. (1995) Mol. Biol. Cell **6**, 151–160
40. Otey, C.A., Pavalko, F.M. and Burridge, K. (1990) J. Cell Biol. **111**, 721–729
41. Pavalko, F.M. and LaRoche, S.M. (1993) J. Immunol. **151**, 3795–3807
42. Otey, C.A., Vasquez, G.B., Burridge, K. and Erickson, B.W. (1993) J. Biol. Chem. **268**, 21193–21197
43. Burridge, K. and Connell, L. (1983) J. Cell Biol. **97**, 359–367
44. Geiger, B., Volk, T. and Volberg, T. (1985) J. Cell Biol. **101**, 1523–1531
45. Kupfer, A., Burn, P. and Singer, S.J.J. (1990) Mol. Cell. Immunol. **4**, 317–325
46. Hemmings, L., Rees, D.J.G., Ohanian, V., Bolton, S.J., Gilmore, A.P., Patel, B., Priddle, H., Trevithick, J.E., Hynes, R.O. and Critchley, D.R. (1996) J. Cell Sci. **109**, 2715–2726
47. Winkler, J., Lunsdorf, H. and Jockusch, B.M. (1997) Eur. J. Biochem. **243**, 430–436
48. McLachlan, A.D., Stewart, M., Hynes, R.O. and Rees, D.J.G. (1994) J. Mol. Biol. **235**, 1278–1290
49. Rees, D.J.G., Ades, S.E., Singer, S.J. and Hynes, R.O. (1990) Nature (London) **347**, 685–689
50. Fox, J.E., Lipfert, L., Clark, E.A., Reynolds, C.C., Austin, C.D. and Brugge, J.S. (1993) J. Biol.Chem. **268**, 25973–25984
51. Nuckolls, G.H., Turner, C.E. and Burridge, K. (1990) J. Cell Biol. **110**, 1635–1644
52. Niggli, V., Kaufmann, S., Goldman, W.H., Weber, T. and Isenberg, G. (1994) Eur. J. Biochem. **227**, 951–957
53. Vaheri, A., Carpen, O., Heiska, L., Helander, T.S., Jaaskelainen, J., Majander-Nordenswan, P., Sainio, M., Timonen, T. and Turunen, O. (1997) Curr. Opin. Cell Biol. **9**, 659–666
54. Marfatia, S.M., Lue, R.A., Branton, D. and Chichti, A.H. (1994) J. Biol. Chem. **269**, 8631–8634
55. Chen, H.C., Appeddu, P.A., Parsons, J.T., Hildebrand, J.D., Schaller, M.D. and Guan, J.L. (1995) J. Biol. Chem. **270**, 16995–16999
56. Gilmore, A.P., Wood, C., Ohanian, V., Jackson, P., Patel, B., Tees, D.J.G., Hynes, R.O. and Critchley, D.R. (1993) J. Cell Biol. **122**, 337–347
57. Bolton, S.J., Barry, S.T., Mosley, H., Patel, B., Jockusch, B.M., Wilkinson, J.M. and Critchley, D.R. (1997) Cell Motil. Cytoskeleton **36**, 363–376
58. Adey, N.B. and Kay, B.K. (1997) Biochem. J. **324**, 523–528
59. Holtzman, D.A., Yang, S. and Drubin, D.G. (1993) J. Cell Biol. **122**, 635–644
60. McCann, R.O. and Craig, S.W. (1997) Proc. Natl. Acad. Sci. U.S.A. **94**, 5679–5684
61. Kaufmann, S., Piekenbrock, T., Goldmann, W.H., Barmann, M. and Isenberg, G. (1991) FEBS Lett. **284**, 187–191
62. DePasquale, J.A. and Izzard, C.S. (1991) J. Cell Biol. **113**, 1351–1359
63. Tsukita, S., Yonemura, S. and Tsukita, S. (1997) Trends Biochem. Sci. **22**, 53–58
64. Hirao, M., Sato, N., Kondo, T., Yonemura, S., Monden, M., Sasaki, T., Takai, Y., Tsukita, S. and Tsukita, S. (1996) J. Cell Biol. **135**, 37–51
65. Kotani, H., Takaihi, K., Sasaki, T. and Takai, Y. (1997) Oncogene **14**, 1705–1713
66. Mackay, D.J.G., Esch, F., Furthmayr, H. and Hall, A. (1997) J. Cell Biol. **138**, 927–938
67. Molony, L., McCaslin, D., Abernethy, J., Paschal, B. and Burridge, K. (1987) J. Biol. Chem. **262**, 7790–7795
68. Burridge, K. and Mangeat, P. (1984) Nature (London) **308**, 744–746
69. Weller, P.A., Ogryzko, E.B., Corben, E.B., Zhidkova, N.I., Patel, B., Price, G.J., Spurr, N.K., Koteliansky, V.E. and Critchley, D.R. (1990) Proc. Natl. Acad. Sci. U.S.A. **87**, 5667–5671
70. Winkler, J., Lunsdorf, H. and Jockusch, B.M. (1996) J. Struct. Biol. **116**, 270–277
71. Price, G.J., Jones, P., Davison, M.D., Patel, B., Bendori, R., Geiger, B. and Critchley, D.R. (1989) Biochem. J. **259**, 453–461
72. Gilmore, A.P., Jackson, P., Waites, G.T. and Critchley, D.R. (1992) J. Cell Sci. **103**, 719–731
73. McGregor, A., Blanchard, A.D., Rowe, A.J. and Critchley, D.R. (1994) Biochem. J. **301**, 225–233

74. Kroemker, M., Rüdiger, A.H., Jockusch, B.M. and Rüdiger, M. (1994) FEBS Lett. **355**, 259–262
75. Turner, C.E., Glenney, Jr., J. and Burridge, K. (1990) J. Cell Biol. **111**, 1059–1068
76. Wood, C.K., Turner, C.E., Jackson, P. and Critchley, D.R. (1994) J. Cell Sci. **107**, 709–717
77. Hüttelmaier, S., Bubeck, P., Rüdiger, M. and Jockusch, B.M. (1997) Eur. J. Biochem. **247**, 1136–1142
78. Brindle, N.P.J., Holt, M.R., Davies, J.E., Price, C.J. and Critchley, D.R. (1996) Biochem. J. **318**, 753–757
79. Reinhard, M., Rüdiger, M., Jockusch, B.M. and Walter, U. (1996) FEBS Lett. **399**, 103–107
80. Lo, S.H. and Chen, L.B. (1994) Cancer Metastasis Rev. **13**, 9–24
81. Hyatt, S.L., Liao, L., Chapline, C. and Jaken, S. (1994) Biochemistry **33**, 1223–1228
82. O'Rourke, F., Matthews, E. and Feinstein, M.B. (1996) Biochem. J. **315**, 1027–1034
83. Koteliansky, V.E., Ogryzko, E.P., Zhidkova, N.I., Weller, P.A., Critchley, D.R., Vancompernolle, K., Vandekerckhove, J., Strasser, P., Way, M., Gimona, M. and Small, J.V. (1992) Eur. J. Biochem. **203**, 767–772
84. Johnson, R.P. and Craig, S.W. (1994) J. Biol. Chem. **269**, 12611–12619
85. Weekes, J., Barry, S.T. and Critchley, D.R. (1996) Biochem. J. **314**, 827–832
86. Johnson, R.P. and Craig, S.W. (1995) Nature (London) **373**, 261–264
87. Gilmore, A.P. and Burridge, K. (1996) Nature (London) **381**, 531–535
88. Tempel, M., Goldmann, W.H., Isenberg, G. and Sackmann, E. (1995) Biophys. J. **69**, 228–241
89. Fukami, K., Endo, T., Imamura, M. and Takenawa, T. (1994) J. Biol. Chem. **269**, 1518–1522
90. Johnson, R.P. and Craig, S.W. (1995) Biochem. Biophys. Res. Commun. **210**, 159–164
91. Herman, B. and Pledger, W.J. (1985) J. Cell Biol. **100**, 1031–1040
92. Blanchard, A., Ohanian, V. and Critchley, D.R. (1989) J. Muscle Res. Cell Motil. **10**, 280–289
93. Matsudaira, P. (1991) Trends Biochem. Sci. **16**, 87–92
94. Winder, S.J. (1996) Biochem. Soc. Trans. **24**, 497–501
95. Flood, G., Kahana, E., Gilmore, A.P., Rowe, A.J., Gratzer, W.B. and Critchley, D.R. (1995) J. Mol. Biol. **252**, 227–234
96. Beggs, A.H., Byers, T.J., Knoll, J.H.M., Boyce, F.M., Bruns, G.A.P. and Kunkel, L.M. (1992) J. Biol. Chem. **267**, 9281–9288
97. Parr, T., Waites, G.T., Patel, B., Millake, D.B. and Critchley, D.R. (1992) Eur. J. Biochem. **210**, 801–809
98. Waites, G.T., Graham, I.R., Jackson, P., Millake, D.B., Patel, B., Blanchard, A.D., Weller, P.A., Eperon, I.C. and Critchley, D.R. (1992) J. Biol. Chem. **267**, 6263–6271
99. Witke, W., Hofmann, A., Koppel, B., Schleicher, M. and Noegel, A.A. (1993) J. Cell Biol. **121**, 599–606
100. Imamura, M., Sakurai, T., Ogawa, Y., Ishikawa, T., Goto, K. and Masaki, T. (1994) Eur. J. Biochem. **223**, 395–401
101. Pavalko, F.M., Walker, D.M., Graham, L., Goheen, M., Doerschuk, C.M. and Kansas, G.S. (1995) J. Cell Biol. **129**, 1155–1164
102. Carpen, O., Pallai, P., Staunton, D.E. and Springer, T.A. (1992) J. Cell Biol. **118**, 1223–1234
103. Heiska, L., Kantor, C., Parr, T., Critchley, D.R., Vilja, P., Gahmberg, C.G. and Carpen, O. (1996) J. Biol. Chem. **271**, 26214–26219
104. Crawford, A.W., Michelson, J.W. and Beckerle, M.C. (1992) J. Cell Biol. **116**, 1391–1393
105. Shibasaki, F., Fukami, K., Fukui, Y. and Takenawa, T. (1994) Biochem. J. **302**, 551–557
106. Mukai, H., Toshimori, M., Shibata, H., Takanaga, H., Kitagawa, M., Miyahara, M., Shimakawa, M. and Ono, Y. (1997) J. Biol. Chem. **272**, 4740–4746
107. Amano, M., Mukai, H., Ono, Y., Chihara, K., Matsui, T., Hamajima, Y., Okawa, K., Iwamatsu, A. and Kaibuchi, K. (1996) Science **271**, 648–650

108. Fukami, K., Furuhashi, K., Inagaki, M., Endo, T., Hatano, S. and Takemawa, T. (1992) Nature (London) **359**, 150–152

109. Reinhard, M., Jouvenal, K., Tripier, D. and Walter, U. (1995) Proc. Natl. Acad. Sci. U.S.A. **92**, 7956–7960

110. Haffner, C., Jarchau, T., Reinhard, M., Hoppe, J., Lohmann, S.M. and Walter, U. (1995) EMBO J. **14**, 19–27

111. Dutartre, H., Davoust, J., Gorvel, J.P. and Chavrier, P. (1996) J. Cell Sci. **109**, 367–377

112. Butt, E., Abel, K., Krieger, M., Palm, D., Hoppe, V., Hoppe, J. and Walter, U. (1994) J. Biol. Chem. **269**, 14509–14517

113. Horstrup, K., Jablonka, B., Honigliedl, P., Just, M., Kochsiek, K. and Walter, U. (1994) Eur. J. Biochem. **225**, 21–27

114. Gertler, F.B., Niebuhr, K., Reinhard, M., Wehland, J. and Soriano, P. (1996) Cell **87**, 227–239

115. Derry, J.M.J., Ochs, H.D. and Francke, U. (1994) Cell **78**, 635–644

116. Symons, M., Derry, J.M.J., Karlak, B., Lemahieu, V., McCormick, F., Francke, U. and Abo, A. (1996) Cell **84**, 723–734

117. Niebuhr, K., Ebel, F., Frank, R., Reinhard, M., Domann, E., Carl, U.D., Walter, U., Gertler, F.B., Wehland, J. and Chakraborty, T. (1997) EMBO J. **16**, 5433–5444

118. Kang, F., Laine, R.O., Bubb, M.R., Southwick, F.S. and Purich, D.L. (1997) Biochemistry **36**, 8384–8392

119. Reinhard, M., Giehl, K., Abel, K., Haffner, C., Jarchau, T., Hoppe, V., Jockusch, B. and Walter, U. (1995) EMBO J. **14**, 1583–1589

120. Nuckolls, G.H., Romer, L.H. and Burridge, K. (1992) J. Cell Sci. **102**, 753–762

121. Albiges-Rizo, C., Frachet, P. and Block, M.R. (1995) J. Cell Sci. **108**, 3317–3329

122. Westmeyer, A., Ruhnau, K., Wegner, A. and Jockusch, B.M. (1990) EMBO J. **9**, 2071–2078

123. Rodriguez Fernandez, J.L., Geiger, B., Salomon, D. and Ben-Zeíev, A. (1993) J. Cell Biol. **122**, 1285–1294

124. Rodriguez Fernandez, J.L., Geiger, B., Salomon, D., Sabanay, I., Zoller, M. and Ben-Zeíev, A. (1992) J. Cell Biol. **119**, 427–438

125. Samuels, M., Ezzell, R.M., Cardozo, T.J., Critchley, D.R., Coll, J.L. and Adamson, E.D. (1993) J. Cell Biol. **121**, 909–921

126. Rodriguez Fernandez, J.L., Geiger, B., Salomon, D. and Ben-Zeíev, A. (1992) Cell Motil. Cytoskeleton **22**, 127–134

127. Alatortsev, V.E., Kramerova, I.A., Frolov, M.V., Lavrov, S.A. and Westphal, E.D. (1997) FEBS Lett. **413**, 197–201

128. Pavalko, F.M. and Burridge, K. (1991) J. Cell Biol. **114**, 481–491

129. Witke, W., Nellen, W. and Noegel, A. (1987) EMBO J. **6**, 4143–4148

130. Witke, W., Schleicher, M. and Noegel, A.A. (1992) Cell **68**, 53–62

131. Roulier, E.M., Fyrberg, C. and Fyrberg, E. (1992) J. Cell Biol. **116**, 911–922

132. Glück, U. and Ben-Zeíev, A. (1994) J. Cell Sci. **107**, 1773–1782

133. Volberg, T., Geiger, B., Kam, Z., Pankov, R., Simcha, I., Sabanay, H., Coll, J.L., Adamson, E. and Ben-Zeíev, A. (1995) J. Cell Sci. **108**, 2253–2260

134. Ridley, A.J. and Hall, A. (1992) Cell **70**, 389–399

135. Ridley, A.J., Paterson, H.F., Johnston, C.L., Diekman, D. and Hall, A. (1992) Cell **70**, 401–410

136. Nobes, C.D. and Hall, A. (1995) Cell **81**, 53–62

137. Kimura, K., Ito, M., Amano, M., Chihara, K., Fukuta, Y., Nakafuka, M., Yamamori, B., Feng, J.H., Nakano, T., Okawa, K., Iwamatsu, A. and Kaibuchi, K. (1996) Science **273**, 245–248

138. Burridge, K., Chrzanowskawodnicka, M. and Zhong, C.L. (1997) Trends Cell Biol. **7**, 342–347

139. Kolanus, W., Nagel, W., Schiller, B., Zeitlmann, L., Godar, S., Stockinger, H. and Seed, B. (1996) Cell **86**, 233–242

140. Meacci, E., Tsai, S.C., Adamik, R., Moss, J. and Vaughan, M. (1997) Proc. Natl. Acad. Sci. U.S.A. **94**, 1745–1748

141. D'Souza-Schorey, C., Boshans, R.L., McDonough, M. and Vaughan, M. (1997) EMBO J. **16**, 5445–5454

142. Serafini, T., Orci, L., Amherdt, M., Brunner, M., Kahn, R.A. and Rothman, J.E. (1991) Cell **67**, 239–253

143. Erickson, J.W., Zhang, C.J., Kahn, R.A., Evans, T. and Cerione, R.A. (1996) J. Biol. Chem. **271**, 26850–26854

Biochem. Soc. Symp. **65**, 101–109
Printed in Great Britain

7

Wnt factors in axonal remodelling and synaptogenesis

Patricia C. Salinas

Developmental Biology Research Centre, The Randall Institute, King's College London, 26–29 Drury Lane, London WC2B 5RL, U.K.

Abstract

'Wiring' of the central nervous system is accomplished by the precise and co-ordinated behaviour of neuronal cells. Proper navigation of axons and formation of synaptic contacts with the correct targets are essential. Although several signalling molecules that control axon guidance, target selection and formation of synapses have been identified, little is known about how these proteins lead to changes in the axonal cytoskeleton. Wnt signalling factors have been shown to induce axonal remodelling in developing neurons. As several components of the Wnt signalling pathway are known, studies on Wnt factors could elucidate the mechanisms by which extracellular molecules regulate the neuronal cytoskeleton. Wnt-7a induces axonal spreading and subsequent increases in synaptic protein levels in mouse cerebellar neurons. These findings suggest a role for Wnt-7a in axon guidance and synapse formation in the developing cerebellum. Based on analyses of the axonal cytoskeleton, a model is proposed in which Wnt-7a induces axonal remodelling by inhibiting glycogen synthase kinase-3β (GSK-3β), a serine/threonine kinase. Inhibition of GSK-3β leads to a decrease in a phosphorylated form of microtubule-associated protein-1B (MAP-1B), a protein involved in microtubule assembly, and a concomitant decrease in the level of stable microtubules. This chapter discusses the novel role of Wnt factors in regulating the axonal cytoskeleton during neuronal development.

Introduction

Axonal extension, turning, retraction, sprouting and the assembly of the synaptic apparatus is achieved by the action of extracellular signals that elicit changes in the neuronal cytoskeleton [1]. Although several guidance molecules, such as the netrin, Eph and semaphorin families of proteins, have been identified, the mechanisms underlying the re-organization of the axonal cytoskeleton are poorly understood [2].

Recently, Wnt signalling molecules, with a well-characterized signalling pathway, have been shown to regulate the neuronal cytoskeleton. Wnts are short-range extracellular glycoproteins involved in early cell-fate decisions [3–5]. Studies in *Drosophila*, *Xenopus* and mammalian cell lines led to the identification of different components of the Wnt signalling pathway. The current model proposes that binding of Wnts to their seven-transmembrane receptor, Frizzled, leads to the activation of casein kinase II and the phosphorylation of Dishevelled, a cytoplasmic protein [6]. Phosphorylation of Dishevelled results in the inactivation of glycogen synthase kinase-3β (GSK-3β), a serine/threonine kinase [7]. In the absence of Wnt, GSK-3β phosphorylates two components of the signalling pathway, β-catenin and adenomatous polyposis coli (APC), resulting in the degradation of β-catenin [8]. Thus Wnt signalling results in an increase in the level of β-catenin through the inactivation of GSK-3β. Subsequently, β-catenin forms a complex with lymphoid-enhancer factor transcription factors, and both are translocated to the nucleus where they activate the transcription of target genes [9]. Studies of Wnts with a well-delineated pathway provide a good opportunity for identifying molecules that will execute cytoskeletal changes during the formation of neuronal connections.

A role for Wnts in regulation of the cytoskeleton was suggested by the findings that components of the Wnt signalling pathway, β-catenin and APC, are associated with the cytoskeleton. β-Catenin may regulate the actin cytoskeleton through the actin-bundling protein fascin [10], while APC binds to β-catenin, GSK-3β [11] and microtubules [12]. In addition, Wnts induce the expression of cadherin proteins that, in turn, induce localized assembly of cytoskeletal components [13]. Cadherins have previously been found to bind to β-catenin and plakoglobin, a β-catenin homologue [14,15], and these complexes are linked to the cytoskeleton through α-catenin, which is also an actin-binding protein [14]. Only recent studies, however, have provided support for the role of Wnts in regulating cytoskeletal organization. In *Caenorhabditis elegans*, mutations in *Wnt* genes and in components of the Wnt pathway lead to re-orientation of the mitotic spindle, and these effects are associated with changes in cell fate [16,17]. However, a more direct piece of evidence for the function of Wnts in microtubule organization comes from analysis of mouse cerebellar neurons exposed to Wnt-7a. In this system, Wnt-7a induces axonal spreading and branching, with a concomitant change in microtubule organization [17a]. These results demonstrate that Wnts regulate the cytoskeleton during both early and later stages of development. In this review, I will discuss evidence for the role of Wnts in cytoskeleton regulation during axonogenesis and synaptogenesis.

Wnt genes are expressed during axonal extension and synaptogenesis

Wnt factors function as developmental switches during early cell-fate decisions [5,8]. However, some members of the Wnt family are expressed after cell differentiation, which suggests that Wnts also function in other cellular processes [18]. Analysis of the pattern of expression of mouse *Wnt* genes shows that *Wnt-3* and *Wnt-7a* are expressed in defined populations of post-mitotic neurons in the postnatal central nervous system [18,19]. In the adult, *Wnt-3* is expressed in Purkinje cells of the cerebellum, the pontine nuclei, the olivary nuclei and the dorsal nuclei of the thalamus [18]. In the postnatal cerebellum, *Wnt-3* is expressed in Purkinje cells during the period of dendritic-tree formation and synaptogenesis [18]. Studies of two mouse mutants in which granule cells fail to mature showed that expression of *Wnt-3* depends on interactions with granule cells [18]. *Wnt-7a*, on the other hand, is expressed in cerebellar granule cells as the cell bodies of these neurons begin their journey to the internal layer of the cerebellar cortex [19]. After cell migration, expression of *Wnt-7a* increases even further, coinciding with the period of synapse formation between granule cells and mossy-fibre axons [19]. These findings suggest a role for Wnt-3 and Wnt-7a in regulating axonal extension and/or synaptogenesis in the developing mouse cerebellum.

Wnt-7a induces axonal remodelling

The hypothesis that Wnt-7a regulates axonal extension and synaptogenesis was tested by exposing cerebellar granule cell neurons to Wnt-7a *in vitro*. In this system, Wnt-7a induces axonal spreading along the entire axon and at the growth cone [19]. Granule cells exposed to Wnt-7a have shorter and more highly branched axons than control neurons, suggesting that axonal spreading and branching occur at the expense of axonal extension [19]. To determine whether Wnt-7a was acting directly on granule cells, lithium was added to cerebellar neurons. Lithium, like Wnts, has been shown to inhibit GSK-3β activity, and therefore mimics Wnt function in several developmental systems [20–22]. Lithium also induces axonal spreading, branching and shortening of the axon length in a similar fashion to Wnt-7a [19]. Although lithium has been shown to affect the inositol pathway [23], axonal remodelling observed in cerebellar granule neurons is independent of inositol turnover [19]. Moreover, lithium increases the levels of β-catenin, strongly suggesting that the effect of lithium is mediated through the inhibition of GSK-3β [19]. These findings indicate that axonal spreading is likely to be due to a direct action of Wnt-7a on granule cell neurons.

Inhibition of GSK-3β by lithium or Wnt-7a leads to microtubule instability and changes in the phosphorylation of MAP-1B

How does the inhibition of GSK-3β change cytoskeletal organization? The shortening of axonal processes in Wnt-7a- or lithium-treated neurons suggests that microtubule organization is affected. Indeed, analysis of stable and dynamic microtubules revealed that GSK-3β inhibition results in the loss of stable microtubules from spread areas and their reduction along the rest of the axon, while dynamic microtubules populate spread areas [17a]. These findings suggest that inhibition of GSK-3β leads to microtubule instability.

The stability of microtubules is regulated by microtubule-associated proteins such as MAP-1B, tau and MAP-2 [24–26]. MAP-1B, a phosphoprotein highly localized to developing axons, is involved in axonal extension [27,28]. The ability of MAP-1B to regulate microtubule organization depends on its phosphorylation state. Phosphorylated MAP-1B binds to microtubules more efficiently than unphosphorylated MAP-1B [24] and may promote axonal outgrowth [29]. As MAP-1B is expressed in cerebellar granule cell axons, changes in MAP-1B phosphorylation were examined in lithium- and Wnt-7a-treated neurons. Using specific antibodies against a phosphorylated isoform of MAP-1B, called MAP-1B-*P*, it was found that lithium induces a dramatic decrease in the levels of MAP-1B-*P* without significantly affecting the total level of MAP-1B [17a]. More importantly, the down-regulation of MAP-1B-*P* occurs within the first 4 h of lithium treatment, while morphological changes are evident after 12 h. These results suggest that lithium inhibits kinase(s) and/or activates phosphatase(s) acting on MAP-1B and that the decrease in the level of phosphorylated MAP-1B is an early event in axonal remodelling.

In vitro kinase assays revealed that GSK-3β phosphorylates MAP-1B directly at a site recognized by the antibody against MAP-1B-*P* [17a]. Therefore changes in the levels of MAP-1B-*P* directly reflect GSK-3β activity in axons. Wnt-7a also decreases the level of MAP-1B-*P* along the axon shaft and

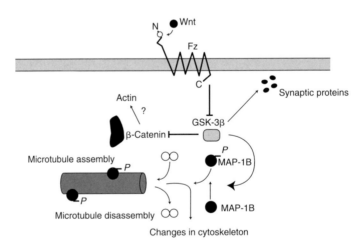

Fig. 1. Proposed Wnt signalling pathway in cerebellar granule cell neurons. Binding of Wnt to its receptor Frizzled (Fz) leads to the inactivation of GSK-3β. This response results in increased levels of β-catenin, which may bind to the actin cytoskeleton. In the absence of Wnt signalling, GSK-3β phosphorylates MAP-1B, resulting in increased microtubule stability. In contrast, inhibition of GSK-3β by Wnt results in a decline in phosphorylated MAP-1B and the loss of stable microtubules, with concomitant axon length shortening and axonal spreading. Inhibition of GSK-3β also increases the levels of synaptic proteins at spread areas of the axon.

in spread areas, as observed in lithium-treated neurons. These results led to the proposal that Wnt-7a inhibits GSK-3β, resulting in a decreased level of MAP-1B-*P*. The loss of phosphorylated MAP-1B may lead to microtubule instability by decreasing microtubule assembly (Fig. 1). This suggestion is in agreement with previous findings showing that phosphorylated MAP-1B has a higher affinity for microtubules than non-phosphorylated MAP-1B, and that phosphorylated MAP-1B is up-regulated during axonal extension [24]. These results also demonstrate that GSK-3β, a kinase highly expressed in the central nervous system [30], could regulate the neuronal cytoarchitecture during the formation of neuronal connections *in vivo*.

How does Wnt-7a induce axonal spreading?

Wnt-7a induces axonal spreading, increases axonal diameter and reduces axonal length [19]. Axon shortening is likely to be due to the loss of stable microtubules [1,31]. However, spreading along the entire axon shaft and at the growth cone cannot be explained solely by changes in microtubule stability. Cell spreading is a process that depends on the actin cytoskeleton and changes in cell adhesion [32]. Thus axonal spreading induced by the inhibition of GSK-3β is probably also associated with increased actin polymerization and changes in cell adhesion. β-Catenin may be involved in this process, as its level increases as a result of GSK-3β inhibition [19]. β-Catenin could affect the actin cytoskeleton by associating with the calcium-dependent adhesion proteins (cadherins) and α-catenin at the plasma membrane [33], or by binding to fascin, an actin-bundling protein, in a cadherin-independent fashion [10]. Thus inhibition of GSK-3β by lithium or Wnt-7a may increase adhesion and actin polymerization at specific areas in the axon through the action of β-catenin. Wnt-7a could also induce axonal spreading by directly increasing cadherin levels, as observed with Wnt factors in other systems [13,34,35]. Further analysis of the actin cytoskeleton and of the distribution of β-catenin and cadherins will lead to a better understanding of Wnt-7a-mediated axonal spreading.

Wnt-7a and synaptogenesis

In the postnatal cerebellum, *Wnt-7a* expression is detected as granule cells begin to migrate to deeper layers of the cerebellar cortex and extend axonal processes. However, the highest levels of *Wnt-7a* expression are observed after neurons have reached their final position and have initiated the period of synaptogenesis [19]. This pattern of expression suggests a role for Wnt-7a in synapse formation in the developing cerebellum. Indeed, granule cells exposed to Wnt-7a express higher levels of synapsin I [19], a protein localized at presynaptic sites and involved in synapse formation and function [36–38]. Synapsin I is particularly localized to spread areas of axons (Fig. 2). These results suggest that Wnt-7a plays a role in the formation of synapses by increasing synapsin I levels in cerebellar neurons [19]. Inhibition of GSK-3β by lithium also increases the levels of synapsin I along the axon, as observed in the

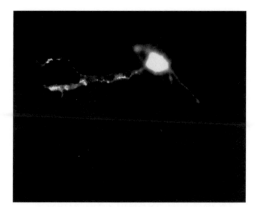

Fig. 2. Wnt-7a induces the localization of synapsin I clusters at spread areas. Granule cell neurons cultured in the presence of Wnt-7a show an increased level of synapsin I at spread areas of the developing axon.

presence of Wnt-7a [19]. Thus Wnt-7a, through GSK-3β, could influence the formation of synapses by increasing the levels of synapsin I at future synaptic sites.

The increase in the level and number of clusters of synapsin I in spread areas of axons following axonal remodelling suggests a possible link between cytoskeletal changes and the regulation of synapsin I levels. Although it has not previously been demonstrated that cytoskeletal re-organization can regulate the level of synaptic proteins, synapsin I is associated with the actin cytoskeleton. Indeed, the function of synapsin I is to link synaptic vesicles carrying neurotransmitters to the actin cytoskeleton, and this process is dependent on the phosphorylation state of synapsin I [39]. Calmodulin has been shown to phosphorylate synapsin I, resulting in the release of the synapsin I bound to synaptic vesicles from the actin cytoskeleton [40]. This process is essential for the localization of synaptic vesicles to the pre-synaptic terminal [41]. It is worth mentioning that spread areas induced by Wnt-7a along the axon are devoid of stable microtubules but enriched in actin filaments and synapsin I (P.C. Salinas, unpublished work). Thus cytoskeletal changes induced by Wnt-7a, through GSK-3β, may allow the accumulation of synapsin I at future synaptic sites. However, these experiments cannot discriminate whether cytoskeletal re-organization is required for synapsin I up-regulation or whether GSK-3β plays a more direct role in the regulation of synapsin I levels [19]. Further experiments are required to determine the mechanism by which GSK-3β controls synapsin I levels and the link with cytoskeleton re-organization.

What is the function of Wnt-7a *in vivo*?

In vitro analyses show that Wnt-7a can act in an autocrine fashion in cerebellar granule cells. However, granule cell axons exposed to Wnt-7a *in vitro* have a different morphology to those observed *in vivo* [19]. As granule cells express Wnt-7a *in vivo*, why then do granule cells lack spread and branched axons *in vivo*? The differences in granule cell morphology *in vivo* and *in vitro* could be explained by a polarized secretion of Wnt-7a *in vivo*. Unfortunately, the *in situ* localization of Wnt-7a cannot be determined due to the lack of specific antibodies against Wnt proteins. Another explanation is that the presence of Wnt-7a antagonists may restrict Wnt-7a actions in the cerebellum. Wnt antagonists have been identified in other developmental systems [42]. These factors, called Frizbees (Fzb), are secreted proteins bearing homology to the extracellular domain of Frizzled, the Wnt receptor [42,43]. Fzb are believed to act by binding to Wnt ligands [44–47]. Although Fzb have not been identified in the postnatal cerebellum, future experiments in this area may elucidate the site of action of Wnt-7a *in vivo*.

Cerebellar granule cells express *Wnt-7a* during axonal extension and synapse formation with their synaptic partners, the mossy-fibre axons and Purkinje cells [19]. As Wnt factors can act in both an autocrine and a paracrine fashion, Wnt-7a could act on mossy-fibre axons, Purkinje cells or any cells that come into close contact with granule cells and express the appropriate receptors. Preliminary results show that Wnt-7a also induces axonal spreading in mossy-fibre axons (F.R. Lucas and P.C. Salinas, unpublished work). Therefore Wnt-7a could act in both a paracrine and an autocrine fashion in the cerebellum.

Wnt-7a null mutant mice have been generated by homologous recombination [48]. Although these mutants exhibit defects in limb patterning, no obvious neurological defects have been described [48]. The lack of neuronal defects suggests that other Wnt factors expressed in the cerebellum may complement Wnt-7a function. Interestingly, *Wnt-7b*, a gene highly homologous to *Wnt-7a*, is expressed in cerebellar granule cell neurons during postnatal development (P.C. Salinas, unpublished work). Detailed analysis of the central nervous system of these mutant mice is required in order to establish the role of Wnt-7a in the formation of neuronal connections *in vivo*.

Conclusions

The mechanisms that regulate the neuronal cytoskeleton have been the centre of numerous studies in recent years. This problem has been tackled by two main approaches. On one hand, analyses of the microtubule and actin cytoskeleton have provided a good understanding of how the cytoskeleton is regulated within the cell. On the other, a number of extracellular signals that influence the behaviour of the cytoskeleton have been identified. However, we have a poor understanding of how extracellular signals are transduced to the cytoplasm to bring about cytoskeletal changes. Studies of the signalling pathways of guidance molecules are beginning to reveal how the extracellular environment modulates the shape and behaviour of cells. Wnt proteins, previously studied as factors that control cell-fate decisions, may also act as neuronal guidance molecules by regulating microtubule organization in developing axons. Studies show that Wnt-7a induces axonal remodelling and subsequently increases the levels of synaptic proteins in primary cerebellar neurons. These responses are mediated through the inhibition of GSK-3β, a serine/threonine kinase highly expressed in developing neurons. Although the relationship between cytoskeletal changes and synaptic protein levels remains unclear, future studies on Wnts may unravel new mechanisms regulating the neuronal cytoskeleton and the formation of synapses in the developing brain.

References

1.   Tanaka, E. and Sabry, J. (1995) Cell **83**, 171–176
2.   TessierLavigne, M. and Goodman, C.S. (1996) Science **274**, 1123–1133
3.   Nusse, R. and Varmus, H.E. (1992) Cell **69**, 1073–1087
4.   Parr, B.A. and McMahon, A.P. (1994) Curr. Opin. Genet. Dev. **4**, 523–528
5.   Nusse, R. (1997) Cell **89**, 321–323
6.   Willert, K., Brink, M., Wodarz, A., Varmus, H. and Nusse, R. (1997) EMBO J. **16**, 3089–3096
7.   Yanagawa, S.I., Van Leeuwen, F., Wodarz, A., Klingensmith, J. and Nusse, R. (1995) Genes Dev. **9**, 1087–1095
8.   Perrimon, N. (1996) Cell **86**, 513–516
9.   Willert, K. and Nusse, R. (1998) Curr. Opin. Genet. Dev. **8**, 95–102
10.   Tao, Y.S., Edwards, R.A., Tubb, B., Wang, S., Bryan, J. and McCrea, P.D. (1996) J. Cell Biol. **134**, 1271–1281
11.   Rubinfeld, B., Albert, I., Porfiri, E., Fiol, C., Munemitsu, S. and Polakis, P. (1996) Science **272**, 1023–1026
12.   Nathke, I.S., Adams, C.L., Polakis, P., Sellin, J.H. and Nelson, W.J. (1996) J. Cell Biol. **134**, 165–179
13.   Hinck, L., Nelson, W.J. and Papkoff, J. (1994) J. Cell Biol. **124**, 729–741
14.   Kemler, R. (1993) Trends Genet. **9**, 317–321
15.   Aberle, H., Schwartz, H. and Kemler, R. (1996) J. Cell. Biochem. **61**, 514–523
16.   Rocheleau, C.E., Downs, W.D., Lin, R.L., Wittmann, C., Bei, Y.X., Cha, Y.H., Ali, M., Priess, J.R. and Mello, C.C. (1997) Cell **90**, 707–716
17.   Thorpe, C.J., Schlesinger, A., Carter, J.C. and Bowerman, B. (1997) Cell **90**, 695–705
17a. Lucas, F.R., Goold, R.G., Gordon-Weeks, P.R. and Salinas, P.C. (1998) J. Cell Sci. **111**, 1351–1361
18.   Salinas, P.C., Fletcher, C., Copeland, N.G., Jenkins, N.A. and Nusse, R. (1994) Development **120**, 1277–1286
19.   Lucas, F.R. and Salinas, P.C. (1997) Dev. Biol. **193**, 31–44
20.   Hedgepeth, C.M., Conrad, L.J., Zhang, J., Huang, H.C., Lee, V.M.Y. and Klein, P.S. (1997) Dev. Biol. **185**, 82–91
21.   Klein, P.S. and Melton, D.A. (1996) Proc. Natl. Acad. Sci. U.S.A. **93**, 8455–8459
22.   Stambolic, V., Ruel, L. and Woodgett, J.R. (1996) Curr. Biol. **6**, 1664–1668
23.   Atack, J.R., Broughton, H.B. and Pollack, S.J. (1995) Trends Neurosci. **18**, 343–349
24.   Gordon-Weeks, P.R. (1997) in Brain Microtubule Associated Proteins: Modifications in Disease (Avila, J., Kosik, K. and Brandt, R., eds.), pp. 53–72, Harwood Academic Publishers, Switzerland
25.   Avila, J., Dominguez, J. and DiazNido, J. (1994) Int. J. Dev. Biol. **38**, 13–25
26.   Maccioni, R.B. and Cambiazo, V. (1995) Physiol. Rev. **75**, 835–864
27.   DiTella, M.C., Feiguin, F., Carri, N., Kosik, K.S. and Caceres, A. (1996) J. Cell Sci. **109**, 467–477
28.   Edelmann, W., Zervas, M., Costello, P. et al. (1996) Proc. Natl. Acad. Sci. U.S.A. **93**, 1270–1275
29.   Ulloa, L., DiazNido, J. and Avila, J. (1993) EMBO J. **12**, 1633–1640
30.   Woodgett, J.R. (1990) EMBO J. **9**, 2431–2438

31. Challacombe, J.F., Snow, D.M. and Letourneau, P.C. (1996) Semin. Neurosci. **8**, 67–80
32. Adams, J.C. (1997) Trends Cell Biol. **7**, 107–110
33. Hoschuetzky, H., Aberle, H. and Kemler, R. (1994) J. Cell Biol. **127**, 1375–1380
34. Shimamura, K., Hirano, S., McMahon, A.P. and Takeichi, M. (1994) Development **120**, 2225–2234
35. Barth, A.I., Nathke, I.S. and Nelson, W.J. (1997) Curr. Opin. Cell Biol. **9**, 683–690
36. De Camilli, P., Benfenati, F., Valtorta, F. and Greengard, P. (1990) Annu. Rev. Cell Biol. **6**, 433–460
37. Ferreira, A., Han, H.Q., Greengard, P. and Kosik, K.S. (1995) Proc. Natl. Acad. Sci. U.S.A. **92**, 9225–9229
38. Ferreira, A., Kosik, K.S., Greengard, P. and Han, H.Q. (1994) Science **264**, 977–979
39. Bahler, M. and Greengard, P. (1987) Nature (London) **326**, 704–707
40. Ceccaldi, P.E., Grohovaz, F., Benfenati, F., Chieregatti, E., Greengard, P. and Valtorta, F. (1995) J. Cell Biol. **128**, 905–912
41. Greengard, P., Valtorta, F., Czernik, A.J. and Benfenati, F. (1993) Science **259**, 780–785
42. Zorn, A.M. (1997) Curr. Biol. **7**, R501–R504
43. Bhanot, P., Brink, M., Harryman Samos, C., Hsieh, J., Wang, Y., Macke, J.P., Andrew, D., Nathans, J. and Nusse, R. (1996) Nature (London) **382**, 225–230
44. Lin, K.M., Wang, S.W., Julius, M.A., Kitajewski, J., Moos, M. and Luyten, F.P. (1997) Proc. Natl. Acad. Sci. U.S.A. **94**, 11196–11200
45. Leyns, L., Bouwmeester, T., Kim, S.H., Piccolo, S. and De Robertis, E.M. (1997) Cell **88**, 747–756
46. Wang, S., Krinks, M., Lin, K., Luyten, F.P. and Moos, Jr., M. (1997) Cell **88**, 757–766
47. Finch, P.W., He, X., Kelley, M.J., Uren, A., Schaudies, R.P., Popescu, N.C., Rudikoff, S., Aaronson, S.A., Varmus, H.E. and Rubin, J.S. (1997) Proc. Natl. Acad. Sci. U.S.A. **94**, 6770–6775
48. Parr, B.A. and McMahon, A.P. (1995) Nature (London) **374**, 350–353

Biochem. Soc. Symp. **65**, 111–123
Printed in Great Britain

8

Rho family proteins and cell migration

Anne J. Ridley*†[1], William E. Allen*‡, Maikel Peppelenbosch* and Gareth E. Jones‡

*The Ludwig Institute for Cancer Research, University College London Branch, 91 Riding House Street, London W1P 8BT, U.K., †Department of Biochemistry and Molecular Biology, University College London, London WC1E 6BT, U.K., and ‡The Randall Institute, King's College London, London WC2B 5RL, U.K.

Abstract

The GTP-binding proteins Rho, Rac and Cdc42 are known to regulate actin organization: Rho induces the assembly of contractile actin-based filaments such as stress fibres, Rac regulates the formation of lamellipodia and membrane ruffles, while Cdc42 is required for filopodium extension. All three proteins can also regulate the assembly of integrin-containing focal adhesion complexes. Cell migration involves co-ordinated and dynamic changes in the actin cytoskeleton and cell adhesion, and we have therefore investigated the roles of Rho family proteins in migration, using two model cell systems. First, in the macrophage cell line Bac1, Rho and Rac were found to be required for colony-stimulating factor-1 (CSF-1)-induced cell migration. In contrast, inhibition of Cdc42 does not prevent macrophages migrating in response to CSF-1, but does prevent recognition of a CSF-1 concentration gradient, so that cells now migrate randomly rather than up the gradient. This implies that Cdc42, and probably filopodia, are required for gradient sensing and cell polarization. Secondly, in the Madin–Darby canine kidney (MDCK) epithelial cell line, Rho and Rac are also essential for migration induced by hepatocyte growth factor/scatter factor. Rac is required for lamellipodium formation and is apparently activated via Ras. Interestingly, however, Rac does not induce lamellipodium formation in unstimulated MDCK cells, indicating that Rac signals differently in epithelial cells compared with fibroblasts or macrophages. Our results point to central roles for Rho, Rac and Cdc42 in co-ordinating cell migration.

[1]To whom correspondence should be addressed, at The Ludwig Institute for Cancer Research.

Introduction

Cell migration is a complex process involving dynamic, co-ordinated changes in cell adhesion and in the cytoskeleton, and also remodelling and degradation of the extracellular matrix to create a passage for cells [1]. In addition, haematopoietic cells need to migrate through endothelial cell barriers to reach their targets in the tissues. Directed migration is stimulated by chemoattractants, and leucocytes are able to detect small differences in the concentrations of these substances [2]. The composition of the extracellular matrix also has a profound influence on the rate of cell migration [1]. The intracellular signalling pathways mediating cell migration therefore have to respond to diverse signals from outside the cell and translate these into very precisely tuned changes within the cell. Of the signalling proteins implicated in cell migration responses, the Rho family of GTPases appear to have the properties that would allow them to act concertedly as co-ordinators of cell migration. Initially identified as regulators of actin organization, they also regulate cell adhesion to the extracellular matrix and to other cells and, in the long term, influence gene expression [3]. They can be activated in response to cytokines, growth factors and extracellular-matrix proteins [3–6]. They can interact with many different target proteins within cells, allowing them to activate a number of downstream signals simultaneously [3,5]. Finally, they show extensive cross-talk with each other [7], which could allow them to co-ordinate different aspects of cell locomotion.

Rho family proteins: structure and function

The Rho family is part of the Ras superfamily of small (around 21 kDa) GTP-binding proteins. The Ras superfamily also includes the Rab family, which regulate intracellular vesicle transport processes, the ADP-ribosylation factor (ARF) family, which are involved both in signal transduction and in vesicle transport, and Ran, which is required for protein transport into the nucleus ([8], and references therein). Members of the Ras superfamily show conserved structural features which reflect their ability to bind guanine nucleotides, although the crystal structures solved so far indicate that there are fine variations between these proteins in the precise mechanisms of binding and hydrolysing GTP [9–11].

So far, 11 mammalian members of the Rho family have been identified: RhoA, RhoB, RhoC, Rac1, Rac2, Cdc42 (two alternatively spliced variants), TC10, RhoD, RhoE, RhoG and TTF [3,12]. Homologues of Cdc42 and Rho are found in yeast, and many Rho family genes have been isolated from higher organisms. For some of these, including *MIG2* from *Caenorhabditis elegans* and *RhoL* from *Drosophila melanogaster*, a mammalian counterpart has not yet been identified [13,14]. Of the mammalian proteins, the best characterized for their ability to regulate actin organization are Rho, Rac and Cdc42. Rho was the first member of this family to be cloned in 1985, followed a few years later by Rac and Cdc42 [8]. As Rho shows homology to Ras, it was at first studied as a potential oncogene, but it was the realization that Rho family proteins are targets for a number of bacterial exoenzymes and toxins that really opened up the field of Rho research [4]. The most frequently used tool for studying Rho function is C3 transferase, an exoenzyme from *Clostridium botulinum*, which ADP-ribosylates and inactivates Rho predominantly, although it can inefficiently modify other members of the family [15]. Treatment of many cell types with C3 transferase induces cell rounding and loss of stress fibres, and this was the first indication that Rho influences the actin cytoskeleton [4]. Subsequently, Rac was also shown to regulate actin organization, and at the same time was independently purified as an essential cofactor for the NADPH oxidase in phagocytic cells [16]. More recently, Rho, Rac and Cdc42 have been implicated in a diverse array of cellular responses, from secretion and endocytosis to transcriptional regulation [3]. In some cases, these effects of Rho family proteins have been reported to be independent of their actions on the actin cytoskeleton.

Proteins interacting with Rho family proteins

All Rho family proteins bind GTP, and the majority have been shown to act as GTPases and cycle between an active, GTP-bound, form and an inactive, GDP-bound, form. An exception to this is RhoE, which does not appear to hydrolyse GTP significantly [17]. Three different types of protein have been found to regulate the cycling of Rho family proteins [18]. First, exchange factors stimulate the release of nucleotide, allowing GTP, which is at a higher concentration than GDP in cells, to bind and thereby activate the protein. Exchange factors for Rho family proteins all contain a homologous 'Dbl' domain, which is sufficient to stimulate exchange [19]. Secondly, the intrinsic GTPase activity of the proteins is enhanced by GTPase-activating proteins (GAPs). The catalytic activity of these GAPs *in vitro* is low, but they certainly have the ability to down-regulate the proteins functionally within cells [20]. Thirdly, Rho, Rac and Cdc42 have all been shown to complex with proteins known as GDIs (guanine nucleotide dissociation inhibitors), which prevent their interaction with other regulatory proteins and keep them complexed in the cytoplasm [8].

It is presumed that incoming signals activate a Rho family protein by increasing the amount of protein bound to GTP. It is known that Ras–GTP levels increase in response to many extracellular signals, but for technical reasons this has not been proven for any member of the Rho family. An alternative possibility is that re-localization is the key step in regulating these proteins. Indeed, translocation of Rho, Rac and Cdc42 to plasma membrane or cytoskeletal fractions is observed following cell stimulation [21,22]. Exchange factor activity for Cdc42 is associated with membrane fractions [23], and in fact the association of Tiam1 (T-lymphoma invasion and metastasis 1), an exchange factor for Rac, with the plasma membrane is regulated by growth factors [24], suggesting another level at which activation of Rho family proteins could be regulated. Finally, GDIs may play an active role in delivering the GTPases to appropriate sites at the plasma membrane, where they can be activated by exchange factors and/or interact with target proteins. For example, RhoGDI can associate in a complex with the transmembrane protein CD44 and the cytoplasmic ERM (ezrin/radixin/moesin) proteins [25]. Rho function is closely linked to ERM proteins: it is required for the association of ERM proteins with the plasma membrane and with CD44, and it can co-localize with ERM proteins [25,26]. In addition, a correct interaction of ERM proteins with actin is required for Rho-induced actin re-organization [27].

Once activated, Rho family proteins can interact with downstream target proteins, stimulating signalling pathways that lead to the observed cellular responses. Many targets for Rho family proteins have been identified, including protein kinases, phosphoinositide kinases and adaptor proteins, which have no enzymic function but have the ability to interact with one or more other proteins [3,5]. A number of targets have the potential to link Rho, Rac or Cdc42 directly with the actin cytoskeleton. For example, IQGAP1 binds to actin filaments and also to Rac and Cdc42 [28,29]. In addition, Rho-associated kinase (ROK) can induce the phosphorylation of myosin II light-chain kinase [3], while the PAK (p21-activated kinase) family of kinases can phosphorylate myosin I heavy chain [30,31]. These kinases are therefore predicted to modulate the activity of the respective myosins and their association with actin filaments [3,32]. Finally, the Rho target p140mDia can bind to profilin [33], an actin-binding protein with the potential to enhance actin polymerization. The precise mechanisms whereby these targets, and undoubtedly others, act to regulate actin re-organization have yet to be elucidated.

Rho family proteins, the actin cytoskeleton and cell adhesion

Swiss 3T3 fibroblasts

The roles of Rho, Rac and Cdc42 in regulating actin organization were first characterized in detail in Swiss 3T3 fibroblasts. These cells have proven to be a good model system for analysing rapid changes in the actin cytoskeleton, as when they are confluent and serum-starved they lose practically all of the two most prominent actin-filament-containing structures found in fibroblasts: stress fibres and lamellipodia. Stress fibres are bundles of actin filaments associated with myosin II filaments and other proteins, forming contractile fibres. They terminate at the plasma membrane in focal adhesions, where transmembrane integrins are clustered and associate both with extracellular-matrix proteins outside the cell and with a large number of proteins inside the cell. Some of these intracellular proteins play a structural role in linking integrins to stress fibres, while others are signal-transducing proteins [32,33]. Lamellipodia are broad, highly dynamic membrane protrusions that extend and retract through a combination of actin polymerization at the plasma membrane, depolymerization within the cytoplasm and myosin-mediated rearward movement of the actin fibres [34].

Constitutively active mutants of Rho and Rac induce the formation of stress fibres and lamellipodia respectively when microinjected into quiescent Swiss 3T3 cells. Conversely, microinjection of C3 transferase to inhibit Rho or of a dominant-negative Rac mutant to inhibit Rac inhibits growth-factor-induced formation of these structures [15,35]. Activated Cdc42 protein induces the extension of filopodia [36,37], which are finger-like plasma membrane protrusions containing actin filament bundles that actively protrude and retract. Under appropriate conditions, Cdc42, Rac and Rho can activate each other sequentially in a cascade: Cdc42 can induce Rac-mediated lamellipodium formation, and Rac can induce Rho-mediated stress fibre formation [35,36]. In fact, in serum-starved, confluent Swiss 3T3 cells the predominant response to activated Cdc42 is stress fibre formation.

Rho, Rac and Cdc42 also regulate the assembly of sites of adhesion to the extracellular matrix in fibroblasts. Rho mediates the formation of focal adhesions, while Rac and Cdc42 induce the formation of smaller adhesion sites to the extracellular matrix, which are located in lamellipodia and at the bases of filopodia [15,36].

Macrophages

To determine how Rho, Rac and Cdc42 act in cell types other than fibroblasts, we have investigated their roles in a variety of cell lines. In particular, we were interested in macrophages as a cell type lacking stress fibres and yet amenable to microinjection. Macrophages originate as monocytes in the circulation, and are stimulated to exit blood vessels in response to inflammatory signals. They then differentiate into macrophages, which can phagocytose invading micro-organisms and cellular debris, as well as releasing cytokines which recruit further leucocytes and other cells to the site of inflammation. The mouse macrophage cell line that we have used, Bac1.2F5, resembles primary macrophages in being dependent upon colony-stimulating factor-1 (CSF-1) for survival and proliferation [38,39].

Constitutively active and dominant-negative mutants of Rho, Rac and Cdc42 were injected into Bac1 macrophages and assessed for their effects on the actin cytoskeleton and on adhesion sites [40]. As in fibroblasts, Rac induces the formation of lamellipodia and membrane ruffles (Fig. 1C), and is required for the formation of these structures in response to CSF-1. Cdc42 induces rapid formation of filopodia (Fig. 1D), and again is required for CSF-1-induced filopodium extension. These cells do not possess stress fibres, but have very fine actin cables within the cytoplasm, running parallel to the plasma membrane and around the nucleus. These cables are not detectable in starved cells, but reappear upon stimulation with CSF-1 after 15–30 min. Rho is required for this response and is activated downstream of Rac. The ability of Rac to activate Rho is thus conserved between fibroblasts and macrophages. Activated Rho also stimulates the formation of these cables in starved cells, and induces cell contraction [40] (Fig. 1B).

Interestingly, Bac1 macrophages have focal complexes which are regulated by Cdc42 acting upstream of Rac [40]. Again, the link between Cdc42 and Rac is present in these cells, as in fibroblasts. These focal complexes contain proteins normally associated with fibroblast focal adhesions, including β_1 integrin, vinculin, focal adhesion kinase and paxillin. In Bac1 cells, Rho does not regulate the formation of focal complexes, suggesting that it does not modulate cell adhesion directly, at least via integrin-containing complexes.

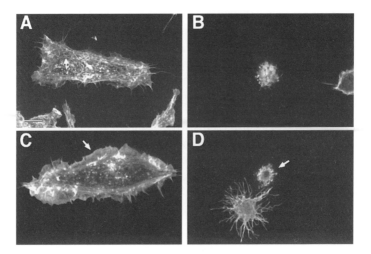

Fig. 1. Changes in actin organization induced by Rho, Rac and Cdc42 in Bac1 macrophages. An example of a polarized Bac1 cell maintained in CSF-1-containing medium is shown in (A). Fine actin cables in the body of the cell are indicated by the arrow. Bac1 cells were starved of CSF-1 for 24 h, then injected with V14-RhoA (B), V12-Rac1 (C) or V12-Cdc42 (D). An injected CSF-1-starved cell is indicated by an arrow in (D). Cells were fixed after 15–30 min and stained with tetramethylrhodamine isothiocyanate–phalloidin to show localization of actin filaments. The bar in (A) represents 10 μm.

Rho family proteins and cell migration

The ability of Rho, Rac and Cdc42 to regulate cell adhesion and actin organization suggests that they could be involved in cell migration responses. A role for Rho in cell motility has been implicated in studies using C3 transferase. For example, when C3 transferase was introduced into neutrophils by electropermeabilization, it was found to inhibit neutrophil migration stimulated by zymosan-activated serum [41]. In addition, a microinjection approach in Swiss 3T3 cells has shown that cell motility is inhibited by either C3 transferase or RhoGDI, which binds to Rho, Rac and Cdc42, as determined by observing phagocytic tracks on colloidal gold [42]. This approach is not particularly quantitative in that it does not measure migration speed, and it does not address whether inhibition of Rho might also affect the ability of cells to take up colloidal gold.

Macrophages

We have investigated the roles of Rho, Rac and Cdc42 proteins in regulating the migration and chemotaxis of Bac1 macrophages. Migration of microinjected cells can be measured directly in a Dunn chamber by time-lapse microscopy (Fig. 2) followed by computer-assisted analysis [43]. In the Dunn chamber, individual cells are observed migrating up a concentration gradient of CSF-1, which is a chemoattractant for Bac1 cells [43]. This allows the immediate effects of microinjecting proteins to be determined, avoiding the less direct

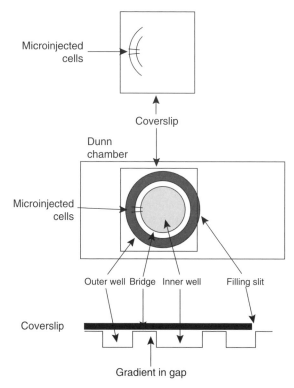

Fig. 2. Use of the Dunn chamber to analyse migration of micro-injected cells. Cells in a marked area of a coverslip are microinjected with protein, then the coverslip is inverted on to the chemotaxis chamber, where both wells are filled with CSF-1-free medium. The marked region is placed over the annular bridge. The coverslip is sealed into position using wax, leaving a slit for draining the outer well. The medium is then drained from this well and replaced with medium containing CSF-1, and the slit is finally sealed with wax. The migration of the microinjected cells in the marked region is recorded, together with that of adjacent uninjected cells.

approach of establishing cell lines, which show altered migratory responses as a result of long-term changes in gene expression induced by the exogenously expressed GTPases.

Using the Dunn chamber, we have observed that activated forms of Rho, Rac and Cdc42 inhibit the migration of macrophages in response to CSF-1 [43a]. This most likely reflects the fact that expression of these proteins dramatically re-organizes the cytoskeleton, such that it is no longer able to be polarized sufficiently to allow migration (Fig. 1) [40]. Inhibition of Rho and Rac, by microinjection of C3 transferase and of the dominant-negative N17-Rac1 protein respectively, prevents the migration of cells. The effect of N17-Rac1 is expected, as lamellipodia are universally observed at the leading edge of migrating cells, and are required for forward protrusion of the membrane. The effect of inhibiting Rho suggests that the contractile actin cables regulated by Rho play an important role in mediating cell migration, probably pulling the body of the cell forward. In the absence of Rho, cells gradually extend dendritic processes, with lamellipodia at the ends. This suggests that the cell is still able to extend its leading edge, but the cell body does not follow and therefore movement is abortive. However, the leading edges keep extending, resembling neurite outgrowth in some respects.

Unlike inhibition of Rho and Rac, which completely abrogates cell migration, inhibition of Cdc42 does not prevent cells moving in response to CSF-1, but actually enhances their migration rate compared with control injected cells [43a]. However, the dominant-negative N17-Cdc42 protein does prevent cells recognizing a gradient, and the cells migrate in random directions, as though they are not exposed to a concentration gradient. This effect of N17-Cdc42 resembles the response of Bac1 cells to tumour necrosis factor-α, which abolishes their ability to detect a gradient of CSF-1 without altering their speed of locomotion [43]. Consistent with a role for Cdc42 in responding to the concentration gradient, tumour necrosis factor-α inhibits CSF-1-induced filopodium formation without affecting lamellipodium formation or membrane ruffling (M. Peppelenbosch and A.J. Ridley, unpublished work).

How do Bac1 cells respond to a gradient of CSF-1, and how does Cdc42 mediate this response? Previous work on Bac1 cells has shown that the CSF-1 receptor is rapidly endocytosed following stimulation with CSF-1, and that the level of CSF-1 receptors on the cell surface after 30 min is very low [38,39]. This makes it unlikely that the cells are continuously sensing and responding to the gradient, a hypothesis supported by the fact that a cell which initially lies directly behind another cell appears to sense the gradient incorrectly and moves persistently away from the source of CSF-1, and does not re-polarize when it is away from the original blocking cell [43]. A possible model for how Bac1 cells polarize is shown in Fig. 3. The earliest response to CSF-1 is extension of filopodia and lamellipodia around the cell, together with membrane ruffling on the upper surface of the cell, followed by cell spreading. It is likely that the gradient is detected during this initial stage, and that the extension of filopodia is necessary for this. The filopodia may act as sensing devices, because they have a high surface area with the potential to carry large numbers of CSF-1 receptors. When they retract back into the cell, they may create a concentrated 'hot spot' of activated receptor which is subsequently endocytosed but is still active. The cell then senses the difference in signal intensity between the front and the back of the cell, and polarizes its actin cytoskeleton accordingly. Possibly, the sites of filopodium retraction act as centres to drive the polarization, recruiting proteins to form the leading lamella. Cdc42, in stimulating the formation of filopodia, thereby initiates the process of gradient detection.

In the absence of Cdc42, the cells still polarize and migrate relatively persistently in one direction, but not in response to the gradient. The extent of polarization is less, however, and the cells remain quite rounded. It is likely that this polarization is due to a stochastic process of increased lamellar growth in one direction, with the cytoskeleton subsequently polarizing towards this extension [1].

One possible explanation for why the cells move faster when Cdc42 is inhibited is that the adhesions to the substratum are weaker, because the inte-

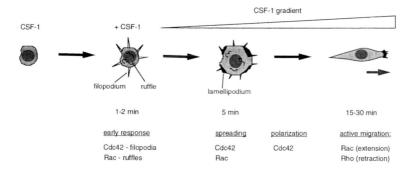

Fig. 3. Model for the roles of Rho, Rac and Cdc42 in macrophage migration and chemotaxis. In the absence of CSF-1, BacI macrophages have a rounded morphology. Following addition of CSF-1, they immediately extend filopodia and form membrane ruffles. Subsequently, they spread via the extension of filopodia and lamellipodia, forming new adhesions to the substratum. In a gradient of CSF-1 the cell then determines which part of the cell has the highest level of signal from activated CSF-1 receptors, and polarizes its actin cytoskeleton such that lamellipodium and filopodium extension become restricted to this region of the cell. This polarization requires Cdc42 activity. Cell migration is subsequently mediated primarily through the actions of Rac, which extends the lamellipodia, and Rho, which causes retraction of the cell body so as to detach the rear of the cell.

grin-containing focal complexes are dispersed. It has been shown that cell migration is maximal at a critical strength of interaction between the cell and the extracellular matrix, dependent on ligand concentration, integrin expression and integrin–ligand affinity [44]. Possibly, due to decreased clustering of integrins, the cells can move faster [1]. Alternatively, the N17-Cdc42-injected cells might move faster because they are more rounded and have broader lamellae.

Cdc42 has also been implicated in generating cell polarity in other systems. In yeast, it is required for polarization of the actin cytoskeleton in response to bud-site selection, and is also a component of the pheromone-activated signalling pathway leading to polarization of cells towards a mating partner, again reflecting changes in the cytoskeleton [45]. In the *D. melanogaster* wing disc epithelium, Cdc42 is required for epithelial cell elongation, which in turn is essential for the generation of apico–basal polarity [46]. Finally, in T-cells, Cdc42 is required for the polarization of the cytoskeleton towards antigen-presenting cells [47]. All of these results suggest that Cdc42 plays a conserved role in mediating the polarization of the actin cytoskeleton.

Epithelial cells

The ability of epithelial cells to migrate involves additional levels of regulation over cells such as fibroblasts and macrophages, because the cells have to detach from each other in order to migrate independently. Sometimes epithelial cells are observed to move as sheets, but in many cases motogens induce loss of cell–cell contacts as well as increased motility.

The Madin–Darby canine kidney (MDCK) epithelial cell line has been used extensively as a model for studying the induction of epithelial cell migration. MDCK cells normally grow as colonies, with adherens junctions, desmosomes and tight junctions between the cells. Hepatocyte growth factor/scatter factor (HGF/SF) induces a biphasic motility response in these cells [48,49]. First, over the first 4 h, the cells extend outwards centripetally but do not physically detach from each other. Secondly, after 4 h the cells begin to separate from one another and migrate independently, similar to fibroblasts, although they still retain the capacity to form some types of junctions and to interact with each other. The migrating cells have few, if any, stress fibres, consistent with the general observation that highly motile cells in culture have far fewer or no stress fibres compared with less motile counterparts [50]. HGF/SF-induced migration is completely prevented by microinjection of activated Rho, which induces stress fibre formation and focal adhesion assembly. How does Rho inhibit migration? It is probable that the dense clustering of integrins in focal adhesions prohibits the retraction stage of cell locomotion [1]. In addition, Rho also prevents HGF/SF-induced lamellipodium extension (A.J. Ridley, unpublished work), possibly because much of the available actin is sequestered in stress fibres. Interestingly, injection of C3 transferase at a concentration sufficient to induce complete loss of stress fibres does not impair HGF/SF-induced migration, but this enzyme does inhibit migration when injected at higher concentrations [49]. Some level of Rho activity is therefore likely to be required for MDCK cell migration, as in macrophages, presumably during the contraction stage of locomotion.

Our studies have also shown that Ras and Rac are required for HGF/SF-induced cell migration, but that activated forms of these proteins are not sufficient to mimic the HGF/SF response [49]. Ras can stimulate lamellipodium extension and cell spreading, but the cells remain attached to each other. Rac by itself does not induce lamellipodium formation, in contrast to its effects in non-epithelial cell types. In fact, it actually appears to enhance accumulation of actin filaments at intercellular junctions, and also inhibits cell migration (A.J. Ridley, unpublished work). This effect may be related to the observation that, in keratinocytes, Rac is required for the formation of cadherin-based adherens junctions induced by increasing extracellular calcium levels [51]. The role of Rac therefore appears to differ depending on the type of cell and also the status of the cell. It will be important to determine how Rac and other related proteins affect the structure of intercellular junctions in MDCK cells, and whether Rac signalling is altered by HGF/SF.

Analysis of cell migration *in vivo*

The ability of Rho family proteins to regulate cell migration *in vitro* is supported by observations on their roles *in vivo* in multicellular organisms. In *D. melanogaster*, axon and dendrite outgrowth is inhibited when dominant-negative inhibitors or activated forms of Rac and Cdc42 inhibit axon and dendrite outgrowth when selectively expressed in neuronal cells [52]. This indicates that they play an important role in axon outgrowth, in accordance with the reported effects of Rac and Cdc42 on neurite outgrowth in mammalian cells [53,54]. In addition, Rac is required for border cell migration during *D. melanogaster* oogenesis [14]. In the mouse, activated Rac arrests both axon outgrowth and dendritic spine morphogenesis in Purkinje cells [55].

In *C. elegans*, a new Rho family member, MIG-2, has been shown to be required specifically for cell migration and for axon guidance [13]. Mutations in *MIG-2* that would be expected to eliminate gene function inhibit the migration of neuroblasts and neurons; the cells migrate more slowly than normal and often fail to reach their destinations. Interestingly, mutations that would be predicted to activate MIG-2 constitutively have a more dramatic phenotype than null mutations, suggesting that MIG-2 might be functionally redundant in most cell types, but is capable of inducing a phenotype when constitutively active.

Conclusions

Our results with macrophages imply that Rho, Rac and Cdc42 act in concert to control different aspects of cell migration. The precise contribution of each protein may well differ depending on the cell type, the chemotactic signal and the composition of the matrix upon which the cells are moving. In addition, it will be of interest to determine whether other members of the Rho family are also involved in regulating cell locomotion. Notably, there are novel Rho family genes in *D. melanogaster* and *C. elegans* that regulate cell migration *in vivo* and for which mammalian counterparts have not yet been identified. Rho family proteins have not so far been implicated in other aspects of cell migration, such as matrix degradation or modification. Particularly interesting in this respect is the recent observation that specific cleavage of extracellular-matrix proteins by metalloproteases is required for optimal migration [56,57]. The signalling pathways regulating metalloprotease activation have yet to be clearly delineated.

There are many different pathological situations where it would be useful to be able to inhibit cell migration selectively. Cancer cell invasion and metastasis, for example, involves migration of cells both through the extracellular matrix and into and out of the vasculature. In chronic inflammatory disorders such as rheumatoid arthritis, tissue damage is caused by migration of macrophages and other cell types into the site of inflammation. Macrophage recruitment also contributes to the development of atherosclerotic plaques. A variety of different approaches can be taken to inhibit cell migration. It can be inhibited from outside the cell, by preventing matrix degradation or interaction of the cell with the extracellular matrix. Alternatively, it can be inhibited from inside the cell, by targeting signal transduction pathways activated by chemoattractants and/or by changes in cell matrix composition. Components of signalling pathways activated by Rho family proteins may be relevant therapeutic targets for inhibiting cell migration.

References

1.   Lauffenburger, D.A. and Horwitz, A.F. (1996) Cell **84**, 359–369
2.   Zigmond, S.H. (1981) J. Cell Biol. **88**, 644–647
3.   Ridley, A.J. (1996) Curr. Biol. **6**, 1256–1264
4.   Machesky, L.M. and Hall, A. (1996) Trends Cell Biol. **6**, 304–310
5.   Tapon, N. and Hall, A. (1997) Curr. Opin. Cell Biol. **9**, 86–92
6.   Barry, S.T., Flinn, H.M., Humphries, M., Critchley, D.R. and Ridley, A.J. (1997) Cell Adhes. Commun. **4**, 387–398
7.   Chant, J. and Stowers, L. (1995) Cell **81**, 1–4
8.   Hall, A. (1994) Annu. Rev. Cell Biol. **10**, 31–54
9.   Schweins, T. and Wittinghofer, A. (1994) Curr. Biol. **4**, 547–550
10.   Hirshberg, M., Stockley, R.W., Dodson, G. and Webb, M.R. (1997) Nature Struct. Biol. **4**, 147–153
11.   Wei, Y., Zhang, Y., Derewenda, U., Liu, X., Minor, W., Nakamoto, R.K., Somlyo, A.V., Somlyo, A.P. and Derewenda, Z.S. (1997) Nature Struct. Biol. **4**, 699–703
12.   Murphy, C., Saffrich, R., Grummt, M., Gournier, H., Rybin, V., Rubino, M., Auvinen, P., Lütcke, A., Parton, R.G. and Zerial, M. (1996) Nature (London) **384**, 427–432
13.   Zipkin, I.D., Kindt, R.M. and Kenyon, C.M. (1997) Cell **90**, 883–894
14.   Murphy, A.M. and Montell, D.J. (1996) J. Cell Biol. **133**, 617–630
15.   Ridley, A.J. and Hall, A. (1992) Cell **70**, 401–410
16.   Ridley, A.J. (1995) Curr. Biol. **5**, 710–712
17.   Foster, R., Hu, K.-Q., Lu, Y., Nolan, K.M., Thissen, J. and Settleman, J. (1996) Mol. Cell. Biol. **16**, 2689–2699
18.   Boguski, M.S. and McCormick, F. (1993) Nature (London) **366**, 643–654
19.   Quillam, L.A., Khosravi-Far, R., Huff, S.Y. and Der, C.J. (1995) BioEssays **17**, 395–404
20.   Lamarche, N. and Hall, A. (1994) Trends Genet. **10**, 436–440
21.   Bockoch, G.M., Bohl, B.P. and Chuang, T.-H. (1994) J. Biol. Chem. **269**, 31674–31679
22.   Dash, D., Aepfelbacher, M. and Siess, W. (1995) J. Biol. Chem. **270**, 17321–17326
23.   Zigmond, S.H., Joyce, M., Borleis, J., Bokoch, G.M. and Devreotes, P.N. (1997) J. Cell Biol. **138**, 363–374
24.   Michiels, F., Stam, J.C., Hordijk, P.L., van der Kammen, R.A., Ruuls-Van Stalle, L., Feltkamp, C.A. and Collard, J.G. (1997) J. Cell Biol. **137**, 387–398
25.   Hirao, M., Sato, N., Kondon, T., Yonemura, S., Monden, M., Sasaki, T., Takai, Y., Tsukita, S. and Tsukita, S. (1996) J. Cell Biol. **135**, 37–51
26.   Kotani, H., Takaishi, K., Sasaki, T. and Takai, Y. (1997) Oncogene **14**, 1705–1713
27.   Mackay, D.J.G., Esch, F., Furthmayr, H. and Hall, A. (1997) J. Cell Biol. **138**, 927–938
28.   Brill, S., Li, S., Lyman, C.W., Church, D.M., Wasmuth, J.J., Weissbach, L., Bernards, A. and Snijders, A. (1996) Mol. Cell. Biol. **16**, 4869–4878
29.   Bashour, A.-M., Fullerton, A.T., Hart, M.J. and Bloom, G.S. (1997) J. Cell Biol. **137**, 1555–1566
30.   Wu, C., Lee, S.-F., Furmaniak-Kazmierczak, E., Côté, G.P., Thomas, D.Y. and Leberer, E. (1996) J. Biol. Chem. **271**, 31787–31790
31.   Brzeska, H., Knaus, U.G., Wang, Z.Y., Bokoch, G.M. and Korn, E.D. (1997) Proc. Natl. Acad. Sci. U.S.A. **94**, 1092–1095
32.   Burridge, K. and Chrzanowska-Wodnicka, M. (1996) Annu. Rev. Cell Dev. Biol. **12**, 463–519
33.   Craig, S.W. and Johnson, R.P. (1996) Curr. Opin. Cell Biol. **8**, 74–85
34.   Welch, M.D., Mallavarapu, A., Rosenblatt, J. and Mitchison, T.J. (1997) Curr. Opin. Cell Biol. **9**, 54–61
35.   Ridley, A.J., Paterson, H.F., Johnston, C.L., Diekmann, D. and Hall, A. (1992) Cell **70**, 401–410
36.   Nobes, C.D. and Hall, A. (1995) Cell **81**, 53–62

37. Kozma, R., Ahmed, S., Best, A. and Lim, L. (1995) Mol. Cell. Biol. **15**, 1942–1952
38. Morgan, C., Pollard, J.W. and Stanley, E.R. (1987) J. Cell. Physiol. **130**, 420–427
39. Boocock, C.A., Jones, G.E., Stanley, E.R. and Pollard, J.W. (1989) J. Cell Sci. **93**, 447–456
40. Allen, W.E., Jones, G.E., Pollard, J.W. and Ridley, A.J. (1997) J. Cell Sci. **110**, 707–720
41. Stasia, M.-J., Jouan, A., Bourmeyster, N., Boquet, P. and Vignais, P.V. (1991) Biochem. Biophys. Res. Commun. **180**, 615–622
42. Takaishi, K., Kikuchi, A., Kuroda, S., Kotani, K., Sasaki, T. and Takai, Y. (1993) Mol. Cell. Biol. **13**, 72–79
43. Webb, S.E., Pollard, J.W. and Jones, G.E. (1996) J. Cell Sci. **109**, 793–803
43a. Allen, W.E., Zicha, D., Ridley, A.J. and Jones, G.E. (1998) J. Cell Biol. **141**, 1147–1157
44. Palacek, S.P., Loftus, J.C., Ginsberg, M.H., Lauffenburger, D.A. and Horwitz, A.F. (1997) Nature (London) **385**, 537–540
45. Chant, J. (1996) Curr. Opin. Cell Biol. **8**, 557–565
46. Eaton, S., Auvinen, P., Luo, L., Jan, Y.N. and Simons, K. (1995) J. Cell Biol. **131**, 151–164
47. Stowers, L., Yelon, D., Berg, L.J. and Chant, J. (1995) Proc. Natl. Acad. Sci. U.S.A. **92**, 5027–5031
48. Gherardi, E. and Stoker, M. (1991) Cancer Cells **3**, 227–232
49. Ridley, A.J., Comoglio, P.M. and Hall, A. (1995) Mol. Cell. Biol. **15**, 1110–1122
50. Burridge, K. (1981) Nature (London) **294**, 691–692
51. Braga, V.M., Machesky, L.M., Hall, A. and Hotchin, N.A. (1997) J. Cell Biol. **137**, 1421–1423
52. Luo, L., Liao, Y.J., Jan, L.Y. and Jan, Y.N. (1994) Genes Dev. **8**, 1787–1802
53. Kozma, R., Sarner, S., Ahmed, S. and Lim, L. (1997) Mol. Cell. Biol. **17**, 1201–1211
54. Lamoureux, P., Altun-Gultekin, S.F., Lin, C., Wagner, J.A. and Heidemann, S.R. (1997) J. Cell Sci. **110**, 635–641
55. Luo, L., Hensch, T.K., Ackerman, L., Barbel, S., Jan, L.Y. and Jan, Y.N. (1996) Nature (London) **379**, 837–840
56. Giannelli, G., Falk-Marzillier, J., Schiraldi, O., Stetler-Stevenson, W.G. and Quaranta, V. (1997) Science **277**, 225–228
57. Pilcher, B.K., Dumin, J.A., Sudbeck, B.D., Krane, S.M., Welgus, H.G. and Parks, W.C. (1997) J. Cell Biol. **137**, 1445–1457

Biochem. Soc. Symp. **65**, 125–146
Printed in Great Britain

9

Rho-like GTPases: their role in cell adhesion and invasion

Frits Michiels and John G. Collard[1]

Division of Cell Biology, The Netherlands Cancer Institute, 121 Plesmanlaan, 1066 CX Amsterdam, The Netherlands

Abstract

Metastasis formation is the leading cause of death in cancer patients. Using an *in vitro* model system, we have identified *Tiam1* (T-lymphoma invasion and metastasis 1) as a gene that can induce invasion by and metastasis of mouse T-lymphoma cells. Subsequent studies showed that Tiam1 is a guanine nucleotide exchange factor for the Rho-like GTPase Rac1, a member of the Ras superfamily of small GTP-binding proteins. Rho-like GTPases play a pivotal role in the orchestration of changes in the actin cytoskeleton in response to receptor stimulation, but have also been shown to be involved in transcriptional activation and cell cycle regulation. Moreover, they can induce oncogenic transformation in fibroblast cells. In this chapter, we first summarize what is known about the signalling pathways that are activated by Tiam1 and Rho-like GTPases, and discuss the putative effectors that may mediate the effects in different cell types. In the latter part, we will more tentatively discuss the role of Tiam1 and Rho-like GTPases in invasion by and metastasis of tumour cells.

Introduction

Tumorigenesis is considered to be a multi-step process in which successive mutations lead to the transformation of normal cells into cells that no longer respond to their environment and which proliferate independently of regulatory signals. Further mutations subsequently may lead to malignant tumour cells which are able to invade the surrounding tissue, enter the bloodstream and extravasate at different sites to establish secondary tumours or metastases. The entire process might take tens of years, but it is generally accepted that environmental factors such as, for instance, diet and smoking, accelerate it. Although metastasis formation is the major cause of death for cancer patients, not much is known about the mechanisms that generate invasive tumour cells.

[1]To whom correspondence should be addressed.

In order to metastasize, tumour cells have to acquire certain properties which their precursor cells did not possess: they have to loosen their normal cell–cell contacts, become motile and produce (or induce the production of) proteolytic enzymes to be able to infiltrate in between the surrounding cells before they can enter the circulation. Extravasation is achieved by cytokine-driven adhesion of the tumour cells to the endothelium or the sub-endothelium of the target organs, followed by the proliferation of the tumour cells at these secondary sites (see Fig. 1). An important factor in this whole process, therefore, is the complex interactions (adhesive as well as de-adhesive) between the tumour cells and the surrounding cells. Some typical examples of adhesion molecules that have been implicated in invasion and metastasis are integrins, cadherins, and the hyaluronic acid receptors CD44 and Rhamm. Other factors that play pivotal roles in metastasis formation are the cytokines, which not only determine the sites of extravasation but are also involved in motility and the recruitment of blood vessels to the primary and secondary tumours. In fact, it has been proposed that metastasis formation of lymphocytes resembles the natural process of extravasation at sites of inflammation.

Since the majority of the studies in the past decades have concentrated on cell proliferation and tumorigenesis, little is known about the processes that determine whether or not a tumour cell will metastasize. Many tumour cell properties have been correlated with metastatic capacity, but direct evidence of a role for specific genes is scarce. Products of oncogenes, such as Ras, can con-

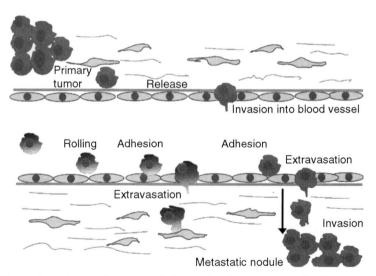

Fig. 1. Invasion and metastasis formation by tumour cells. For metastasis formation, tumour cells have to be released from the primary tumour, invade the surrounding tissue and enter blood or lymphatic vessels for transport to other parts of the body (upper part). Extravasation at ectopic sites may lead to the formation of a metastatic nodule (lower right). The process of extravasation resembles the extravasation of normal lymphoid cells at, for instance, sites of inflammation (lower left). Tumour cells are shown in dark grey; normal lymphoid cells are shaded.

fer metastatic capacity (see [1]) and so do specific splice variants of CD44, the expression of which correlates with metastasis in model systems and certain human tumours [2,3]. Other gene products, such as Nm23, which shows homology with nucleoside diphosphate kinases [4], and KAI1, which encodes a glycosylated transmembrane protein [5], can suppress metastasis [6,7]. In contrast, the *mta1* gene, another metastasis-associated gene, is overexpressed in mammary carcinoma cells [8]. Proteins that influence the invasive properties of cells strongly influence metastasis formation, as has been shown for proteases and their inhibitors (see [9]), and for adhesion molecules which play a role in cell–cell and cell–matrix interactions. Loss of E-cadherin is essential for detachment of carcinoma cells from the primary tumour [10], whereas integrins such as very late antigen 2 (VLA-2) [11] and lymphocyte function-associated antigen 1 (LFA-1) [12] are required for formation of metastases.

In a search for genes that can induce invasion and metastasis formation by T-lymphoma cells, we identified an activator or guanine nucleotide exchange factor (GEF) of the Rho-like GTPase Rac1. Activation of this gene, called *Tiam1* (T-lymphoma invasion and metastasis 1), allowed the T-lymphoma cells to infiltrate a monolayer of fibroblast cells, and to metastasize after injection into the tail vein of syngeneic mice [13]. Subsequent studies showed that a dominant active mutant of Rac1, G12V-Rac1, as well as a dominant active mutant of another member of the Rho-like GTPases, G12V-Cdc42, can induce invasiveness [14, 14a]. In this chapter, we review some of our results and discuss the role of Rho-like GTPases in the process of invasion and metastasis of lymphoid and non-lymphoid tumour cells.

Model system

Fibroblast invasion assay

To study the genetic control of invasion and metastasis, we use an *in vitro* invasion model, which is based on the capacity of T-lymphoma cells to infiltrate a monolayer of rat embryonic fibroblasts (Fig. 2). The parental cell line, BW5147, does not infiltrate the fibroblast monolayer spontaneously, but invasion can be induced by various experimental manipulations. Moreover, non-infiltrating cells can be washed away, and infrequent infiltrating variants can be isolated from the monolayer by trypsinization and replating the cells on a larger surface. After attachment of the fibroblasts, the infiltrating lymphoma cells can be propagated and subjected to further selection cycles in order to isolate invasive variants. The major advantage of this assay is that invasive cells can be selected with very high efficacy. As shown previously, one single invasive cell admixed to 1×10^6 non-invasive cells can be isolated within three selection rounds [15]. This powerful *in vitro* selection system allows the selection of infrequent invasive variants induced by various experimental manipulations.

In this way we have found that T-cell hybridomas, generated by fusion of non-invasive BW5147 T-lymphoma cells with inherently invasive activated normal T-cells, acquired a highly invasive phenotype [15,16]. Further analyses of invasive interspecies hybrids revealed the presence of an invasion-inducing

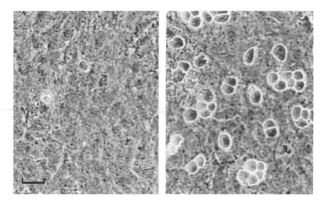

Fig. 2. Fibroblast invasion assay. The left panel shows a fibroblast mono-layer that has been incubated overnight with non-invasive T-lymphoma cells. All T-lymphoma cells have been removed during the subsequent washing pro-cedure. The right panel shows a similar monolayer that has been incubated with invasive variants. After washing of the non-invasive cells, the invaded T-lymphoma variants can be clearly seen as round, refractile cells in between the fibroblasts.

gene on human chromosome 7 [15]. Segregation of this chromosome from the hybrids resulted in loss of invasive properties. By the generation of deletion hybrids, we sublocalized the invasion-relevant region to chromosome 7p12-cen [17]. Invasive variants can also be isolated at low frequency from 5-azacytidine-treated BW5147 cells, suggesting that invasion-inducing genes can be activated by DNA hypomethylation [18]. Furthermore, we have shown that overexpression of an exogenous dominant active mutant of H-Ras, G12V-Ras, leads to invasive T-lymphoma cells [19]. In all these cases, the capacity of the T-lymphoma cells to infiltrate the fibroblast monolayer was correlated with the formation of experimental metastases after injection of the cells into the tail veins of syngeneic mice.

Identification of *Tiam1* by retroviral insertional mutagenesis

To isolate invasion-relevant genes, we performed a retroviral insertional mutagenesis screen. Separate populations of non-invasive BW5147 T-lym-phoma cells were infected with wild-type Moloney murine leukaemia virus, resulting in, on average, 15 proviral insertions per cell. From these populations, invasive cell variants were selected *in vitro* on monolayers of fibroblasts (Fig. 3). A panel of independently selected invasive variants was subsequently analysed for common proviral integration sites in order to identify genes responsible for the induction of invasiveness [13]. Retrovirally induced DNA rearrangements in a locus that we termed *Tiam1* were detected in 12 out 30 (40%) independently selected invasive cell clones [13]. No insertions in *Tiam1* were found in the remaining invasive cell clones, suggesting that other genes must be involved in the acquisition of invasiveness by these cells.

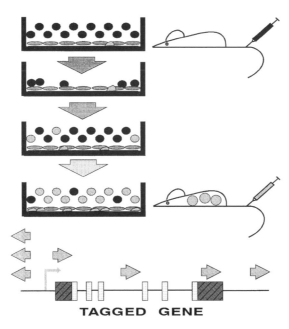

Fig. 3. Retroviral insertional mutagenesis screen for invasive variants. To isolate genes that induce invasion, we infected non-invasive BW5147 T-lymphoma cells (depicted as round, black cells) with Moloney murine leukaemia virus [13]. Independent pools of infected cells were subjected to several rounds of selection for invasive cells on a fibroblast monolayer (depicted as dark grey, flat cells), which resulted in the isolation of invasive variants (depicted as light grey, round cells). In these invasive variants, the integration of a provirus in the vicinity of an invasion-relevant gene must have led to the activation of that gene, leading to the invasive phenotype (bottom panel). This approach has led to the identification of Tiam1 [13]. Injection of the parental, non-invasive cells into the tail vein of syngeneic mice does not lead to metastasis formation (upper right), while injection of the invasive variants results in multiple metastases (lower right).

The proviral insertions are found within coding exons of the *Tiam1* gene, resulting in truncated 5′- and 3′-end transcripts which give rise to elevated levels of truncated N- and C-terminal Tiam1 protein fragments respectively. In one invasive variant, an amplification of the *Tiam1* locus was observed, with a concomitant increase in the amount of normal Tiam1 protein [13]. This suggests that invasion can be induced either by increased amounts of the normal Tiam1 protein or by protein truncation. Indeed, transfection of truncated Tiam1 cDNAs into non-invasive T-lymphoma cells rendered these cells invasive, establishing the invasion-inducing capacity of Tiam1 [13].

Tiam I is an activator of the Rho-like GTPase Rac

Conserved domains in Tiam I

Tiam1 has been highly conserved during evolution. Mouse and human Tiam1 proteins are 95% identical at the amino acid level [20]. Mouse *Tiam1* maps to the distal end of chromosome 16. The human homologue of *Tiam1* maps to 21q22, thus excluding the possibility that *Tiam1* is the invasion-inducing gene on human chromosome 7 [21]. The gene is highly expressed in brain and testis, and at low levels in most other mouse and human tissues [13,20]. Tiam1 mRNA is present in virtually all tumour cell lines tested, including B- and T-cell lymphomas, melanomas, neuroblastomas and carcinomas of the breast, lung, ovary, bladder and pancreas [20]. Whether Tiam1 expression levels correlate with the invasive and metastatic capacity of human tumours is currently being investigated.

Tiam1 encodes a protein of 1591 amino acid residues which contains different domains that show identity with other proteins ([13]; and see Fig. 4). At the N-terminus, a myristoylation site is present that could anchor the protein to membranes (Figure 5). This sequence is followed by two PEST domains, sequence motifs that in other proteins have been associated with protein insta-

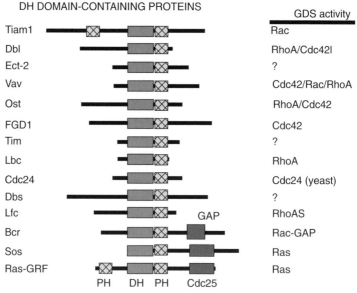

Fig. 4. Dbl family of GEFs for Rho-like GTPases. Shown is the conserved catalytic Dbl homology (DH) domain (grey) in combination with the adjacent pleckstrin homology (PH) domain (hatched). Tiam I is one of several proteins that contain an additional PH domain in the N-terminal region. The specificity of the GEFs is indicated on the right. Bcr contains a DH–PH module, but functions as a Rac-specific GAP by virtue of an additional Rho-GAP domain. Similarly, Sos and Ras-GRF (guanine nucleotide release factor) contain an additional Cdc25 domain and function as Ras-specific GEFs. GDS, guanine nucleotide exchange factor.

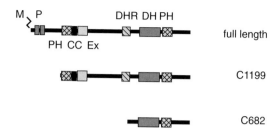

Fig. 5. Conserved domains in Tiam1. The full-length Tiam1 protein contains a consensus myristoylation sequence (M) at the N-terminus, which might aid in targeting Tiam1 to the plasma membrane. Also at the N-terminus are two PEST regions (P), which are generally thought to mediate protein instability. Tiam1 contains two pleckstrin homology (PH) domains. The N-terminal PH domain is part of a larger complex, consisting of an adjacent putative coiled-coil (CC) domain and an extended region (Ex), which is required for serum-inducible membrane localization of Tiam1. The function of the discs large homology region (DHR), or PDZ domain, a protein–protein interaction motif, is currently unknown. The Dbl homology region (DH) represents the catalytic domain for exchange activity of Tiam1. This domain is followed by a PH domain, as found in all other GEFs for Rho-like GTPases. The large (C1199) and small (C682) N-terminally truncated mutant Tiam1 proteins are also indicated.

bility. In the C-terminal half of the protein, Tiam1 contains a Dbl homology (DH) domain. This DH domain is shared with a number of structurally different gene products, including Dbl, Bcr, Cdc24, Vav and Ect-2 ([1,22]; see Fig. 4). Most of these gene products act as GEFs for Rho-like GTPases and exhibit transforming properties after N-terminal truncation. Tiam1 also contains two pleckstrin homology (PH) domains (Fig. 5), a recently identified protein module which is present in many signalling molecules and is thought to play a role either in specific protein–protein interactions, comparable to the role of SH2 and SH3 domains [23], or in interactions with phosphatidylinositol derivatives [24]. One of the PH domains of Tiam1 is located C-terminally adjacent to the DH domain, as in all other GEFs for Rho-like GTPases, whereas the second PH domain is more N-terminally located. Adjacent to the N-terminal PH domain (PHn), we identified two further regions, a putative coiled-coil (CC) domain and the extended region (Ex), which are required for Tiam1 functioning (see below). Furthermore, Tiam1 contains a so-called PDZ or discs large homology region, which is also thought to mediate protein–protein interactions [25]. Based on these sequence identities, we proposed that Tiam1 may act as an activator of Rho-like small GTP-binding proteins which transduce extracellular signals towards the cytoskeleton [13,14].

Rho-like GTPases

Small GTP-binding proteins of the Rho subfamily, including Cdc42, Rac1 and RhoA, are key control molecules in the re-organization of the actin cytoskeleton in response to stimulation of growth factor receptors [26]. Like Ras GTPases, Rho-like GTPases cycle between the active GTP-bound state and the inactive GDP-bound state [27]. Their activity is influenced by three different classes of proteins [28]. Activation occurs by interaction with GEFs. Inactivation occurs by interactions with GTPase-activating proteins (GAPs), whereas guanine nucleotide dissociation inhibitors (GDIs) are thought to prevent the activation and/or inactivation of small GTPases.

Cdc42, Rac1 and RhoA participate in the formation of distinct patterns of actin re-organization after stimulation of cell surface receptors. In the yeast *Saccharomyces cerevisiae*, Cdc42 is required for polarized cell growth [29]. A human homologue, Cdc42HS, regulates the formation of actin-containing microspikes or filopodia in Swiss 3T3 fibroblasts, and might be activated after treatment with bradykinin and interleukin 1 [30,31]. Growth factors such as platelet-derived growth factor (PDGF), epidermal growth factor (EGF) and insulin rapidly induce the formation of lamellipodia and membrane ruffles in certain fibroblast cell lines, a process controlled by Rac1 [32]. Introduction of constitutively active Rac1 (V12-Rac1) into fibroblasts increases the amount of cortical F-actin and induces the formation of membrane ruffles and pinocytotic vesicles [32]. Both V12-Rac and V12-Cdc42 also induce the formation of focal complexes, which contain integrins and certain signalling molecules but which differ from focal contacts because they are not obviously attached to actin stress fibres [33]. Activation of RhoA causes the formation of actin filament bundles (stress fibres) and the assembly of focal contacts in cells [34]. Treatment of cells with C3 transferase, which inactivates RhoA [35], results in the dissociation of actin stress fibres [36,37]. The assembly of focal contacts and stress fibres in serum-starved fibroblasts is also seen after exposure of the cells to the serum component lysophosphatidic acid (LPA), and this can be inhibited by inactivation of RhoA [34].

Studies using microinjection of Rho-like GTPases in Swiss 3T3 cells suggested that there is a hierarchy in the way in which these GTPases regulate each other's activities. Expression of activated Cdc42 leads to activation of Rac, while V12-Rac can stimulate RhoA activity [33]. Similarly, injection of V12-Ras also leads to activation of Rac, resulting in the induction of membrane ruffling [38]. Little is known, however, about the mechanisms responsible for this cross-talk between the GTPases. A molecule that may interconnect or stimulate both Ras and Rac signalling pathways is phosphoinositide 3-kinase (PI 3-kinase). This kinase is rapidly activated after stimulation of various cell surface receptors, as well as by oncogenic Ras [39,40]. PI 3-kinase stimulates the synthesis of PtdIns(3,4,5)P_3 from PtdIns(4,5)P_2 in target cells. Although PI 3-kinase is considered to be an effector of Ras, it is also able to activate Rac [41]. In addition, activation of PI 3-kinase by growth factor receptors leads to an activation of Rac, probably via the activation of a GEF [40]. A possible mechanism for the activation of RhoA by Rac1 is suggested by the finding that activation of Rac1 leads to the production of leukotrienes, which in turn activate RhoA [42].

Tiam1 activates Rac1

We were able to show that Tiam1 can act as a GEF for the Rho-like GTPases RhoA, Rac1 and Cdc42 *in vitro* [14]. However, fibroblast cells that overexpress Tiam1, or an N-terminally truncated form of Tiam1, resemble V12-Rac-expressing cells, in that they show extensive membrane ruffling and contain numerous pinocytotic vesicles. Furthermore, they are flat and round, more like pancakes, and have an epithelial-like phenotype due to enhanced cell–cell interactions [14]. This Tiam1-induced phenotype can be inhibited by co-expression of dominant-negative N17-Rac1, but not by co-expression of N17-Cdc42 or N19-RhoA, or by inhibition of RhoA by treatment with C3 transferase [1,14]. Also, in other cell types, e.g. N1E-115 neuroblastoma cells and epithelial Madin–Darby canine kidney (MDCK) cells, we found that the phenotype of Tiam1-expressing cells resembles that of V12-Rac-expressing cells, and that this phenotype can be inhibited by co-expression of N17-Rac only [43,44]. Thus it seems reasonable to assume that Tiam1 acts as a specific activator of Rac1 *in vivo*.

Downstream effectors of Rho-like GTPases

Several molecules that bind to activated forms of the Rho-like GTPases have recently been identified. These include kinases (mostly serine/threonine kinases), lipid kinases, GAPs and many molecules without any known function. No conclusive function for any of these proteins can yet be given. For the sake of this article, we will describe only the signalling pathways downstream of the Rho-like GTPases that are potentially involved in invasion and metastasis of tumour cells. These include transcriptional activation and regulation of the cytoskeleton.

Effectors of Rac/Cdc42

In contrast to Ras, which activates the Raf/mitogen-activated protein kinase (MAPK) pathway, Rho-like GTPases stimulate other kinase pathways that lead to the activation of p38 MAPK and JNK [Jun N-terminal kinase; also termed stress-activated protein kinase (SAPK)]. It has been demonstrated that constitutively activated Cdc42 and Rac1, but not RhoA, activate JNK [45,46], as do GEF proteins for both Cdc42 and RhoA, and Tiam1, an activator of Rac [46–48]. Rac-mediated activation of JNK is blocked by kinase-defective mutants of PAK65 (p21-activated kinase of 65kDa), a serine/threonine kinase that binds to activated Cdc42 and Rac1, and by dominant-negative mutants of SEK1 (SAPK kinase) and MEK1 (MAPK kinase), kinases that act upstream of JNK [31,49,50]. This suggests that Cdc42 and Rac activate a kinase cascade, similar to the activation of Raf and its downstream effectors by Ras. It is unclear at the moment whether PAK65 is involved in the stimulation of JNK activity [51]. Other candidates are the mixed-lineage kinases (MLKs), of which MLK3 has been shown to bind to activated Rac and Cdc42 [52]. Both Ras and Rac can also induce activation of the transcription factor nuclear factor κB, probably mediated by the generation of oxygen radicals [53]. The production of reactive oxygen species may be functionally similar to phagocytic oxidase production in neutrophils, which is regulated by Rac2, a close relative of Rac1, and reactive oxygen species might be used as second messengers in other cell types.

Potential downstream effectors of Rac and Cdc42 involved in the re-organization of the cytoskeleton are Por1, WASP (Wiskott–Aldrich syndrome protein), the Rac-specific GAP N-chimaerin, and IQGAP, a protein with identity to Ras-related GAPs [54–58]. Por1 was found in a yeast two-hybrid screen to interact only with activated Rac1. Truncated versions of Por1 interfere with the induction of membrane ruffles by activated Rac1, and evidence is provided that Por1 is acting in a signalling complex downstream of Rac. Unfortunately, Por1 does not show any identity with other proteins [54]. WASP, a protein that is affected in Wiskott–Aldrich syndrome immunodeficiency patients, binds to activated Cdc42 and, to a lesser extent, to activated Rac [56]. Wiskott–Aldrich syndrome patients show a severe disorganization of the cytoskeleton in T-cells and platelets. The introduction of WASP into fibroblasts causes a clumping of the F-actin, which can be inhibited by N17-Cdc42. However, WASP does not mediate filopodia formation, because mutants of Cdc42 that do not bind to WASP *in vitro* can still induce filopodia [59]. Overexpression of N-chimaerin, a Rac/Cdc42-specific GAP, induces ruffling in fibroblast cells and the outgrowth of neurites in N1E-115 neuroblastoma cells, both of which are characteristic of activation of Rac [58]. These effects are inhibited by co-expression of dominant-negative N17-Rac or N17-Cdc42, indicating that endogenous Rac and/or Cdc42 is required. This suggests that the GAP, or the Rac/Cdc42–GAP complex, provides the signal to the cytoskeleton [58]. The IQGAPs, which are related to Ras-GAPs, do not show GAP activity towards Rac or Cdc42, but bind to the activated forms of these GTPases and in fact inhibit the intrinsic GTPase activity [57,60,61]. The localization of IQGAPs within cells is consistent with a function downstream of Rac/Cdc42 [62]. Other effectors that have been shown to bind to Rac and/or Cdc42 are PtdIns4*P* 5-kinase [63], PI 3-kinase [64,65] (note that this enzyme has also been proven to function upstream of Rac; see above) and tubulin [66]. It is, however, not clear at the moment whether these proteins have a function as effectors of these GTPases.

Effectors of Rho

Activation of RhoA by serum or LPA leads to activation of the serum response factor [67]. The roles of the recently identified protein kinase N [68,69], its close relative protein kinase C-related kinase 2 [70], and several other proteins with unknown function that all bind to activated RhoA, such as citron [71], rhophilin [68], rhotekin [72] and p116(Rip) [73], are not known at the moment. An important family of kinases that bind to and are activated by GTP-bound RhoA are the Rho kinases p160Rock [74], Rho kinase [75] and Rok (α and β) [76,77]. Rho kinase phosphorylates the catalytic subunit of myosin light chain phosphatase, thereby inhibiting its activity [78]. This leads to an enhanced phosphorylation of myosin light chain, resulting in the activation of myosin and the formation of stress fibres [79]. Rho kinase can also phosphorylate myosin light chain at the same serine residue as calmodulin-dependent myosin light chain kinase, resulting in smooth muscle contraction [80]. Injection of the catalytic domain of Rho kinase induces stress fibre formation and focal contacts in fibroblast cells [79]. These results strongly suggest that Rho kinase is the downstream target of Rho that mediates the organization of the cytoskeleton.

Role of Rho-like proteins in oncogenic transformation

We and others have shown that V12-Rac1, V12-Cdc42 and, to a lesser extent, V14-RhoA induce an oncogenic phenotype when expressed in NIH-3T3 cells or Rat1 fibroblasts. The efficiency, however, is much lower than that of V12-Ras [81,82]. Moreover, most activators (GEFs) of Rho-like GTPases, such as Dbl, Ect-2, Vav, Tim, Lbc, Lfc, Dbs and Ost, have been identified as 5'-end truncated proto-oncogenes in NIH-3T3 transformation assays [1,22]. It appears that N-terminal truncation activates the oncogenic potential of these proteins. N-terminal truncation of Tiam1 also leads to an oncogenic phenotype in NIH-3T3 cells [81]. Interestingly, different tumour cell morphologies were obtained with distinct N-terminally truncated Tiam1 proteins. A large C-terminal Tiam1 fragment induced large, rather flat cells, while tumour cells that expressed a shorter C-terminal Tiam1 protein were much smaller and not much different from normal NIH-3T3 cells [81]. This suggests that transcriptional activation, rather than morphological transformation, is required for oncogenic transformation. These kinds of studies substantiate the findings that activation of Rac and Rho signalling pathways induces an oncogenic phenotype in cells.

Dominant-negative mutants of Cdc42, Rac1 and RhoA inhibit focus formation induced by V12-Ras in NIH-3T3 fibroblasts, but not that induced by an activated downstream target of Ras, RafCAAX. It has also been demonstrated that V12-Rac1, V12-Cdc42 and V14-RhoA synergize with RafCAAX in focus formation assays [83–86]. These data indicate that Ras and Rho-like GTPases act in concert in cell transformation and that oncogenic V12-Ras drives both the Raf/MAPK and the Rac/Rho/Cdc42 pathways. The finding that V12-Cdc42, V12-Rac1 and V14-RhoA cause progression of cells through S-phase [87], as well as the finding that overexpression of V12-Cdc42 and V12-Rac1 results in multinucleated cells [14,81,85], support the concept that Rho-like GTPases are involved in cell cycle regulation [88]. De-regulation of Rho-like GTPases could thus indeed be important for malignant transformation.

Properties of Tiam1

The induction of a tumorigenic phenotype in NIH-3T3 cells can be accomplished by using two distinct N-terminally truncated Tiam1 mutants: a large C-terminal Tiam1 fragment that includes PHn and the PDZ domain (C1199 Tiam1), and a small C-terminal fragment that contains only the DH domain and the C-terminal PH domain (C682 Tiam1) (see Fig. 5). Further examination of these cells showed that the cells expressing C1199 Tiam1 had the typical flat, pancake-like morphology and contained membrane ruffles and pinocytotic vesicles (Figs. 6A and 6B), while the cells expressing C682 Tiam1 looked like normal fibroblasts [48,81]. Immunoelectron microscopy indicated that the C682 Tiam1 protein was localized exclusively in the cytoplasm, while the C1199 Tiam1 protein was also present at the plasma membrane [48]. A series of deletion mutants showed us that PHn, the adjacent putative CC region and a neighbouring region of about 200 amino acids (Ex; see Fig. 5) are all required for membrane localization of Tiam1 and the subsequent induction

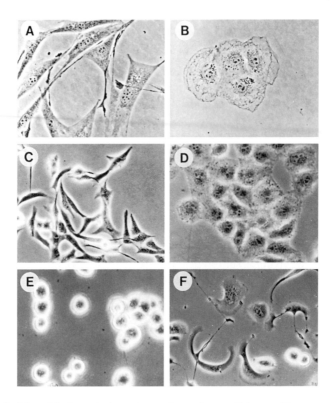

Fig. 6. Tiam1-induced phenotypes in various cell types. Phase-contrast images showing the morphological phenotypes of mock-transfected control cells (A, C, E), and the phenotypes of cells that have been transfected with a Tiam1-expressing construct (B, D, F). Shown are normal NIH-3T3 cells (A) and C1199-Tiam1-expressing NIH-3T3 cells (B); fibroblastoid V12-Ras-transformed MDCKf3 cells (C) and C1199-Tiam1-expressing MDCKf3 cells (D); and normal (E) and C1199-Tiam1-expressing (F) NIE-115 neuroblastoma cells. Note the epithelial phenotypes, induced by enhanced cadherin binding, in (D). The cells in (E) and (F) were grown in the presence of serum on laminin-coated dishes. Compare the absence of neurite-like extensions in control cells (E) with the neurite-bearing Tiam1-expressing cells (F).

of membrane ruffling [48,89]. Deletions in any of these regions abolished the presence of Tiam1 at the membrane [48,89]. This membrane targeting domain of Tiam1 could be functionally replaced by the N-terminal region of c-Src, which contains a myristoylation sequence. For reasons that are unknown at the moment, the myristoylation sequence of Tiam1 is insufficient for membrane localization [48]. Membrane localization of Tiam1 was also found to be important for the induction of JNK activity by Tiam1: only mutants of Tiam1 that localized at the plasma membrane and induced the formation of membrane ruffles were capable of activating JNK [48]. The fact that C682 Tiam1, which does not contain the PHn–CC–Ex region and did not activate JNK, caused onco-

genic transformation of NIH-3T3 cells makes it unlikely that activation of
JNK is involved in oncogenic transformation.

The localization of Tiam1 at the plasma membrane can be induced by
serum [48]. Serum-starved fibroblast cells that express C1199 Tiam1 hardly
contain any membrane ruffles, and immunoelectron microscopy showed that
under these conditions the amount of Tiam1 that is present at the plasma membrane is greatly reduced. The addition of serum restored both the membrane
localization of C1199 Tiam1 and the presence of membrane ruffles within a few
hours. None of PDGF, EGF or insulin could substitute for serum. Although
these growth factors are known to mediate lamellae formation in these cells
through activation of Rac within 5 min, their addition did not lead to the reappearance of the typical Tiam1-dependent ruffling [48]. To our surprise, LPA
was found to be required for the reconstitution of the Tiam1 phenotype.
Within 4 h after addition of LPA, and after a transient induction of stress fibres
by LPA, the Tiam1-expressing cells again showed their flat, epithelial-like
morphology and contained numerous membrane ruffles. Thus, although serum
and LPA are known to activate RhoA, the downstream signalling pathways are
somehow required for Tiam1-mediated activation of Rac. An alternative explanation might be that serum starvation deprives the cells of the ability to support
membrane ruffling. However, we found that cells that express a chimaeric protein in which the myristoylated c-Src sequences were fused to a short
C-terminal Tiam1 protein (Ms-C580 Tiam1) still showed membrane ruffling
after serum starvation [48]. Furthermore, serum-starved cells that are injected
with a V12-Rac1-expressing construct also show the induction of membrane
ruffling [32]. Therefore we believe that serum (LPA) is somehow required for
the localization of Tiam1 at the plasma membrane, mediated by PHn–CC–Ex.
The mechanisms by which this is achieved are unknown at the moment.
Preliminary results suggest a role for RhoA and PI 3-kinase, since inhibitors
such as C3 transferase and wortmannin (which blocks PI 3-kinase activity)
interfere with the LPA-induced reappearance of the Tiam1 phenotype. The latter result would fit with data from Cantley's group that PHn of Tiam1 binds
more effectively to PtdInsP_3 than to PtdInsP_2 [90].

Cell motility and adhesion

Phagokinesis and wounding migration assays have been used to investigate potential effects of Tiam1 on cell motility. NIH-3T3 cells transfected with
full-length or C1199 Tiam1 constructs, which both induce a flat, pancake-like
morphology with many membrane ruffles, are less motile than untransfected
controls. Similarly, V12-Rac1-expressing cells show decreased motility compared with control cells. Thus, although these cells display numerous
membrane ruffles, which has been supposed to be a hallmark of motile cells,
they are strongly impaired in their motility. The motility of V14-RhoA-
expressing cells, which contain numerous stress fibres, is somewhat reduced
compared with that of control cells. However, C3 transferase treatment inhibited the motility of all cells studied, indicating that RhoA activity is required
for motility. Interestingly, the small cells derived from tumours induced by

C682 Tiam1 (see above) are rather motile (almost as much as control cells). Since cells expressing C682 Tiam1 do not show any apparent activation of Rac1, as Tiam1 is not localized at the plasma membrane because of the lack of the PHn–CC–Ex region, it is possible that this cytoplasmic Tiam1 protein activates other signalling pathways that can induce motility, e.g. activation of the recently identified cytoplasmic Rac3 [91].

The low degree of motility (20% of control values) of full-length or C1199 Tiam1 transfectants might be explained by (1) increased adhesion of these cells to the substrate, possibly by activation of integrins; (2) increased homotypic interactions, as these cells exhibit a more epithelial-like phenotype compared with normal cells; or (3) a loss of cell polarity which prevents directed motility. All of these processes are most probably regulated by Rho-like proteins, and we have found evidence for the first two possibilities in other cell types.

It has been known for some time that RhoA regulates the neuronal morphology of N1E-115 neuroblastoma cells [92]. In the absence of serum, these cells produce neurite-like extensions. Thrombin and LPA both induce cell rounding and neurite retraction of differentiated N1E-115 cells, which can be inhibited by treatment with C3 transferase or dominant-negative N19-RhoA [73]. We have found recently that Tiam1 and active mutant forms of Cdc42 and Rac1 can induce neurite outgrowth even in the presence of serum, provided that the cells are plated on laminin-coated dishes [44] (Figs. 6E and 6F). Neuronal differentiation was prevented if the cells were plated on plastic or fibronectin-coated dishes, even in the absence of serum. Neurite formation on laminin was enhanced by co-expression of N19-RhoA or p190 Rho-GAP, and inhibited by V14-RhoA [44]. This suggests that Rac and RhoA have opposing roles in the regulation of neuronal morphology, and that activation of Rac somehow leads to the down-modulation of RhoA in these cells. The Tiam1- and V12-Rac-expressing cells showed a significant increase in adhesion to laminin compared with control cells, as well as redistribution of $\alpha_6\beta_1$ integrin, a laminin receptor, to the distal end of advancing lamellae. No co-localization was found between these integrins and components of focal contacts such as paxillin and vinculin [44]. Taken together, this study shows that Tiam1 and Rac promote the integrin-mediated adhesion and spreading of N1E-115 cells on laminin.

A different effect of Tiam1 and Rac was observed in epithelial MDCK cells. E-cadherin is a transmembrane protein that provides homotypic interactions between neighbouring cells. Together with cytoplasmically associated proteins such as α- and β-catenins, E-cadherin forms adhesive complexes at sites of cell–cell contacts which are stabilized by connections with cortical actin via molecules such as α-actinin [93]. Hepatocyte growth factor (HGF)/scatter factor (SF) induces scattering of these cells because it reduces the E-cadherin-mediated cell–cell adhesion and stimulates motility [93]. It has been suggested that Rac is involved in E-cadherin-mediated adhesion. Overexpression of dominant-negative N17-Rac1 inhibits HGF/SF-induced scattering, while V12-Rac promotes actin polymerization at sites of cell–cell contacts and leads to an enhanced staining of components of E-cadherin complexes at the adherens junctions [94,95]. We have recently shown that overexpression of V12-Rac, as well as of Tiam1, inhibits HGF/SF-induced scattering by potentiating E-cadherin-mediated cell–cell adhesions, probably by enhancing interactions between the E-cadherin complexes and cortical F-actin [43]. If E-cadherin interactions are blocked, by adding anti-E-cadherin antibodies or by growing cells in low Ca^{2+}, the Tiam1- or V12-Rac-expressing cells respond normally to HGF/SF. This is in contrast with N17-Rac, which rather seems to prevent cell motility. Ras-transformed MDCKf3 cells have a fibroblastoid phenotype and do not grow in colonies, as a result of reduced E-cadherin-mediated cell–cell adhesion [10]. Overexpression of Tiam1 or V12-Rac in these cells leads to enhanced cell–cell contacts and completely restores the epithelial phenotype ([43]; Figs. 6C and 6D). Furthermore, invasion by MDCKf3 in a collagen matrix is completely blocked by Tiam1 or V12-Rac, because of the restoration of E-cadherin-mediated adhesion [43].

We have provided evidence that Tiam1 and Rac are important determinants of cell–cell as well as cell–matrix interactions. As already mentioned, NIH-3T3-derived tumour cells that express Tiam1 or V12-Rac display a more epithelial-like phenotype. Since NIH-3T3 cells express N-cadherin rather than E-cadherin, this suggests that Tiam1/Rac also influence N-cadherin-mediated interactions. Whether or not integrin-mediated adhesion is enhanced is currently under investigation. The reduced motility of Tiam1- or V12-Rac1-expressing NIH-3T3 cells might thus be explained by the enhanced cell–cell and/or cell–matrix interactions. The mechanism by which this is achieved is largely unknown, but probably involves the formation of cortical F-actin, which provides excessive anchoring sites for integrin and E-cadherin complexes. In this respect, it is noteworthy that only Tiam1 proteins that contain the PHn–CC–Ex region and that localize to the plasma membrane (see next section) are capable of enhancing both integrin-mediated adhesion of N1E-115 cells to laminin and E-cadherin-mediated interactions in MDCK cells. The phenotype that is induced in these cells by PHn–CC–Ex-containing Tiam1 proteins is more pronounced than that of V12-Rac-expressing cells, which might reflect the specific localization of Tiam1 at the cell periphery, compared with the more cytoplasmic localization of V12-Rac. We have not been able so far to demonstrate any direct interactions between Tiam1 and components of the E-cadherin complex. We are still searching for proteins that interact with the PHn–CC–Ex region and which might play a role in the specific localization of Tiam1. Since PtdInsP_3 might also be involved in the membrane localization of Tiam1 (see previous section), an attractive model would be that the Tiam1-mediated activation of Rac at the plasma membrane results in the recruitment of PI 3-kinase, leading to the local production of PtdInsP_3. This, in turn, could recruit additional Tiam1 molecules etc., which would lead to an amplification of the signal. In this model, PI 3-kinase would be both upstream and downstream of Rac1. The validation of this model, however, awaits further examination.

Invasiveness of lymphoid cells

Role of Tiam1 and Rho-like GTPases

Tiam1 was identified in a retroviral insertional mutagenesis screen for invasive variants of the parental non-invasive BW5147 T-lymphoma cell line. In several independently isolated invasive variants, retroviral insertions led to the generation of truncated small N-terminal and large C-terminal Tiam1 fragments. To verify that overexpression of these fragments was the cause of the invasive phenotype, we transfected different plasmids into the BW5147 cells, and indeed were able to isolate invasive variants that overexpressed the truncated proteins [13]. In this way, we found that a large C-terminal fragment which contains the PHn–CC–Ex region (C1199), and a smaller C-terminal fragment that lacks the PHn–CC–Ex region (C682), were both able to generate invasive variants. However, the cell line expressing the C682 Tiam1 protein in particular was rather unstable, probably due to the high overexpression of this protein which was required for invasiveness. Using a retroviral transduction method, we found that only Tiam1 mutants that contain an intact PHn–CC–Ex region are able to generate invasive cells. These results suggest that, at moderate expression levels, membrane localization of Tiam1 through interactions with the PHn–CC–Ex region is also required for the induction of invasiveness.

We have shown previously that BW5147 T-lymphoma cells which overexpress V12-H-Ras or V12-Rac become invasive on a fibroblast monolayer [14,19], similar to T-lymphoma cells that overexpress full-length or truncated Tiam1 proteins. We have found recently that V12-Cdc42 can also induce invasion by these cells [14a]. In contrast, BW cells that overexpress V14-RhoA did not acquire an invasive phenotype [14]. To identify which of the downstream pathways might be involved, we made use of effector mutants of Rac and Cdc42 (kindly provided by Cathy Lamarche and Alan Hall, UCL, London, U.K.) which were shown to be defective in one or other signalling pathway [59]. Effector mutations were made in an L61 background, which is a GTPase-deficient, dominant active mutation, similar to V12. L61/A37-Rac1 still binds to and activates PAK, and stimulates JNK activation, but does not induce membrane ruffling and focal complexes. L61/C40-Rac1, on the other hand, does not bind to PAK and cannot stimulate JNK activity, but still induces membrane ruffling and focal complex formation [59]. We have found that L61/C40-Rac1, but not L61/A37-Rac1, induces invasion by BW cells. This suggests that the Rac-mediated rearrangement of the cytoskeleton, but not the activation of PAK or JNK, is the cause of the invasive phenotype. Effector mutations in Cdc42 were also tested, but the results were less clear. L61/A37-Cdc42 does not stimulate PAK or JNK activity and does not lead to the activation of Rac, but induces filopodia formation, while L61/C40-Cdc42 is defective in the activation of PAK and JNK, but signals to Rac and also induces filopodia formation [59]. Both mutants were found to be capable of generating invasive variants. This suggests that invasion involves a mutual downstream signalling pathway, which might be the alteration of the cytoskeleton, and that Cdc42 can induce invasion independently of Rac1.

Although V14-RhoA cannot induce invasion, we have found that treatment of Tiam1- or V12-Rac-expressing cells with C3 transferase inhibits invasion by these cells, suggesting that RhoA activity is required but not sufficient for invasion. In line with these results is the fact that serum is absolutely required for invasion; this can be replaced, however, by LPA or sphingosine 1-phosphate. Since these factors are known to stimulate RhoA, we generated cells that expressed both V12-Rac and V14-RhoA. Although these cells showed an enhanced invasive phenotype when compared with cells that express only V12-Rac1, they were still dependent on the presence of LPA or sphingosine 1-phosphate for invasiveness. This suggests that LPA (or sphingosine 1-phosphate) stimulates additional, Rho-independent pathways that are required for invasion [14a]. It is known from other cell types that LPA can stimulate the MAPK pathway through a pertussis-toxin-sensitive G_q-protein, and can activate phospholipase C through a pertussis-toxin-insensitive G_i-protein [92]. Therefore we tested whether any of these pathways is required for invasion. Invasion by Tiam1- or V12-Rac1-expressing cells is not inhibited by pertussis toxin and, in line with these results, an activated mutant of Raf (RafCAAX) did not lead to serum-independent invasion by cells that co-expressed RafCAAX and V12-Rac1. In contrast, U73122, an inhibitor of phospholipase C, abolished invasion by Tiam1- or V12-Rac-expressing cells almost completely, as had also been found for other T-lymphoma variants. These results suggest that at least three activities, Rac, Rho and phospholipase C, are required for invasion by lymphoma cells.

Role of integrin-mediated adhesion

We have recently started to analyse whether other properties, such as changes in adhesion, could be correlated with the acquisition of an invasive phenotype by T-lymphoma cells. We have shown that overexpression of Tiam1 or V12-Rac1 in N1E-115 neuroblastoma cells enhanced the adhesion of these cells to laminin [44]. Therefore we tested whether invasive T-lymphoma variants could adhere to certain substrates, in contrast with non-invasive cells. Of all substrates tested, kalinine (laminin 5) was found to be the most informative. Only invasive cells adhered readily to kalinine and spread within minutes on this substrate. Adhesion could be efficiently blocked with antibodies against α_6 or β_1 integrin, which together form the kalinine receptor. However, invasion into the REF monolayer by Tiam1- or V12-Rac1-expressing cells is only partially blocked by antibodies against α_6 or β_1 integrins, suggesting that other integrins might be involved. Further evidence for a role for integrins in the acquisition of an invasive phenotype comes from the human erythroleukaemia cell line K562. As for mouse T-lymphoma cells, K562 variants that overexpress Tiam1 or V12-Rac1 invade into the fibroblast monolayer and can be selected from non-invasive cells. K562 cells express only the $\alpha_5\beta_1$ integrin, which is a fibronectin receptor. This receptor is normally not active on these cells, but can be activated by phorbol esters such as PMA. We found that invasive K562 cells which overexpress Tiam1 or V12-Rac1 adhere strongly to fibronectin. Furthermore, both adhesion and invasion could be blocked by antibodies against the $\alpha_5\beta_1$ integrin, suggesting that the activation of this integrin by Tiam1/Rac is involved in the acquisition of an invasive phenotype in this cell type.

Towards a model

It is clear from the presented data that Tiam1 and Rho-like GTPases play pivotal roles in diverse signalling pathways. On top of this, the activation of these pathways will lead to different results, depending on the cell type that is investigated. We have shown that overexpression of Tiam1/V12-Rac1/V12-Cdc42 leads to an invasive phenotype in lymphoid cells such as BW5147 T-lymphoma cells and K562 cells, while overexpression of the same molecules in epithelial cells leads to inhibition of a V12-Ras-induced invasive phenotype (Fig. 7). What is the common denominator in these processes? A clue to this might come from the fact that these phenomena are solely induced by Tiam1/Rac1 mutants that affect cytoskeletal organization. The main difference that we observed between normal MDCK cells and Tiam1/V12-Rac1-expressing MDCK cells is that the latter display enhanced staining of F-actin at the periphery of the cells, which might stabilize the E-cadherin complexes to the cytoskeleton. Stabilization of E-cadherin complexes might also explain the phenotypic reversion of V12-Ras-transformed MDCKf3 cells by Tiam1 or V12-Rac. A similar concentration of cortical F-actin was found in the N1E-115 neuroblastoma cells, which display enhanced $\alpha_6\beta_1$-integrin-mediated binding to laminin after overexpression of Tiam1/V12-Rac1. This suggests that integrin

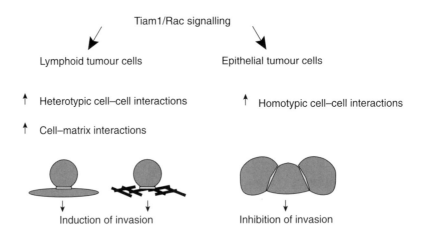

Fig. 7. Different effects of Tiam1/Rac signalling on invasion. Activation of Tiam1/Rac in T-lymphoma cells induces invasion by these cells, possibly due to increased heterotypic cell–cell and cell–matrix interactions. Activation of Tiam1/Rac in epithelial tumour cells reduces the invasive potential of these cells due to enhanced E-cadherin-mediated homotypic cell–cell interactions. A common denominator might be that activation of Tiam1/Rac leads to an enhancement of the cortical F-actin, providing extra anchorage sites for both integrin and cadherin complexes.

complexes (which did not contain other proteins that are commonly found in focal complexes) might also be stabilized by the cortical F-actin. Thus, although there are several questions to be answered, our current model proposes an important role for the actin cytoskeleton.

How, then, do tumour cells invade? To answer this question, we have to discriminate between lymphoid cells and adherent carcinoma cells. We have shown that at least three activities are important for invasion by T-lymphoma cells: Rac, RhoA and phospholipase C. As discussed, activation of Rac might lead to a stabilization of integrin complexes, promoting the adhesion of the lymphoid cells to the fibroblast monolayer. RhoA and phospholipase C might subsequently be necessary for motility and for the contractile forces that are required for the lymphoid cells to squeeze in between the monolayer cells. In the latter processes the actin cytoskeleton again plays a crucial role.

A different picture emerges if we deal with carcinoma cells. These cells typically show cadherin-mediated homotypic interactions, which have to be released before the cells can escape from their normal environment and invade into the surrounding tissue. In these cells, the stabilization of the cadherin complexes by Tiam1/Rac would lead to a reduction of the invasive potential.

Whether other signalling pathways downstream from Rho-like GTPases, e.g. transcriptional activation, are involved in the process of tumour formation and metastasis remains to be elucidated. Transcriptional activation could lead to the activation of proteolytic enzymes, or the production of growth factors or their receptors, which would be important factors in this process. However, further research is required to establish this aspect.

The proposed model is far from complete, and will become more complex when further players come into the picture. The data discussed here strongly suggest that the orchestration of changes in the actin cytoskeleton by Rho-like GTPases and their regulators, leading to alterations in cell–cell and cell–matrix interactions, is an important aspect in the processes of invasion and metastasis of tumour cells.

We thank our colleagues Frank van Leeuwen, Peter Hordijk, Jord Stam, Rob van der Kammen and Eva Sander for stimulating discussions and careful reading of this manuscript. The work in J.C.'s lab is supported by grants from the Dutch Cancer Foundation and the Netherlands Organization for Scientific Research.

References

1. Collard, J.G. (1996) Int. J. Oncol. **8**, 131–138
2. Hofmann, M., Ruby, W., Zoller, M., Tolg, C., Ponta, H., Herrlich, P. and Gunthert, U. (1991) Cancer Res. **51**, 5292–5297
3. Wielenga, V.J.M., Heider, K.H., Offerhaus, G.J.A., Adolf, G.R., Vandenberg, F.M., Ponta, H., Herrlich, P. and Pals, S.T. (1993) Cancer Res. **53**, 4754–4756
4. Leone, A., Flatow, U., King, C.R., Sandeen, M.A., Margulies, I.M., Liotta, L.A. and Steeg, P.S. (1991) Cell **65**, 25–35
5. Dong, J., Lamb, P.W., Rinker-Schaeffer, C.W., Vulkanovic, J., Ichikawa, T., Isaacs, J.T. and Barrett, J.C. (1995) Science **268**, 884–886
6. Bevilacqua, G., Sobel, M.E., Liotta, L.A. and Steeg, P.S. (1989) Cancer Res. **49**, 5185–5190

7.   Dong, J.T., Suzuki, H., Pin, S.S., Bova, G.S., Schalken, J.A., Isaacs, W.B., Barrett, J.C. and
       Isaacs, J.T. (1996) Cancer Res. **56**, 4387–4390
8.   Toh, Y., Pencil, S.D. and Nicolson, G.L. (1995) Gene **159**, 97–104
9.   Liotta, L.A. and Stetler Stevenson, W.G. (1991) Cancer Res. **51**, 5054–5059
10.   Vleminckx, K., Vakaet, Jr., L., Mareel, M., Fiers, W. and Van Roy, F. (1991) Cell **66**,
       107–119
11.   Chan, B.M.C., Matsuura, N., Takada, Y., Zetter, B.R. and Hemler, M.E. (1991) Science **251**,
       1600–1602
12.   Roossien, F.F., de Rijk, D., Bikker, A. and Roos, E. (1989) J. Cell Biol. **108**, 1979–1985
13.   Habets, G.G.M., Scholtes, E.H.M., Zuydgeest, D., Vanderkammen, R.A., Stam, J.C., Berns,
       A. and Collard, J.G. (1994) Cell **77**, 537–549
14.   Michiels, F., Habets, G.G.M., Stam, J.C., Vanderkammen, R.A. and Collard, J.G. (1995)
       Nature (London) **375**, 338–340
14a.   Stam, J.C., Michiels, F., Van der Kammen, R.A., Moolenaar, W.H. and Collard, J.G. (1998)
       EMBO J., in the press
15.   Collard, J.G., Van de Poll, M., Scheffer, A., Roos, E., Hopman, A.H.M., Geurts van Kessel,
       A.H.M. and Van Dongen, J.J.M. (1987) Cancer Res. **47**, 6666–6670
16.   Roos, E., La Rivière, G., Collard, J.G., Stukart, M.J. and de Baetselier, P. (1985) Cancer Res.
       45, 6238–6243
17.   Habets, G.G.M., Van der Kammen, R.A., Willemsen, V., Balemans, M., Wiegant, J. and
       Collard, J.G. (1992) Cytogenet. Cell Genet. **60**, 200–205
18.   Habets, G.G.M., Van der Kammen, R., Scholtes, W.H.M. and Collard, J.G. (1990) Clin.
       Exp. Metastasis **8**, 567–577
19.   Collard, J.G., Schijven, J.F. and Roos, E. (1987) Cancer Res. **47**, 754–759
20.   Habets, G.G.M., Van der Kammen, R.A., Stam, J.C., Michiels, F. and Collard, J.G. (1995)
       Oncogene **10**, 1371–1376
21.   Habets, G.G.M., Vanderkammen, R.A., Jenkins, N.A., Gilbert, D.J., Copeland, N.G.,
       Hagemeijer, A. and Collard, J.G. (1995) Cytogenet. Cell Genet. **70**, 48–51
22.   Cerione, R.A. and Zheng, Y. (1996) Curr. Opin. Cell Biol. **8**, 216–222
23.   Gibson, T.J., Hyvonen, M., Birney, E., Musacchio, A. and Saraste, M. (1994) Trends
       Biochem. Sci. **19**, 349–353
24.   Harlan, J.E., Hajduk, P.J., Yoon, H.S. and Fesik, S.W. (1994) Nature (London) **371**, 168–170
25.   Kennedy, M.B. (1995) Trends Biochem. Sci. **20**, 350
26.   Tapon, N. and Hall, A. (1997) Curr. Opin. Cell Biol. **9**, 86–92
27.   Bourne, H.R., Sanders, D.A. and McCormick, F. (1991) Nature (London) **349**, 117–127
28.   Boguski, M.S. and McCormick, F. (1993) Nature (London) **366**, 643–654
29.   Johnson, D.I. and Pringle, J.R. (1990) J. Cell Biol. **111**, 143–152
30.   Kozma, R., Ahmed, S., Best, A. and Lim, L. (1995) Mol. Cell. Biol. **15**, 1942–1952
31.   Bagrodia, S., Derijard, B., Davis, R.J. and Cerione, R.A. (1995) J. Biol. Chem. **270**,
       27995–27998
32.   Ridley, A.J., Paterson, H.F., Johnston, C.L., Diekmann, D. and Hall, A. (1992) Cell **70**,
       401–410
33.   Nobes, C.D. and Hall, A. (1995) Cell **81**, 53–62
34.   Ridley, A.J. and Hall, A. (1992) Cell **70**, 389–399
35.   Sekine, A., Fujiwara, M. and Narumiya, S. (1989) J. Biol. Chem. **264**, 8602–8605
36.   Chardin, P., Boquet, P., Madaule, P., Popoff, M.R., Rubin, E.J. and Gill, D.M. (1989)
       EMBO J. **8**, 1087–1092
37.   Paterson, H.F., Self, A.J., Garrett, M.D., Just, I., Aktories, K. and Hall, A. (1990) J. Cell
       Biol. **111**, 1001–1007
38.   Ridley, A.J. (1994) BioEssays **16**, 321–327
39.   Rodriguezviciana, P., Warne, P.H., Khwaja, A., Marte, B.M., Pappin, D., Das, P.,
       Waterfield, M.D., Ridley, A. and Downward, J. (1997) Cell **89**, 457–467

40. Hawkins, P.T., Eguinoa, A., Qiu, R.G., Stokoe, D., Cooke, F.T., Walters, R., Wennstrom, S., Claessonwelsh, L., Evans, T., Symons, M. and Stephens, L. (1995) Curr. Biol. **5**, 393–403

41. Reif, K., Nobes, C.D., Thomas, G., Hall, A. and Cantrell, D.A. (1996) Curr. Biol. **6**, 1445–1455

42. Peppelenbosch, M.P., Qiu, R.G., Devriessmits, A.M.M., Tertoolen, L.G.J., Delaat, S.W., McCormick, F., Hall, A., Symons, M.H. and Bos, J.L. (1995) Cell **81**, 849–856

43. Hordijk, P.L., ten Klooster, J.P., Van der Kammen, R.A., Michiels, F., Oomen, L.C.J.M. and Collard, J.G. (1997) Science **278**, 1464–1466

44. van Leeuwen, F.N., Kain, H.E.T., Van der Kammen, R.A., Michiels, F., Kranenburg, O. and Collard, J.G. (1997) J. Cell Biol. **139**, 797–807

45. Minden, A., Lin, A.N., Claret, F.X., Abo, A. and Karin, M. (1995) Cell **81**, 1147–1157

46. Coso, O.A., Chiariello, M., Yu, J.C., Teramoto, H., Crespo, P., Xu, N.G., Miki, T. and Gutkind, J.S. (1995) Cell **81**, 1137–1146

47. Crespo, P., Bustelo, X.R., Aaronson, D.S., Coso, O.A., Lopezbarahona, M., Barbacid, M. and Gutkind, J.S. (1996) Oncogene **13**, 455–460

48. Michiels, F., Stam, J.C., Hordijk, P.L., Vanderkammen, R.A., Ruulsvanstalle, L., Feltkamp, C.A. and Collard, J.G. (1997) J. Cell Biol. **137**, 387–398

49. Brown, J.L., Stowers, L., Baer, M., Trejo, J., Coughlin, S. and Chant, J. (1996) Curr. Biol. **6**, 598–605

50. Fanger, G.R., Johnson, N.L. and Johnson, G.L. (1997) EMBO J. **16**, 4961–4972

51. Westwick, J.K., Lambert, Q.T., Clark, G.J., Symons, M., Vanaelst, L., Pestell, R.G. and Der, C.J. (1997) Mol. Cell. Biol. **17**, 1324–1335

52. Burbelo, P.D., Drechsel, D. and Hall, A. (1995) J. Biol. Chem. **270**, 29071–29074

53. Irani, K., Xia, Y., Zweier, J.L., Sollott, S.J., Der, C.J., Fearon, E.R., Sundaresan, M., Finkel, T. and Goldschmidt Clermont, P.J. (1997) Science **275**, 1649–1652

54. van Aelst, L., Joneson, T. and Bar-Sagi, D. (1997) EMBO J. **15**, 3778–3786

55. Dsouzaschorey, C., Boshans, R.L., McDonough, M., Stahl, P.D. and Vanaelst, L. (1997) EMBO J. **16**, 5445–5454

56. Symons, M., Derry, J.M.J., Karlak, B., Jiang, S., Lemahieu, V., McCormick, F., Francke, U. and Abo, A. (1996) Cell **84**, 723–734

57. Hart, M.J., Callow, M.G., Souza, B. and Polakis, P. (1996) EMBO J. **15**, 2997–3005

58. Kozma, R., Ahmed, S., Best, A. and Lim, L. (1996) Mol. Cell. Biol. **16**, 5069–5080

59. Lamarche, N., Tapon, N., Stowers, L., Burbelo, P.D., Aspenstrom, P., Bridges, T., Chant, J. and Hall, A. (1996) Cell **87**, 519–529

60. Brill, S., Li, S.H., Lyman, C.W., Church, D.M., Wasmuth, J.J., Weissbach, L., Bernards, A. and Snijders, A.J. (1996) Mol. Cell. Biol. **16**, 4869–4878

61. Kuroda, S., Fukata, M., Kobayashi, K., Nakafuku, M., Nomura, N., Iwamatsu, A. and Kaibuchi, K. (1996) J. Biol. Chem. **271**, 23363–23367

62. Bashour, A.M., Fullerton, A.T., Hart, M.J. and Bloom, G.S. (1997) J. Cell Biol. **137**, 1555–1566

63. Ren, X.D., Bokoch, G.M., Traynorkaplan, A., Jenkins, G.H., Anderson, R.A. and Schwartz, M.A. (1996) Mol. Biol. Cell **7**, 435–442

64. Zheng, Y., Bagrodia, S. and Cerione, R.A. (1994) J. Biol. Chem. **269**, 18727–18730

65. Bokoch, G.M., Vlahos, C.J., Wang, Y., Knaus, U.G. and Traynorkaplan, A.E. (1996) Biochem. J. **315**, 775–779

66. Best, A., Ahmed, S., Kozma, R. and Lim, L. (1996) J. Biol. Chem. **271**, 3756–3762

67. Narumiya, S. (1996) J. Biochem. (Tokyo) **120**, 215–228

68. Watanabe, G., Saito, Y., Madaule, P., Ishizaki, T., Fujisawa, K., Morii, N., Mukai, H., Ono, Y., Kakizuka, A. and Narumiya, S. (1996) Science **271**, 645–648

69. Amano, M., Mukai, H., Ono, Y., Chihara, K., Matsui, T., Hamajima, Y., Okawa, K., Iwamatsu, A. and Kaibuchi, K. (1996) Science **271**, 648–650

70. Vincent, S. and Settleman, J. (1997) Mol. Cell. Biol. **17**, 2247–2256

71. Madaule, P., Furuyashiki, T., Reid, T., Ishizaki, T., Watanabe, G., Morii, N. and Narumiya, S. (1995) FEBS Lett. **377**, 243–248

72. Reid, T., Furayashiki, T., Ishizaki, T., Watanabe, G., Watanabe, N., Fujisawa, K., Morii, N., Madaule, P. and Narumiya, S. (1996) J. Biol. Chem. **271**, 13556–13560

73. Gebbink, M.F.B.G., Kranenburg, O., Poland, M., Vanhorck, F.P.G., Houssa, B. and Moolenaar, W.H. (1997) J. Cell Biol. **137**, 1603–1613

74. Ishizaki, T., Maekawa, M., Fujisawa, K., Okawa, K., Iwamatsu, A., Fujita, A., Watanabe, N., Saito, Y., Kakizuka, A., Morii, N. and Narumiya, S. (1996) EMBO J. **15**, 1885–1893

75. Amano, M., Ito, M., Kimura, K., Fukata, Y., Chihara, K., Nakano, T., Matsuura, Y. and Kaibuchi, K. (1996) J. Biol. Chem. **271**, 20246–20249

76. Leung, T., Manser, E., Tan, L. and Lim, L. (1995) J. Biol. Chem. **270**, 29051–29054

77. Leung, T., Chen, X.Q., Manser, E. and Lim, L. (1996) Mol. Cell. Biol. **16**, 5313–5327

78. Kimura, K., Ito, M., Amano, M., Chihara, K., Fukata, Y., Nakafuku, M., Yamamori, B., Feng, J., Nakano, T., Okawa, K., et al. (1996) Science **273**, 245–248

79. Amano, M., Chihara, K., Kimura, K., Fukata, Y., Nakamura, N., Matsuura, Y. and Kaibuchi, K. (1997) Science **275**, 1308–1311

80. Kureishi, Y., Kobayashi, S., Amano, M., Kimura, K., Kanaide, H., Nakano, T., Kaibuchi, K. and Ito, M. (1997) J. Biol. Chem. **272**, 12257–12260

81. Vanleeuwen, F.N., Vanderkammen, R.A., Habets, G.G.M. and Collard, J.G. (1995) Oncogene **11**, 2215–2221

82. Symons, M. (1996) Trends Biochem. Sci. **21**, 178–181

83. Qiu, R.G., Chen, J., Kirn, D., McCormick, F. and Symons, M. (1995) Nature (London) **374**, 457–459

84. Qiu, R.G., Chen, J., McCormick, F. and Symons, M. (1995) Proc. Natl. Acad. Sci. U.S.A. **92**, 11781–11785

85. Qiu, R.G., Abo, A., McCormick, F. and Symons, M. (1997) Mol. Cell. Biol. **17**, 3449–3458

86. Khosravi-Far, R., Solski, P.A., Clark, G.J., Kinch, M.S. and Der, C.J. (1995) Mol. Cell. Biol. **15**, 6443–6453

87. Olson, M.F., Ashworth, A. and Hall, A. (1995) Science **269**, 1270–1272

88. Hirai, A., Nakamura, S., Noguchi, Y., Yasuda, T., Kitagawa, M., Tatsuno, I., Oeda, T., Tahara, K., Terano, T., Narumiya, S., et al. (1997) J. Biol. Chem. **272**, 13–16

89. Stam, J.C., Sander, E.E., Michiels, F., van Leeuwen, F.N., Kain, H.E.T., Van der Kammen, R.A. and Collard, J.G. (1997) J. Biol. Chem. **272**, 28447–28454

90. Rameh, L.E., Arvidsson, A.K., Carraway, K.L., Couvillon, A.D., Rathbun, G., Crompton, A., Vanrenterghem, B., Czech, M.P., Ravichandran, K.S., Burakoff, S.J., et al. (1997) J. Biol. Chem. **272**, 22059–22066

91. Haataja, L., Groffen, J. and Heisterkamp, N. (1997) J. Biol. Chem. **272**, 20384–20388

92. Moolenaar, W.H., Kranenburg, O., Postma, F.R. and Zondag, G.C.M. (1997) Curr. Opin. Cell Biol. **9**, 168–173

93. Takeichi, M. (1991) Science **251**, 1451–1455

94. Ridley, A.J., Comoglio, P.M. and Hall, A. (1995) Mol. Cell. Biol. **15**, 1110–1122

95. Braga, V.M.M., Machesky, L.M., Hall, A. and Hotchin, N.A. (1997) J. Cell Biol. **137**, 1421–1431

Biochem. Soc. Symp. **65**, 147–172
Printed in Great Britain

10

Microtubule involvement in regulating cell contractility and adhesion-dependent signalling: a possible mechanism for polarization of cell motility

M. Elbaum*, A. Chausovsky†, E. T. Levy†, M. Shtutman† and A. D. Bershadsky†[1]

*Department of Materials and Interfaces, and †Department of Molecular Cell Biology, Weizmann Institute of Science, P.O.B. 26, Rehovot 76100, Israel

Abstract

The dynamic shape of an isolated cell results from an interplay between protrusion, adhesion and contraction activities. These are most closely associated with the actin cytoskeleton. In many cell types, microtubules have been shown to be involved in the development of morphological polarity required for directional migration. This suggests a role for the microtubule system in regulating both the actin cytoskeleton and the formation of cell–substrate adhesions. The most prominent role of microtubules in the cell is in transport of vesicles and organelles. Disruption of the microtubules, on the other hand, leads to a significant increase in actomyosin-driven contractility. This suggests the involvement of microtubules in the control of forces produced by the cell against the points at which it contacts the substrate or extracellular matrix. We show that microtubule disruption also activates an adhesion-dependent signal transduction cascade and promotes the formation of focal adhesions and associated actin microfilament bundles. Using overexpression of caldesmon, a regulatory protein which inhibits the interaction between actin and myosin, we show that these effects of microtubule disruption depend on the activation of contractility. Formation of focal adhesions induced by the small GTPase Rho is also blocked by the caldesmon inhibition of contractility. We infer that there is a step in the adhesion-dependent signalling pathway that requires mechanical tension applied to cell–substrate contacts. Although the experimental data are

[1]To whom correspondence should be addressed.

based on complete microtubule disruption, we suggest that a similar effect occurs locally following depolymerization of individual microtubules. We speculate that the interplay among microtubule dynamics, actomyosin contractility and adhesion-dependent signalling can produce a mechanism for the determination of cell polarity and direction of migration. In essence, microtubule depolymerization would create a local increase in contractile force, testing and promoting the maturation of nearby cell–substrate adhesions.

Introduction

The shape and locomotion of cells attached to a substrate depend on three basic parameters: protrusional activity at the periphery, adhesion to the extracellular matrix, and contraction throughout the body of the cell [1–3]. Cell protrusions such as filopodia and lamellipodia may form adhesive contacts to the underlying substrate, mediated by interactions of specific receptors with the extracellular matrix (ECM). In most cell types, the major adhesion receptors are known as integrins [4,5]. The points of adhesion are placed in tension [6–9,176] by contraction of structures within the cell [8–10]. Release of adhesion leads to cell rounding [11]. The interplay between protrusional activity, internal contractility and adhesion may result in spreading, development of asymmetrical shape and/or directional motility along the substrate. Motility requires the formation of new protrusions and new adhesions, plus detachment of the cell from old adhesion sites [3].

The main function of lamellipodial extensions appears to be adhesive, both in making new membrane–ECM contacts and in attaching these to the cytoskeleton [12,13]. The molecular complexes making up these contacts are called focal adhesions, or focal contacts. The forward edge of the lamellipodium of a typical motile cell is studded by many small, dot-like contacts. Firm anchoring of the cell coincides with the maturation of these contacts into larger elongated patches [14,15]. Cell motility is most rapid at an optimal intermediate level of adhesion [16].

A necessary prerequisite for directional motility is a breaking of symmetry at the cell periphery. Protrusional activity associated with the forward direction of cell movement is concentrated at the leading edge. This process is known as polarization of the cell shape [3,17]. Note that we refer here to a property of the individual cell, and not to the apical–basal axis of epithelial and endothelial cells, which depend on lateral cell–cell interactions [18,19]. This polarization is often, as in the case of cultured fibroblasts, associated with a fan-like shape with a broad lamella at the leading edge and a narrower trailing edge anchored by old adhesions. Quantitative morphological indices describing the cell outline may then be a good measure of the polarization [20,21]. In other cases, e.g. fish keratocytes [22], segregation of the cell edge into protrusionally active and inactive parts is not accompanied by strong asymmetry in the cell outline. In addition to morphology, a number of specific proteins have been discovered to concentrate preferentially at lamellar extensions. In higher eukaryotic cells these include β-actin [23], profilin [24], VASP (vasodilator-stimulated phosphoprotein) and its homologue, Mena [25,94], Arp 2/3 complex

[94, 177], cofilin [178] and proteins of the ERM (ezrin/radixin/moesin) family [26], particularly ezrin [27]. These are probably involved in the regulation of polymerization, cross-linking and membrane attachment of actin filaments.

Besides the actin cytoskeleton mentioned above, the microtubule system often plays a role in the determination and maintenance of cell polarity. This was first noted in the experiments of Vasiliev and co-workers on fibroblasts exposed to colcemide in order to depolymerize the microtubules [28]. The treated cells lost the segregation of their peripheries into active and stable regions, as well as their ability to migrate directionally [28,29]. Other agents that depolymerize microtubules (nocodazole, vinblastine, colchicine) have a similar effect. More recent studies showed that artificial microtubule stabliza-tion also reduces cell polarization and motility. This is achieved by the microtubule-stabilizing drug taxol [30,31] or by extremely low doses of depolymerizing agents [32,33], which paradoxically also have a stabilizing effect [34,35]. Therefore it appears that the proper dynamics of microtubules, and not simply their existence, plays a role in the regulation of cell polarity.

On the other hand, some cell types can move directionally after complete disruption of their microtubules (reviewed in [36]). For example, fish kerato-cytes and even fragments of these cells lacking microtubules entirely preserve the capacity for directional movement [37]. Neoplastically transformed fibro-blasts which adhere particularly weakly to the substrate can be polarized in a microtubule-independent manner [38]. Moreover, chicken fibroblasts in pri-mary culture can initially polarize their shape independently of microtubules, but lose this ability after several generations [39].

Contractility is another important factor responsible for polarization of cell shape during locomotion. Myosin II is particularly implicated in overall cell contraction [11]. In some cell systems, such as *Dictyostelium* amoebae, directional motility is still possible in a myosin II knockout strain [40]. Myosin II is required for forward motion when cell adhesion to the substrate is increased [41]. In cultured fibroblasts, myosin II is a component of both the submembranous cortex and the contractile bundles of actin filaments [42] asso-ciated with focal adhesions. It is this interplay between contractility and cell adhesion on which we will focus attention. We will discuss the evidence that tension developed by myosin is necessary for the maturation of focal contacts. Furthermore, the microtubule system may regulate adhesion via control of myosin II-driven contractility.

In this review we aim to study the role of microtubules as a co-ordinator of motility-related events. We begin with a discussion of microtubules them-selves, particularly with regard to their mechanical characteristics and connections to other cell components. We look next at the actin cytoskeleton in its generation of cell protrusion and especially myosin-dependent contrac-tion. We then discuss in detail the evidence for regulation of these actin cytoskeleton functions by microtubules. We present original data concerning the interplay between microtubules and the contractile actin network in the formation and maturation of focal adhesions. Finally, we propose a mechanism by which dynamic microtubules participate in the establishment of cell polarity.

Microtubules *in vitro* and *in vivo*

Microtubules are polymers of heterodimeric α- and β-tubulin. They form as a hollow cylindrical structure, 25 nm in diameter, with an intrinsic polarity based on the directionality of the α- and β-subunits. The two dissimilar ends are known as plus and minus (reviewed in [43,179]). The polymerization/depolarization of microtubules is not an equilibrium thermodynamic process. Rather, it is associated with hydrolysis of GTP associated with the β-subunit. This energy-dependence leads to a rich behaviour of assembly and disassembly (reviewed in [43–45, 62]). Over a certain range of temperature and concentration, comparable with physiological levels, *in vitro* polymerization displays a dynamic instability consisting of periods of assembly at a constant rate interspersed with rapid, catastrophic depolymerization events [46,47]. Growth rates are different at the two ends [43–45]. Current models suggest that growing ends contain a cap in which the tubulin is associated with GTP, while loss of this cap due to stochastic GTP hydrolysis leads to disassembly [43–46]. Recent work has also suggested a temporarily stable state, particularly at the minus end [48].

In spite of their unconventional dynamics, microtubules are nonetheless elastic solid bodies with well-defined mechanical properties. These properties have been measured by a variety of methods. For example, microtubules were formed inside phospholipid vesicles and then the vesicles were subjected to micropipette aspiration. Measurement of the suction required for buckling of a single microtubule or of a bundle of microtubules, as well as for the contraction of a circumferential ring of microtubules, was used to extract the microtubule bending rigidity [49,50]. Values from these and other measurements [49–53] differ by small factors, but converge within an order of magnitude to give a bending rigidity of 2×10^{-19} N·cm^2 (2×10^{-14} dyn·cm^2). This corresponds to a persistence length of roughly 5 mm. This means that, at shorter length scales, the microtubule maintains its overall straightness against Brownian fluctuations.

The polymerization of microtubules can apply a force against an opposing obstacle, such as a lipid membrane. Microtubules have been shown to draw out tubular structures in cell extracts [54], as well as in artificial membranes as described above [49,50]. The scale of this extensive force is limited by the bending rigidity; the microtubule can support a compression which depends on its length. For a segment of length 10 μm, for example, the critical force required to induce buckling is 2 pN, which is somewhat less than the force produced by individual motor proteins [the critical force required to buckle a rod compressed at its ends is $\pi^2 \kappa / L^2$, where κ is the bending rigidity and L is length]. Depolymerizing microtubules can also exert a tensile force. This is seen most dramatically in the poleward movement of chromosomes during mitosis, an effect that has been duplicated *in vitro* [54–57]. Depolymerization can induce inward tubulation of the plasma membrane [58]. These tensile effects of depolymerization depend on links created by other proteins between the microtubule ends and the external structures.

In the typical undifferentiated cell in culture, microtubules form a radial system, with the minus ends nucleated at the centrosome and the plus ends growing towards the periphery. Their total number is of the order of hundreds. They display a dynamic instability *in vivo* (reviewed in [44,59]), with a growth rate of about 5 μm/min measured in newt lung epithelial cells [59]. This is comparable with the rate measured for purified tubulin *in vitro* under physiological conditions [47]. The disassembly rate *in vivo* is approx. 15 μm/min in the same cells. This is half the rate measured *in vitro*.

In some cell types, a fraction of the microtubules can be observed to be disconnected from the centrosome, e.g. the microtubules in the processes of nerve cells (reviewed in [44]). It was found that some microtubules in epithelial cells nucleate at the centrosome, disconnect and move away from it autonomously [60]. Further experiments will undoubtedly find other dynamic behaviours in various cell types.

A peculiar type of behaviour called 'treadmilling' has also been registered *in vivo*, in which the plus end of the microtubule lengthens while the minus end shrinks, inducing movement of the microtubule as a whole [61]. It is interesting that although treadmilling was originally suggested based on observations *in vitro* (reviewed in [62,180]), a more recent measurement using purified tubulin and analysis based on real-time dynamic observations did not find evidence for this regime [63]. A likely conclusion is that the effect *in vivo* is due to co-operation with the many other associated proteins in the cell. Among the hundreds of proteins which associate with microtubules in the cytoplasm, one notable class is that of motor proteins. The motor proteins belong to two protein families, kinesins and dyneins, which hydrolyse ATP to move in directions determined with respect to microtubule polarity. Dyneins move from the plus end towards the minus end, while the majority of kinesins move from minus to plus [64–67]. These motor proteins move vesicles and membranous organelles as transport cargo [67,68]. Perhaps they also 'piggyback' mRNAs and soluble proteins. Some motor forms can induce sliding of microtubules one against the other. In the presence of cross-linkers, sliding can be transformed into the characteristic bending motion, as in flagella and cilia. Kinesin motors are also involved in the generation of tractile forces by microtubule depolymerization [69].

The second class of microtubule-associated proteins (MAPs), the non-motor MAPs [70], have less well-defined functions. They may stabilize microtubules, and also cross-link or associate them with other cytoskeletal elements. Phosphorylation of the MAPs by specific kinases can interrupt the stabilizing effect, and provide the cell with a regulatory mechanism for microtubule stabilization or disruption [71,72]. MAP binding to microtubules can also be regulated by special modulator proteins [73]. Other auxiliary proteins may bind the tubulin dimers in solution, thereby skewing the microtubule dynamics towards depolymerization [74,75]. Some members of the kinesin family also participate in regulation of microtubule dynamics, provoking catastrophic depolymerization [76]. Some MAPs bind to microtubule ends, and may participate in the anchoring of microtubules to peripheral cytoplasmic structures [181]. Yet another group of proteins can sever microtubules [77–79]. These may be involved in the separation of microtubules from the centrosome.

Proteins normally associated with many aspects of signal transduction have also been found in association with microtubules. Among these are mitogen-activated protein kinase (MAPK) [80,81], phosphatase 2A [82], phosphoinositide 3-kinase [83] and the tumour-suppressor protein adenomatous polyposis coli [84,85]. In general the physiological relevance of the association of these proteins with microtubules is not understood. Two possibilities may be suggested: as other MAPs, they may be involved in regulation of the microtubules themselves or, more intriguingly, they may make use of the microtubule network to localize their sensory and effectory functions.

Our attention here is focused on the possible role of microtubules in generating the polarization needed for directed locomotion of the cell as a whole. To this end, we should examine microtubule interactions with the actomyosin contractile system.

Contraction and protrusion: elementary actin-dependent motile processes

The shape of a cell may be very complex. Nonetheless, the complexity can be built on the basis of elementary processes: local protrusion, and both local and global cell contraction. These depend on the actin cytoskeleton. The formation of actin-filled membrane extensions such as filopodia and lamellipodia has been discussed in many excellent reviews [1,3,22,86–88]. Apparently, actin polymerization is an essential component in the physical deformation of the cell boundary. This is seen by analogy with the transport process of *Listeria* and several other bacterial types, which move within the cell by means of actin polymerization at their tail ends [89–91]. Indeed, it appears that actin polymerization alone can generate sufficient force to deform the plasma membrane [92,182]. Mechanisms of induction and regulation of actin polymerization are under active investigation, and have been reviewed by other authors [86,93–95,183].

Bud formation in yeast could represent another parallel with pseudopod formation in higher eukaryotes [96,97]. In budding yeast the rapid extensive growth depends on the availability of construction material near the protrusion site. For pseudopod extension in vertebrate cells, incorporation of new membrane material is well documented [98,99,184]. Furthermore, disruption of the normal delivery of β-actin mRNA to the lamellipodia of fibroblasts inhibits cell polarization [100,101].

Beyond simple polymerization, actin cross-linking is likely to be a crucial element in the formation of cell protrusions. For example, the Arp 2/3 complex nucleates formation of actin filaments, but furthermore connects pointed filament ends to the sides of other filaments, thereby creating branching networks [185,186]. Both *Dictyostelium* amoebae [102] and human melanoma cells [103] that lack actin-cross-linking proteins are deficient in protrusional activity. Re-expression of these proteins restores the ability of the cell to form protrusions. Interestingly, ectopic expression of MAP2c, which is seen to cross-link actin *in vitro*, also restores protrusional activity [104]. This example points to a possible functional link between the microtubule and actin networks.

The many forms of cellular contraction are generated by the sliding of actin filaments driven by myosin motor proteins. The myosin superfamily includes at least 13 classes [105,106], but the predominant role in contractility is played by the myosin II class, also known as the conventional myosins. This class appears ubiquitously in cell types ranging from amoebae to striated muscle of higher vertebrates [105]. Its distinctive feature is the ability to form bipolar filaments, along which actin filaments are drawn in opposite directions [107]. The character of contraction developed depends on the nature of the actin arrays on which the myosin acts.

The exact mode of action of the majority of unconventional myosins is less clear. Members of the myosin I class are often found in association with lipid membranes and therefore may be involved in vesicular transport, as well as in anchoring actin filaments to the plasma membrane [105,106]. Myosin V is also probably involved in transport processes [108], sometimes in concert with microtubule motors [187]. In particular, this myosin may provide materials that are specifically required for the filopodial extension of neuronal growth cones [109]. It is possible, of course, that some of the unconventional myosins will be shown to be involved in contractility.

Myosin II-based contractility takes many diverse forms. Besides its classical function in striated and smooth muscle, genetic mutation data show that it is involved in the contraction of the cleavage furrow during cytokinesis [110,111], and also that it drives closure of the dorsal cleft in *Drosophila* development [112]. The contraction of actin ring structures in epithelial monolayers is a mechanism of wound healing based on a purse-string model [113].

Many types of cells in culture (perhaps all naturally adherent ones) are contractile to some degree. The typical test is the deformation of an elastic or gel substrate [6–10,176]. The degree of contractility varies significantly, however, both among cell types and in response to external conditions. For example, keratocytes generate much weaker forces than fibroblasts [8,114,115]. Release of adherent cells by proteolytic enzymes (e.g. trypsin) usually causes them to detach from the substrate and to round into rough spheres, suggesting the presence of internally generated tensile forces. Permeabilization of the cell membrane and addition of ATP has a similar effect. It has been shown that the detachment of permeabilized cells can be blocked by a peptide which interferes specifically with the actin–myosin interaction [11,116].

Organized cell contraction is an obvious element in the protrusion, attachment and retraction model of cell motility. Evidence for the involvement of myosin II in crawling locomotion comes mainly from experiments using null mutants of *Dictyostelium* amoebae. The first experiments found surprisingly little effect of the loss of myosin, primarily that the knockouts moved more slowly [40]. Later studies showed that more significant differences appear in cells plated on to susbtrates made especially adhesive by coating with poly-lysine [41]. Myosin II was found to concentrate mainly at the receding edge in these cells, suggesting a role in retraction of the trailing edge [117]. It also appears transiently in the tips of retracting pseudopods [117]. The motility of typical tissue culture cells derived from higher animals, e.g. fibroblasts, resembles much more closely that of amoeboid motion on adherent substrates than in its natural mode. Experiments using microinjection of anti-(myosin II) anti-bodies showed marked changes in cell shape, particularly an elongation of the trailing edge, which was not effectively retracted [118]. Similar observations have been made following other means of myosin blocking [119].

Myosin II filaments organize into superstructures. The most prominent are seen in the sarcomeres of striated muscle. A less periodic arrangement exists in smooth muscle as well, with shorter, thinner filaments. In fibroblasts, myosin filaments aggregate into small ribbons which form within and give structure to the actin network [42]. A recent suggestion based on these structural observations is that the ribbons organize the actin filaments into bundles which traverse the cell. These heavy bundles are a dominant actin structure in many types of cultured cells. They normally terminate in focal adhesions [120], and are the main source of tension applied to the culture substrate via adhesion sites. In keratocytes, myosin II forms discrete clusters of bipolar minifilaments in lamellipodia that assemble into actomyosin bundles, contraction of which moves the cell body forward [186] (see Chapter 12 in this volume). The common feature among the various types of myosin superstructures is the bipolar, 'back-to-back' organization of the myosin molecules. Contractility results from the oppositely directed tension generated in associated actin filaments. The more diffuse structure in non-muscle cells permits a more flexible and dynamic local organization of the basic contractile elements.

Stimulation of motile activities leads to regulation of contraction and protrusion. A central role in integrating the stimuli into controlled organization of the actin cytoskeleton is played by three small GTPases of the Rho family: Cdc42, Rac and Rho itself [121,122]. These act as switches, being active in the GTP-bound form and inactive in the GDP-bound form. Specific mutations may keep them in the constitutively activated GTP-bound form. Microinjection or ectopic expression of the activated protein variants shows that each induces specific changes in the actin cytoskeleton. In the case of Cdc42, exaggerated filipodia are formed in several cell types [123,124]. In a similar fashion, Rac induces large and active lamellipodia [125], while Rho induces formation of stress fibres and cell–substrate focal adhesions [126]. The changes are most apparent in cells preincubated in serum-free medium, under which conditions these features are normally suppressed. Specific components of serum, as well as certain hormones, can selectively trigger activation of the Rho switches. For example, lysophosphatidic acid activates Rho [126], growth factors such as platelet-derived growth factor, epidermal growth factor and insulin were shown to activate Rac [125], and bradykinin activates Cdc42 [123,124].

Of course, the activation is not direct, and many stages are likely to be discovered between ligand–receptor binding and consequent GTPase activation. There is also evidence for a synergistic activation among the Rho-family GTPases. Some degree of hierarchy has already been revealed: Cdc42 activation can lead to activation of Rac, and likewise Rac can activate Rho [121–124]. It is possible, however, that inhibitory actions also exist, and perhaps Cdc42 activation actually depresses the formation of stress fibres normally associated with Rho activation [123]. The level at which these stimulatory or inhibitory effects occur is not clear at present. It may even be impossible to separate entirely the 'upstream' GTPase switches from the cytoskeletal re-organizations that they induce, if the latter themselves generate feedback on their putative regulators. This would suggest an integrated system for spatio–temporal control of the basic protrusion and contraction elements involved in cell motility. Returning to a comparison introduced earlier, it is striking to note the similarities in the ways in which Rho-family proteins act in the regulation of cell protrusions in higher eukaryotes and in budding processes in yeast [127].

The process of lamellipodia formation depends on actin polymerization, so it is not surprising that Rac should be involved in its initiation [128]. Rho, on the other hand, has been shown to be involved in the regulation of contractility. In the following, we expand on the control mechanisms for myosin superstructure assembly and its force-producing interaction with actin.

The main feature of myosin II regulation in smooth muscle and non-muscle cells alike is the phosphorylation of the regulatory light chain at Ser-19 [129]. On the one hand, this modification triggers actin-dependent ATPase activity driving movement of actin filaments along the myosin heads. On the other, it induces the formation of organized myosin filaments. Phosphorylation of the neighbouring Thr-18 enhances these effects. Myosin light-chain kinase is a Ca^{2+}/calmodulin-dependent enzyme that performs these phosphorylations [129]. This shows an important link between Ca^{2+} and regulation of contractility in smooth muscle and non-muscle cells. Recently it was shown that GTP-Rho activates another kinase, named Rho-associated kinase (ROK), one of whose substrates is the same Ser-19 on the myosin light chain [130]. Dephosphorylation at this site is carried out by myosin light-chain phosphatase, which is itself under the double control of Rho: GTP-Rho can bind to its myosin-binding subunit, while ROK can inactivate it by phosphorylation [131]. We can infer from this that some background level of Rho is necessary as a counter to the myosin phosphatase. Furthermore, an increase in Rho can stimulate contractility in a Ca^{2+}-independent manner by activation of ROK. Indeed, these effects can be seen *in vivo*. Activation of Rho by lysophosphatidic acid or microinjection of the constitutively active form increases contractility [132], while inactivation by the C3 toxin of *Botulinum* was shown to block contractility in some cell systems [133]. Microinjection of ROK mimics the effect of Rho activation on stress fibre formation [134], providing additional evidence that myosin activation by ROK is involved in this cytoskeletal rearrangement.

A distinct control mechanism for myosin function is based on the regulation of its interaction with actin filaments. This is achieved in smooth muscle and non-muscle cells by the regulatory proteins calponin and caldesmon. Our attention here is focused on the latter [135]. *In vitro*, caldesmon can bind actin filaments and myosin, reducing the actin-dependent myosin ATPase activity and motor function. Another actin-binding protein, tropomyosin, enhances the inhibitory effect of caldesmon [136]. Moreover, caldesmon can also bind Ca^{2+}/calmodulin, and this binding suppresses its inhibitory effect [135]. This provides one more means for the regulation of contractility by Ca^{2+}, this time independently of myosin modification. This mechanism functions in a manner similar to that of the troponin/tropomyosin-based regulation of contractility in striated muscle. Phosphorylation of caldesmon also overturns its inhibitory action [137].

We have discussed here the principles of regulation governing basic elements in the generation of cell locomotion, specifically protrusional and contractile activity. The involvement of microtubules in these phenomena is less obvious than that of the actin cytoskeleton. Yet, at least in some cell types, they play an essential role, to which we turn our attention in the sections below.

Microtubule-dependent regulation of the actin cytoskeleton *in vivo*

Disruption of microtubules leads to a series of changes in cell shape, based on a gradual re-organization of the actin cytoskeleton. Most striking, though, is a sharp and immediate increase in contractility. By culturing fibroblast cells on deformable elastic substrates, as developed by Harris [6,7], Danowski demonstrated this effect using a number of microtubule-depolymerizing drugs [138]. A quantitative study measuring the isometric tension generated by similar cells cultured in a deformable collagen gel also showed an increase in tension following the application of nocodazole [9]. As measured in the gel system, the tension approximately doubled. Taxol, a microtubule stabilizer, had no effect [9,138], although it does cause a slower re-organization of the microtubule system, including bundling and the loss of radial character [30,31]. Furthermore, the cell contractility induced by microtubule disruption is accompanied by an increase in phosphorylation of the myosin regulatory light chain [139]. The connection between microtubule disruption and contractility is present in other experimental systems [140,141], and in fact this mechanism operates in live animals as well. In the heart, one type of contractile failure is associated with an excess of cytoplasmic microtubules [142,143], while disruption of microtubules causes an increase in the beating rate of cardiac myocytes [144]. In neural cells, microtubule disruption causes retraction of the neurites [145], whereas in *Xenopus* oocytes it induces cortical contraction [188].

How does microtubule disruption induce contractility of the actin network? Perhaps the intact microtubules interact mechanically with the actin cytoskeleton. In favour of this hypothesis, a number of proteins may make a physical link between the two systems. Some types of MAPs [104] and plectin [146,147] may interact with both microtubules and actin filaments. Plectin may also interact with myosin [147]. Intermediate filaments can attach to both microtubules and actin filaments, further cross-linking these two cytoskeletal systems [148–150].

Microtubules are the most rigid of the cytoplasmic filaments. It has been proposed that actomyosin-generated contraction places the microtubules under compression, and furthermore that the overall cell shape is determined by this force balance. In fact, two types of balance are possible, either from compression of microtubules within the cell or from tension balanced externally by the substratum through points of adhesive contact [151,152]. As mentioned earlier, recent experiments *in vitro* have determined the rigidity of single microtubules quantitatively. It is instructive to compare the scale of forces involved in contraction with those relevant to microtubule buckling.

The most recent data on the tension produced by cell contractions point to whole-cell forces of hundreds of nN [115], i.e. of the same order as the force for cell detachment [16]. In the collagen gel experiments, disruption of microtubules caused an approximate doubling of the contractile force [9]. Therefore we can take these numbers as estimates of the resting forces as well. The critical buckling compression for a microtubule of 10 μm length is approx. 2 pN, so something of the order of 100000 microtubules would be required to counteract the resting-level cell contraction mechanically, assuming that each microtubule acts individually. Clearly there are not enough individual microtubules to counteract the whole-cell contraction in the simplest picture of force balance. A recent work using 'optical tweezers' found that microtubule-associated proteins may increase the rigidity of microtubules significantly, though not by orders of magnitude [189]. In contrast with individual microtubules, a laterally bound bundle should have a rigidity proportional to the square of their number, which would still require hundreds of microtubules in such bound bundles to reach the required force. In cultured fibroblasts overexpressing MAP2 and tau, even very heavy bundles of microtubules were not able to make projections unless the cortical cytoskeleton was weakened by cytochalasin treatment [153]. When weakened, on the other hand, the formation of straight projections resembled the corresponding distortions made in synthetic lipid vesicles. It was recently shown that in migrating epithelial cells, microtubules in the lamella move centripetally with actomyosin-driven retrograde flow [190]. Taken together, these considerations make it highly improbable that microtubules alone can resist the actomyosin-generated tension mechanically as compressively loaded struts. This leaves the cell–substrate adhesions as the primary load-bearing elements setting the cell shape in competition with internal tension. We cannot exclude the possibility that linking of microtubules to other cytoskeletal elements could produce a network of appropriate strength. This possibility awaits experimental evidence.

The microtubule would then be under tension on at least part of its length. It is also possible that the transport function of microtubules is involved in the active removal of some components involved in contractility. Loss of the microtubules would leave a higher concentration of these elements near the cell periphery. Similarly, microtubules may sequester proteins which interact both with microtubules and with actin filaments. Again, loss of the microtubules would cause a sudden increase in the concentration of these proteins and might lead to the observed contractile effect.

There is evidence for association of the microtubules with proteins that are involved in a variety of signal transduction pathways [80–85]. Some of these pathways may be related to the regulation of contractility. Depolymerization of microtubules would lead to a sudden increase in their availability for other targets. Rho is one more candidate, because it binds to kinectin [155] (a receptor for kinesin [154]) on kinesin-transported vesicles. Many other mechanisms can be suggested.

Protrusional activity, which like contractility depends primarily on the actin cytoskeleton, may also be regulated by microtubules. Depolymerization of microtubules suppresses the normal protrusion/retraction activity of pseudopods at the leading edge of fibroblasts [156,191]. At the same time, it may permit formation of pseudopods in normally stable regions of the cell periphery [28,29]. In fibroblasts, long-term treatment with microtubule-disrupting agents leads to loss of the characteristic fan-like shape. The cells become discoid, with pseudopodial activity dispersed around the edge.

The function of microtubules in vesicular transport has led to the suggestion that their role in pseudopod formation is based on the delivery of fresh components to the extending membrane. For example, new membrane has been shown to be incorporated first at the leading edge [98]. Moreover, disruption of intracellular trafficking by brefeldin A induces changes in pseudopodial activity and cell shape which are almost identical to those caused by microtubule disruption [157]. Microinjection of an anti-kinesin antibody has a similar effect [158]. Beyond membrane lipids and associated components, some proteins may rely on microtubule-based transport to reach the regions of protrusion. In fission yeast, the protein Tea1 is carried to the outer ends of the microtubules, where they reach the growing tips [159]. Tea1 has about 40% similarity with the ERM-family proteins, whose localization to the growth cone of neurons was also shown to depend on microtubules [160]. The tumour-suppressor protein adenomatous polyposis coli is found localized to microtubule ends, adjacent to regions of protrusion formation [85].

The transport hypothesis can represent only part of the regulation of protrusional activity. It is obvious to suggest that some aspects of contractility induced by microtubule disruption can affect cell protrusions. For example, the increase in contractility could strengthen or accelerate the retraction of pseudopods. In general, the whole mechanical construct in which the microtubules and actin cytoskeleton play a part must be appreciated in the context of cell spreading on a substrate, including the specific interactions of membrane receptors with the extracellular matrix.

Actin cytoskeleton, microtubules and focal adhesion formation

Cells adhere to a substrate using transmembrane receptors. The best studied class are the integrins [4]. Members of this family of heterodimers interact with specific ligands of the ECM proteins, including fibronectin, collagen, vitronectin, laminin, etc. The cytoplasmic part of the β-integrin subunit associates with a large number of proteins, connecting eventually to the actin cytoskeleton (reviewed in [5,161–163,193]). The focal adhesion sites of cell–ECM contact are normally discrete and isolated. They may mature into streak-shaped patches of several μm in length. The complex of molecules forming the cytoplasmic part is called the focal adhesion plaque. Some of these molecules, such as talin and α-actinin, which interact *in vitro* both with actin and with the cytoplasmic β-integrin tail, appear to play a primarily structural role as connectors [161–163]. Vinculin can bind talin and α-actinin to each other, and also to actin, stabilizing the assembly [192]. The focal plaque is also rich in kinases, such as focal adhesion kinases, their targets, and adaptor proteins (e.g. tensin and paxillin) [161–163]. Although associated with signalling functions, this certainly does not exclude the possibility that such proteins participate directly or indirectly in mechanical linking. Based on evidence presented below, we will suggest that these functions are in fact not distinct, and that the adhesion plaque as a whole can act as a tension-sensing device.

Adhesion of cells is accompanied by a number of signalling events that lead eventually to downstream activation of cyclin-dependent kinases and cell proliferation [193]. In fibroblasts, adhesion deprivation leads to cell-cycle arrest, whereas epithelial cells tend to undergo apoptosis [194]. At least in normal (i.e. non-cancerous) fibroblasts, the formation of focal adhesions is a necessary step in this signalling pathway. Without them, progression through the cell cycle is halted [164]. Maturation of focal adhesion complexes also involves some signal transduction pathways. This includes some steps that are dependent on tyrosine phosphorylation, and these can be blocked by specific inhibitors such as genistein and herbimycin [161–163,165]. Signalling that is dependent on activated Rho is also an indispensable step in the formation of focal adhesions, as described above [121,122]. In serum-free medium, Rho is not active, and many cell types are unable to form stable focal adhesions in spite of the presence of the correct extracellular matrix.

Maturation of focal adhesions from dot-like contacts to streak-like patches is accompanied by the formation of actin stress fibres which traverse the cell and whose ends are buried within the focal adhesion plaques. We have discussed these actin structures above in the context of contractility. Even partial disruption of the actin cytoskeleton with cytochalasin prevents maturation of focal adhesions and stress fibres [161–163]. Furthermore, it blocks the adhesion-dependent signalling associated with tyrosine phosphorylation and downstream events, including MAPK activation. At the same time, blocking of tyrosine phosphorylation is sufficient to prevent actin stress fibre formation [161–163,165]. Therefore the processes of focal adhesion maturation and actin cytoskeletal rearrangement are closely related through the signalling proteins.

Perhaps it is by now not a surprise that microtubules also appear to be involved in the regulation of focal adhesions. Disruption of microtubules leads to the formation of strong focal adhesions associated with stress fibres in fibroblasts cultured in serum-free medium on ECM-coated substrates [166] (Fig. 1). This process depends on activation of the entire adhesion-dependent signal transduction pathway, since it cannot occur in the absense of a proper ECM. It is accompanied by the typical adhesion-dependent signalling events also induced during attachment of suspended cells to the ECM, or by Rho activation in serum-starved, ECM-attached cells. These include tyrosine phosphorylation of focal adhesion kinase, paxillin [166] and the adaptor protein SHC, as well as MAPK activation (Fig. 2). Inhibitors of tyrosine phosphorylation, as well as the Rho inhibitor C3 transferase, suppress the effects of microtubule disruption [166, 167, 195]. Even in Madin–Darby canine kidney (MDCK) epithelial cells, which normally do not form mature focal adhesions and stress fibres, disruption of microtubules induces formation of

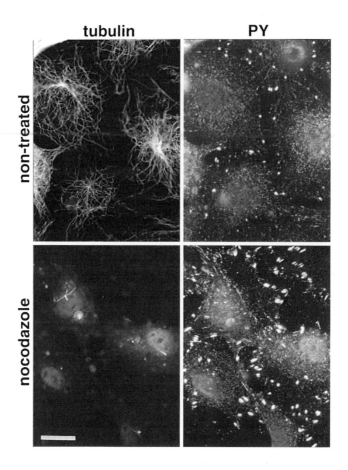

Fig. 1. Effect of microtubule disruption on the formation of focal adhesions in Swiss 3T3 fibroblasts. Cells were serum-starved for 24 h. Control cells retain a well-developed radial system of microtubules, visualized using an anti-tubulin antibody. The focal adhesions, stained with an antibody against phosphotyrosine (PY), are sparse and very small under these conditions. At 30 min after addition of nocodazole, the microtubule network has completely disappeared, while focal adhesions have become numerous and large. Staining with antibodies against the specific focal adhesion markers vinculin and paxillin reveals the same pattern [166]. Scale bar = 20 μm.

these structures very efficiently (Fig. 3). Maturation of focal adhesions induced by microtubule disruption was observed in a variety of fibroblastic cells [196].

What is the common link between these two processes that induce maturation of focal adhesions, namely microtubule disruption and activation of Rho? As we have already discussed in detail, both are powerful stimulators of myosin II-driven contractility. Contractility induced in cells attached to an immobilized extracellular matrix produces tension at the points of adhesion. We propose that it is this tension, sensed by nascent focal adhesion complexes, which acts as the signal that induces the downstream events of actin cytoskel-

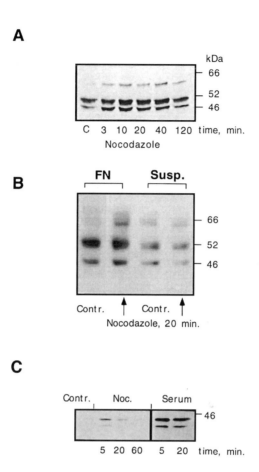

Fig. 2. Some aspects of signal transduction induced by microtubule disruption. (A) Serum-starved Swiss 3T3 cells were lysed at the time points indicated following nocodazole treatment. The lysate was immunoprecipitated with anti-phosphotyrosine antibody. After gel electrophoresis, the proteins were blotted and stained with an antibody against the adaptor protein SHC. The level of phosphorylation of all SHC isoforms was increased. C denotes control. (B) SHC phosphorylation induced by microtubule disruption was compared in cells adherent to fibronectin-coated substrates (FN) and in suspension (Susp.). The increase in phosphorylation seen at 20 min, particularly evident for the 66 kDa isoform, occurred only in the fibronectin-attached cells. Contr. denotes control. In our previous work [166], it was shown that similar adhesion-dependent phosphorylation of focal adhesion kinase and paxillin occurs following disruption of microtubules. (C) Total cell lysate was also immunoblotted using an antibody against the activated form of MAPK (antibody kindly provided by Rony Zeger, Weizmann Institute of Science). A transient activation followed nocodazole (Noc.) treatment within 5 min, while serum addition caused a more pronounced activation. Contr. denotes control.

non-treated   **nocodazole**

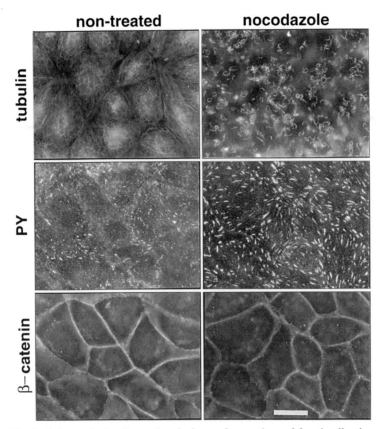

Fig. 3. Microtubule disruption induces formation of focal adhesions in MDCK cells, without disturbing cell–cell junctions. The cells were serum-starved for 12 h before treatment. In the top row, tubulin staining is shown before and 30 min after nocodazole addition. The microtubule system is completely disrupted, aside from a few curved remnants. In the middle row, phosphotyrosine (PY) staining shows the increase in number and strengthening of focal adhesions, similar to the effect shown in Fig. 1. In the bottom row, staining with an antibody against β-catenin shows that cell–cell adherens junctions are not affected by nocodazole treatment. The middle and bottom rows show the same field of cells, though at different focal planes. Scale bar = 20 μm.

eton re-organization and the cascade of tyrosine phosphorylation, probably culminating in transcription activation leading to progression through the cell cycle.

Evidence supporting this hypothesis begins with observations obtained using chemical inhibitors of myosin light-chain phosphorylation or myosin–actin interaction. These chemical agents prevent the formation of focal adhesions and stress fibres, as well as inhibiting the associated signalling events induced by either microtubule disruption or Rho activation [132,166]. It is difficult to evaluate the specificity of simple chemical inhibitors, however, and therefore these data were confirmed using a molecular genetic tool. We have shown (Helfman, D., Levy, E.T., Berthier, C., Shtutman, M., Elbaum, M. and Bershedsky, A.D., unpublished work) that overexpression of caldesmon in cultured fibroblasts efficiently inhibits the contractility of these cells, and prevents the formation of focal adhesions and stress fibres by serum addition, Rho activation or microtubule disruption (Figs. 4 and 5).

Caldesmon interferes specifically with the interaction of actin and myosin. According to *in vitro* evidence, it does not affect the integrity or polymerization of the actin filaments [135]. In cells its action is Ca^{2+}-sensitive, which points to a specific interaction *in vivo* as well. Truncated variants of caldesmon are not effective in preventing focal adhesions and stress fibres. The caldesmon experiments, together with those using chemical inhibitors, show that generation of tension at cell–ECM contacts is an essential part of the process of maturation of focal adhesions and the associated formation of actin stress fibres.

Two other types of experiment have led to a similar conclusion regarding the tension-signalling role of cell–ECM adhesions. In one category, fibroblasts and epithelial cells were cultured on very pliable substrates or in loose gels. They were unable to develop contractile tension because the substrate was too easily deformable, and could not form mature focal adhesions and stress fibres even in serum-containing medium [168]. On the other hand, cells cultured on elastic substrates which can be stretched externally showed an increase in the tyrosine phosphorylation of focal adhesion kinase and in other signalling events of the adhesion-dependent signalling pathway [169]. Another approach is to address the cell contractility apparatus locally, by means of an ECM–ligand-coated bead applied to the cell surface. Such beads are transported from the cell periphery towards the centre, suggesting that the cell applies a contractile force at the point of contact [170]. In a recent experiment, 'optical tweezers' were used to apply a counterforce to hold the bead in place [171]. The cells responded by increasing the internally generated force on the bead until it could no longer be held in the laser trap. Apparently, external resistance leads to signalling for the recruitment and activation of contractile or adhesive elements at the adhesion site.

The suggestion that mechanical tension can lead to phosphorylation events has already been made in the context of mitotic checkpoint regulation. The separation of mitotic chromosome pairs is delayed until attachment of all the chromosomes to the spindle microtubules via the kinetochores is completed. The cell tests the attachments mechanically, and can be tricked to proceed if tension is applied using a microneedle [172,173]. In this case the tension induces dephosphorylation of kinetochore components [173].

Returning to focal adhesions, the activity of some proteins depends on large-scale conformational changes. Vinculin, for example, can exist in a relatively inactive closed conformation in which the active sites on head and tail regions are mutually hidden, and in an active open conformation in which it can bind talin and α-actinin at the head, and actin at the tail [165,192]. One can imagine that this opening of the molecule is governed by mechanical forces.

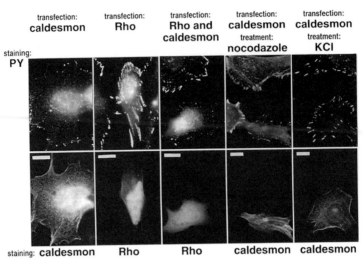

Fig. 4. Effect of caldesmon transfection on the formation of focal adhesions in SV80 fibroblasts. The upper and lower rows show identical fields. Focal adhesions are visualized in the upper row by anti-phosphotyrosine (PY) staining. Cells were transiently transfected with constructs encoding non-muscle caldesmon, activated Rho, or both. The constructs contain either HA (haemagglutinin) or VSV (vesicular stomatitis virus) epitope tags that make it possible to visualize the corresponding proteins using anti-tag antibodies. Beginning from the left-hand side, a cell transfected with caldesmon alone has few and small focal adhesions, even in serum-containing medium (first column). Transfection by activated Rho induces formation of very large and bright focal adhesions, which are much larger than those of neighbouring non-transfected cells, again in serum-containing medium. In the third column, cells from a culture simultaneously transfected with caldesmon and Rho are shown. The lower cell clearly expresses Rho, but contains focal adhesions of the caldesmon type. The focal adhesions are in this case weaker than those of neighbouring non-transfected cells in serum-containing medium. Thus caldesmon interferes with the Rho-dependent induction of focal adhesion formation. In the fourth column, nocodazole treatment induces the formation of focal adhesions in serum-starved cells, as seen in Figs. 1 and 3. Large focal adhesions are absent from the caldesmon-positive cell (lower panel). In the last column, focal adhesions do form in a caldesmon-positive cell if the intracellular Ca^{2+} level is increased by depolarization of the membrane potential using KCl. This is in accordance with *in vitro* evidence that caldesmon interferes with the actin–myosin interaction in a Ca^{2+}-sensitive manner. Scale bars = 10 μm.

Microtubule involvement in cell motility

The evidence for a role for microtubules in regulating the basic processes related to cell motility comes primarily from experiments involving major disruption of the cytoplasmic microtubule network. In the normal life of a cell, such a situation occurs only for a short period immediately preceding formation of the mitotic spindle [174]. This is accompanied in the majority of cell types by strong contractions and cell rounding. It is possible that the consider-

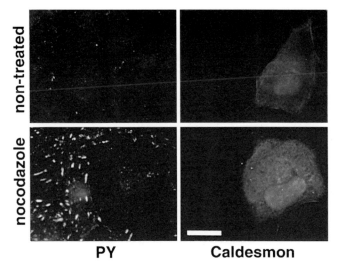

Fig. 5. Caldesmon interferes with the formation of focal adhesions induced by microtubule disruption in serum-starved MDCK cells. The two rows correspond to the same field visualized using anti-phosphotyrosine antibody (PY; left column) and an anti-tag antibody for caldesmon (right column). Note in the lower row that focal adhesions are absent from the area corresponding to the caldesmon-positive cell following nocodazole treatment.

ations that we have described above apply to this period, and that the collapse of cytoplasmic microtubules is accompanied by a burst of signalling activity that assists the cell's progression into the mitotic phase [175].

During normal interphase, the microtubule network is stable or slowly changing as a whole, but individual microtubule ends are constantly extending and retracting, as discussed earlier. Here we suggest that the local effect of microtubule collapse on the actin cytoskeleton network and focal adhesion sites nearby is similar to that of global microtubule disruption on the whole cell. The collapse of a microtubule end would induce a local increase in myosin II-generated contractile tension. This tension would be communicated to the nascent cell–ECM contacts via the actin network, as depicted schematically in Fig. 6. If it breaks, its components would be reabsorbed and dispersed by the cell. If it holds, however, the ensuing signals would act as a positive feedback, leading to increased build up and eventually to maturation of a focal adhesion plaque.

The suggested mechanism involving positive feedback creates an intrinsic instability in the strength of adhesions. The testing of adhesions based on microtubule collapse inducing contractility would provide a means for the cell to simultaneously check the nature of its adhesive contacts all along the periphery. A slight advantage in adhesion due to gradients in ECM density, to chemotaxis or even to random fluctuations would be sensed and amplified. This could help to explain the microtubule dependence of the development of cell polarization in the direction of externally imposed gradients, at least in the

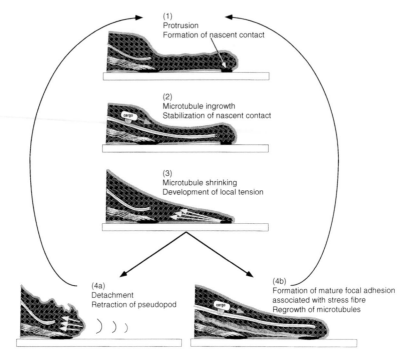

Fig. 6. Scheme to illustrate our model of the role of microtubules in the establishment of cell polarity. Microtubules can assist in the delivery of components for new cell protrusions and cell–ECM contacts. On the other hand, depolymerization of microtubules leads to an increase in cell contractility and consequent tension against the nascent substrate contacts. If weak, these contacts will break and the cell will retract (step 4a). If they are strong enough, the increase in tension activates adhesion-dependent signalling pathways which lead to assembly of focal adhesions and associated actin microfilament bundles (step 4b). The evidence for this model comes from experiments involving global destabilization of the microtubule network. We propose that depolymerization of individual microtubules in the course of dynamic instability may have a similar local effect. Cell polarization would develop in the direction corresponding to the strongest adhesion-dependent signalling. The positive feedback creates an intrinsic instability which could explain the development of polarization in the absence of externally imposed gradients.

case of fibroblasts. In the absence of obvious gradients, these cells nonetheless polarize, although in random directions. The positive feedback makes the symmetrical cell shape unstable to small fluctuations.

Some cell types do not depend on microtubules for the development of polarity. One notable example is that of fish keratocytes [37]. These, as well as many other types of poorly adherent and highly motile cells, including neoplastically transformed fibroblasts [38], require less contractile activity in order to release old adhesion sites at their trailing edges. For cancer cells in particular, it is reasonable to suggest that adhesion-dependent signalling is constitutively activated, which of course restricts its regulatory capacities. We can suggest by analogy that, in other types of weakly adherent and weakly contractile cells, the response of adhesion-dependent signalling to the relevant tensile stimulation is also likely to be small. Further experiments will be needed to clarify this issue.

The suggested role of microtubule depolymerization in regulating local contractility complements the more traditional point of view, according to which microtubule-based transport enhances the formation of cell protrusions and initial cell–ECM contacts by delivery of new membrane and other components [32,98,157,158]. It has been noticed that growing microtubules are often targeted towards the sites of nascent focal adhesions [197]. It is possible that dynamic microtubules assist in the creation of new contacts in their growing phase, and then check them by inducing contractility in conjunction with their collapse.

It is important to stress that the model of local regulation of contractility based on microtubule depolymerization depends on the presence of dynamic microtubule ends. This is in agreement with the observation that taxol-treated fibroblasts are non-polarized [30,31]. Another method of stabilizing microtubules against dynamic instability is the application of very low doses of depolymerizing agents, such as vinblastine or nocodazole [35,36]. Studies have shown that the microtubule-stabilizing effect in cultured cells is accompanied by a loss of polarization and motility [32,33]. The microtubule-based outward transport is not likely to be inhibited by these treatments.

Summary

In summary, we have shown that microtubule disruption affects cell adhesion via its effect on cell contractility. The same and other evidence points to a dependence of adhesion signalling on the tension generated at cell–ECM contacts. This suggests the possibility that microtubule dynamics are involved in the local regulation of focal adhesion complexes and the associated contractile stress fibres. We have discussed how this model can be applied to adhesion-dependent signalling as well as phenomena such as the polarization of cell locomotion in various situations.

Supported in part by grants from the Minerva Foundation, The Israel Sciences Foundation, and The Israel Ministry of Science to A.D. Bershadsky, from The Israel Sciences Foundation to M. Elbaum, and from the Crown Endowment Fund to A.D. Bershadsky and M. Elbaum.

References

1. Bray, D. (1992) Cell Movements, Garland Publishing, New York and London
2. Sheetz, M.P., Felsenfeld, D.P. and Galbraith, C.G. (1998) Trends Cell Biol. 8, 52–54
3. Lauffenburger, D.A. and Horwitz, A.F. (1996) Cell 84, 359–369
4. Hynes, R.O. (1992) Cell 69, 11–25
5. Schwartz, M.A., Schaller, M.D. and Ginsberg, M.H. (1995) Annu. Rev. Cell Dev. Biol. 11, 549–599
6. Harris, A.K., Wild, P. and Stopak, D. (1980) Science 208, 177–179
7. Harris, A.K. (1982) in Cell Behaviour (Bellairs, R., Curtis, A. and Dunn, G., eds.), pp. 3–20, Associated Scientific Publishers, Amsterdam

8.   Oliver, T., Lee, J. and Jacobson, K. (1994) Semin. Cell Biol. **5**, 139–147

9.   Kolodney, M.S. and Wysolmerski, R.B. (1992) J. Cell Biol. **117**, 73–82

10.   Kolodney, M.S. and Elson, E.L. (1993) J. Biol. Chem. **268**, 23850–23855

11.   Sims, J.R., Karp, S. and Ingber, D. (1992) J.Cell Sci. **103**, 1215–1222

12.   Vasiliev, J.M. (1982) in Cell Behaviour (Bellairs, R., Curtis, A. and Dunn, G., eds.), pp. 135–158, Associated Scientific Publishers, Amsterdam

13.   Kucik, D.F., Kuo, S.C., Elson, E.L. and Sheetz, M.P. (1991) J. Cell Biol. **114**, 1029–1036

14.   Bershadsky, A.D., Tint, I.S., Neyfakh, A.A. and Vasiliev, J.M. (1985) Exp. Cell. Res. **158**, 433–444

15.   Rinnerthaler, G., Geiger, B. and Small, J.V. (1988) J. Cell Biol. **106**, 747–760

16.   Palecek, S.P., Loftus, J.C., Ginsberg, M.H., Lauffenburger, D.A. and Horwitz, A.F. (1997) Nature (London) **385**, 537–540 [erratum published in Nature (London) **388**, 210]

17.   Vasiliev, J.M. (1991) J. Cell Sci. **98**, 1–4

18.   Mays, R.W., Nelson, W.J. and Marrs, J.A. (1995) Cold Spring Harbor Symp. Quant. Biol. **60**, 763–773

19.   Eaton, S. and Simons, K. (1995) Cell **82**, 5–8

20.   Dunn, G.A. and Brown, A.F. (1986) J. Cell Sci. **83**, 313–340

21.   Dunn, G.A. and Brown, A.F. (1990) Lect. Notes Biomath. **89**, 10–34

22.   Lee, J., Ishihara, A. and Jacobson, K. (1993) Symp. Soc. Exp. Biol. **47**, 73–89

23.   Herman, I.M. (1993) Curr. Opin. Cell Biol. **5**, 48–55

24.   Buss, F., Temm-Grove, C., Henning, S. and Jockusch, B.M. (1992) Cell Motil. Cytoskeleton **22**, 51–61

25.   Gertler, F.B., Niebuhr, K., Reinhard, M., Wehland, J. and Soriano, P. (1996) Cell **87**, 227–239

26.   Sato, N., Funayama, N., Nagafuchi, A., Yonemura, S., Tsukita, S. and Tsukita, S. (1992) J. Cell Sci. **103**, 131–143

27.   Lamb, R.F., Ozanne, B.W., Roy, C., McGarry, L., Stipp, C., Mangeat, P. and Jay, D.G. (1997) Curr. Biol. **7**, 682–688

28.   Vasiliev, J.M., Gelfand, I.M., Domnina, L.V., Ivanova, O.Y., Komm, S.G. and Olshevskaya, L.V. (1970) J. Embryol. Exp. Morphol. **24**, 625–640

29.   Vasiliev, J.M. and Gelfand, I.M. (1976) Cold Spring Harbor Conf. Cell Proliferation **3**, 279–304

30.   Schiff, P.B. and Horwitz S.B. (1980) Proc. Natl. Acad. Sci. U.S.A. **77**, 1561–1565

31.   Pletjushkina, O.J., Ivanova, O.J., Kaverina, I.N. and Vasiliev, J.M. (1994) Exp. Cell Res. **212**, 201–208

32.   Bershadsky, A.D. and Vasiliev, J.M. (1993) Symp. Soc. Exp. Biol. **47**, 353–373

33.   Liao, G., Nagasaki, T. and Gundersen, G.G. (1995) J. Cell Sci. **108**, 3473–3483

34.   Dhamodharan, R., Jordan, M.A., Thrower, D., Wilson, L. and Wadsworth, P. (1995) Mol. Biol. Cell **6**, 1215–1229

35.   Vasquez, R.J., Howell, B., Yvon, A.-M.C., Wadsworth, P. and Cassimeris, L. (1997) Mol. Biol. Cell **8**, 973–985

36.   Schliwa, M. and Höner, B. (1993) Trends Cell Biol. **3**, 377–380

37.   Euteneur, U. and Schliwa, M. (1984) Nature (London) **310**, 58–61

38.   Ivanova, O.Y., Svitkina, T.M., Vasiliev, J.M. and Gelfand, I.M. (1980) Exp. Cell Res. **128**, 457–461

39.   Middleton, C.A., Brown, R.F., Brown, R.M., Karavanova, I.D., Roberts, D.J.H. and Vasiliev, J.M. (1989) J. Cell Sci. **94**, 25–32

40.   Wessels, D., Soll, D.R., Knecht, D., Loomis, W.F., De Lozanne, A. and Spudich, J. (1988) Dev. Biol. **128**, 164–177

41.   Jay, P.Y., Pham, P.A., Wong, S.A. and Elson, E.L. (1995) J. Cell Sci. **108**, 387–393

42.   Verkhovsky, A.B., Svitkina, T.M. and Borisy, G.G. (1995) J. Cell Biol. **131**, 989–1002

43.   Wade, R.H. and Hyman, A.A. (1997) Curr. Opin. Cell Biol. **9**, 12–17

44. Gelfand, V.I. and Bershadsky, A.D. (1991) Annu. Rev. Cell Biol. 7, 93–116
45. Inoué, S. and Salmon, E.D. (1995) Mol. Biol. Cell 6, 1619–1640
46. Mitchison, T. and Kirschner, M. (1984) Nature (London) 312, 237–242
47. Fygenson, D.K., Braun, E. and Libchaber, A. (1994) Phys. Rev. E Stat. Phys. Plasmas Fluids Relat. Interdiscip. Top. 50, 1579–1588
48. Tran, P.T., Walker, R.A. and Salmon, E.D. (1997) J. Cell Biol. 138, 105–117
49. Elbaum, M., Fygenson, D.K. and Libchaber, A. (1996) Phys. Rev. Lett. 76, 4078–4081
50. Fygenson, D.K., Elbaum, M., Shraiman, B. and Libchaber, A. (1997) Phys. Rev. E Stat. Phys. Plasmas Fluids Relat. Interdiscip. Top. 55, 850–859
51. Felgner, H., Frank, R. and Schliwa, M. (1996) J. Cell Sci. 109, 509–516
52. Gittes, F., Mickey, B., Nettleton, J. and Howard, J. (1993) J. Cell Biol. 120, 923–934
53. Venier, P., Maggs, A.C., Carlier, M.-F. and Pantaloni, D. (1994) J. Biol. Chem. 269, 13353–13360 (erratum published in J. Biol. Chem. 270, 17056)
54. Waterman-Storer, C.M., Gregory, J., Parsons, S.F. and Salmon, E.D. (1995) J. Cell Biol. 130, 1161–1169
55. Coue, M., Lombillo, V.A. and McIntosh, J.R. (1991) J. Cell Biol. 112, 1165–1175
56. Waters, J.C., Mitchison, T.J., Rieder, C.L. and Salmon, E.D. (1996) Mol. Biol. Cell 7, 1547–1558
57. Inoue, S. (1996) Cell Struct. Funct. 21, 375–379
58. van Deurs, B., von Bulow, F., Vilhardt, F., Holm, P.K. and Sandvig, K. (1996) J. Cell Sci. 109, 1655–1665
59. Cassimeris, L. (1993) Cell Motil. Cytoskeleton 26, 275–281
60. Keating, T.J., Peloquin, J.G., Rodionov, V.I., Momcilovic, D. and Borisy, G.G. (1997) Proc. Natl. Acad. Sci. U.S.A. 94, 5078–5083
61. Rodionov, V.I. and Borisy, G.G. (1997) Science 275, 215–218
62. Waterman-Storer, C.M. and Salmon, E.D. (1997) Curr. Biol. 7, R369–R372
63. Fygenson, D.K. (1995) Ph.D. Thesis, Princeton University, Princeton, NJ
64. Holzbaur, E.L.F. and Vallee, R.B. (1994) Annu. Rev. Cell Biol. 10, 339–372
65. Bloom, G.S. and Endow, S. (1994) Protein Profile 1, 1059–1116
66. Cole, D.G. and Scholey, J.M. (1995) Trends Cell Biol. 5, 259–262
67. Hirokawa, N. (1998) Science 279, 519–526
68. Sheetz, M.P. (1996) Cell Struct. Funct. 21, 369–373
69. Desai, A. and Mitchison, T.J. (1995) J. Cell Biol. 128, 1–4
70. Lee, G. (1993) Curr. Opin. Cell Biol. 5, 88–94
71. Drewes, G., Trinczek, B., Illenberger, S., Biernat, J., Schmitt-Ulms, G., Meyer, H.E., Mandelkow, E.-M. and Mandelkow, E. (1995) J. Biol. Chem. 270, 7679–7688
72. Drewes, G., Ebneth, A., Preuss, U., Mandelkow, E.M. and Mandelkow, E. (1997) Cell 89, 297–308
73. Ulitzur, N., Humbert, M. and Pfeffer, S.R. (1997) Proc. Natl. Acad. Sci. U.S.A. 94, 5084–5089
74. Belmont, L.D. and Mitchison, T.J. (1996) Cell 84, 623–631
75. Marklund, U., Larsson, N., Gradin, H.M., Brattsand, G. and Gullberg, M. (1996) EMBO J. 15, 5290–5298
76. Walczak, C.E., Mitchison, T.J. and Desai, A. (1996) Cell 84, 37–47
77. McNally, F.J., Okawa, K., Iwamatsu, A. and Vale, R.D. (1996) J. Cell Sci. 109, 561–567
78. Salisbury, J.L. (1995) Curr. Opin. Cell Biol. 7, 39–45
79. Shiina, N., Gotoh, Y., Kubomura, N., Iwamatsu, A. and Nishida, E. (1994) Science 266, 282–285
80. Reszka, A.A., Seger, R., Diltz, C.D., Krebs, E.G. and Fischer, E.H. (1995) Proc. Natl. Acad. Sci. U.S.A. 92, 8881–8885
81. Morishima-Kawashima, M. and Kosik, K.S. (1996) Mol. Biol. Cell 7, 893–905

82. Sontag, E., Nunbhakdi-Craig, V., Bloom, G.S. and Mumby, M.C. (1995) J. Cell Biol. **128**, 1131–1144

83. Kapeller, R., Toker, A., Cantley, L.C. and Carpenter, C.L. (1995) J. Biol. Chem. **270**, 25985–25991

84. Smith, K.J., Levy, D.B., Maupin, P., Pollard, T.D., Vogelstein, B. and Kinzler, K.W. (1994) Cancer Res. **54**, 3672–3675

85. Näthke, I.S., Adams, C.L., Polakis, P., Sellin, J.H. and Nelson, W.J. (1996) J. Cell Biol. **134**, 165–180

86. Condeelis, J. (1993) Annu. Rev. Cell Biol. **9**, 411–444

87. Small, J.V. (1994) Semin. Cell Biol. **5**, 157–163

88. Mitchison, T.J. and Cramer, L.P. (1996) Cell **84**, 371–379

89. Tilney, L.G. and Tilney, M.S. (1993) Trends Microbiol. **1**, 25–31

90. Southwick, F.S. and Purich, D.L. (1994) BioEssays **16**, 885–891

91. Lasa, I. and Cossart, P. (1996) Trends Cell Biol. **6**, 109–114

92. Mogilner, A. and Oster, G. (1996) Biophys. J. **71**, 3501–3510

93. Barkalow, K. and Hartwig, J.H. (1995) Curr. Biol. **5**, 1000–1002

94. Machesky, L. (1997) Curr. Biol. **7**, R164–R167

95. Theriot, J.A. (1997) J. Cell Biol. **136**, 1165–1168

96. Chant, J. and Pringle, J.R. (1991) Curr. Opin. Genet. Dev. **1**, 342–350

97. Chant, J. (1994) Trends Genet. **10**, 328–333

98. Singer, S.J. and Kupfer, A. (1986) Annu. Rev. Cell Biol. **2**, 337–365

99. Craig, A.M., Wyborski, R.J. and Banker, G. (1996) Nature (London) **375**, 592–594

100. Kislauskis, E.H., Zhu, X.-C. and Singer, R.H. (1994) J. Cell Biol. **127**, 441–451

101. Kislauskis, E.H., Zhu, X.-C. and Singer, R.H. (1997) J. Cell Biol. **136**, 1263–1270

102. Cox, D., Condeelis, J., Wessels, D., Soll, D., Kern, H. and Knecht, D. (1992) J. Cell Biol. **116**, 943–955

103. Cunningham, C., Gorlin, J., Kwiatkowsky, D., Hartwig, J., Janmey, P. and Stossel, T. (1992) Science **255**, 325–327

104. Cunningham, C.C., Leclerc, N., Flanagan, L.A., Lu, M., Janmey, P.A. and Kosik, K.S. (1997) J. Cell Biol. **136**, 845–857

105. Mooseker, M.S. and Cheney, R.E. (1995) Annu. Rev. Cell Dev. Biol. **11**, 633–675

106. Titus, M.A. (1997) Trends Cell Biol. **7**, 119–123

107. Warrick, H.M. and Spudich, J.A. (1987) Annu. Rev. Cell Biol. **3**, 379–421

108. Titus, M.A. (1997) Curr. Biol. **7**, R301–R304

109. Wang, F.S., Wolenski, J.S., Cheney, R.E., Mooseker, M.S. and Jay, D.G. (1996) Science **273**, 660–663

110. Spudich, J.A. (1989) Cell Regul. **1**, 1–11

111. Karess, R.E., Chang, X.-J., Edwards, K.A., Kulkarni, S., Aguilera, I. and Kiehart, D.P. (1991) Cell **65**, 1177–1189

112. Young, P.E., Richman, A.M., Ketchum, A.S. and Kiehart, D.P. (1993) Genes Dev. **7**, 29–41

113. Bement, W.M., Forscher, P. and Mooseker, M.S. (1993) J. Cell Biol. **121**, 565–578

114. Lee, J., Leonard, M., Oliver, T., Ishihara, A. and Jacobson, K. (1994) J. Cell Biol. **127**, 1957–1964

115. Burton, K. and Taylor, D.L. (1997) Nature (London) **385**, 450–454

116. Crowley, E. and Horwitz, A.F. (1995) J. Cell Biol. **131**, 525–537

117. Moores, S.L., Sabry, J.H. and Spudich, J.A. (1996) Proc. Natl. Acad. Sci. U.S.A. **93**, 443–446

118. Höner, B., Citi, S., Kendrick-Jones, J. and Jockusch, B. (1988) J. Cell Biol. **107**, 2181–2189

119. Volberg, T., Geiger, B., Citi, S. and Bershadsky, A.D. (1994) Cell Motil. Cytoskeleton **29**, 321–338

120. Heath, J.P. and Dunn, G.A. (1978) J. Cell Sci. **29**, 197–212

121. Hall, A. (1998) Science **279**, 509–514

122. Tapon, N. and Hall, A. (1997) Curr. Opin. Cell Biol. **9**, 86–92

123. Kozma, R., Ahmed, S., Best, A. and Lim, L. (1995) Mol. Cell. Biol. **15**, 1942–1952
124. Nobes, C.D. and Hall, A. (1995) Cell **81**, 53–62
125. Ridley, A.J. (1994) BioEssays **16**, 321–327
126. Ridley, A.J. and Hall, A. (1992) Cell **70**, 389–399
127. Chant, J. and Stowers, L. (1995) Cell **81**, 1–4
128. Hartwig, J.H., Bokoch, G.M., Carpenter, C.L., Janmey, P.A., Taylor, L.A., Toker, A. and Stossel, T.P. (1995) Cell **82**, 643–653
129. Tan, J.L., Ravid, S. and Spudich, J.A. (1992) Annu. Rev. Biochem. **61**, 721–759
130. Amano, M., Ito, M., Kimura, K., Fukata, Y., Chihara, K., Nakano, T., Matsuura, Y. and Kaibuchi, K. (1996) J. Biol. Chem. **271**, 20246–20249
131. Kimura, K., Ito, M., Amano, M., Chihara, K., Fukata, Y., Nakafuku, M., Yamamori, B., Feng, J., Nakano, T., Okawa, K., et al. (1996) Science **273**, 245–248
132. Chrzanowska-Wodnicka, M. and Burridge, K. (1996) J. Cell Biol. **133**, 1403–1415
133. Jalink, K., van Corven, E.J., Hengeveld, T., Morii, N., Narumiya, S. and Moolenaar, W.H. (1994) J. Cell Biol. **126**, 801–810
134. Amano, M., Chihara, K., Kimura, K., Fukata, Y., Nakamura, N., Matsuura, Y. and Kaibuchi, K. (1997) Science **275**, 1308–1311
135. Matsumura, F. and Yamashiro, S. (1993) Curr. Opin. Cell Biol. **5**, 70–76
136. Fraser, I.D. and Marston, S.B. (1995) J. Biol. Chem. **270**, 19688–19693
137. Yamashiro, S., Yamakita, Y., Hosoya, H. and Matsumura, F. (1991) Nature (London) **349**, 169–172
138. Danowski, B. (1989) J. Cell Sci. **93**, 255–266
139. Kolodney, M.S. and Elson, E.L. (1995) Proc. Natl. Acad. Sci. U.S.A. **92**, 10252–10256
140. Brown, R.A., Talas, G., Porter, R.A., McGrouther, D.A. and Eastwood, M. (1996) J. Cell Physiol. **169**, 439–447
141. Sheridan, B.C., McIntyre, Jr., R.C., Meldrum, D.R., Cleveland, Jr., J.C., Agrafojo, J., Banerjee, A., Harken, A.H. and Fullerton D.A. (1996) J. Surg. Res. **62**, 284–287
142. Tstsui, H., Ishihara, K. and Cooper, G. (1993) Science **260**, 682–687
143. Yoshida, T., Urabe, Y., Sugimachi, M. and Takeshita, A. (1996) Am. J. Physiol. **271**, H1978–H1987
144. Lampidis, T.J., Kolonias, D., Savaraj, N. and Rubin, R.W. (1992) Proc. Natl. Acad. Sci. U.S.A. **89**, 1256–1261
145. Heidemann, S.R. (1966) Int. Rev. Cytol. **165**, 235–296
146. Koszka, C., Leichtfried, F.E. and Wiche, G. (1985) Eur. J. Cell Biol. **38**, 149–156
147. Svitkina, T.M., Verkhovsky, A.B. and Borisy, G.G. (1996) J. Cell Biol. **135**, 991–1007
148. Tint, I.S., Hollenbeck, P.J., Verkhovsky, A.B., Surgucheva, I.G. and Bershadsky, A.D. (1991) J. Cell Sci. **98**, 375–384
149. Cary, R.B., Klymkowsky, M.W., Evans, R.M., Domingo, A., Dent J.A. and Backhus, L.E. (1994) J. Cell Sci. **107**, 1609–1622
150. Fuchs, E. and Cleveland, D.W. (1998) Science **279**, 514–519
151. Ingber, D.E., Dike L., Hansen, L., Karp, S., Liley, H., Maniotis, A., McNamee, H., Mooney, D., Plopper, G., Sims, J. and Wang, N. (1994) Int. Rev. Cytol. **150**, 173–224
152. Chicurel, M.E., Chen, C.S. and Ingber, D.E. (1998) Curr. Opin. Cell Biol. **10**, 232–239
153. Weisshaar, B. and Matus, A.J. (1993) Neurocytol. **22**, 727–734
154. Toyoshima, I., Yu, H., Steuer, E.R. and Sheetz, M.P. (1992) J. Cell Biol. **118**, 1121–1131
155. Hotta, K., Tanaka, K., Mino, A., Kohno, H. and Takai, Y. (1996) Biochem. Biophys. Res. Commun. **225**, 69–74
156. Bershadsky, A.D., Vaisberg, E.A. and Vasiliev, J.M. (1991) Cell Motil. Cytoskel. **19**, 152–158
157. Bershadsky, A.D. and Futerman, A.H. (1994) Proc. Natl. Acad. Sci. U.S.A. **91**, 5686–5689
158. Rodionov, V.I., Gyoeva, F.K., Tanaka, E., Bershadsky, A.D., Vasiliev, J.M. and Gelfand V.I. (1993) J. Cell Biol. **123**, 1811–1820

159. Mata, J. and Nurse, P. (1997) Cell **89**, 939–949
160. Vega, L.R. and Solomon, F. (1997) Cell **89**, 825–828
161. Geiger, B., Yehuda-Levenberg, S. and Bershadsky, A.D. (1995) Acta Anat. **154**, 46–62
162. Yamada, K.M. and Geiger, B. (1997) Curr. Opin. Cell Biol. **9**, 76–85
163. Ben-Ze'ev, A. and Bershadsky, A.D. (1997) Adv. Mol. Cell Biol. **24**, 125–163
164. O'Neill, C., Jordan P., Riddle, P. and Ireland, G. (1990) J. Cell Sci. **95**, 577–586
165. Burridge, K., Chrzanowska-Wodnicka, M. and Zhong, C. (1997) Trends Cell Biol. **7**, 342–347
166. Bershadsky, A.D., Chausovsky, A., Becker, E., Lyubimova, A. and Geiger, B. (1996) Curr. Biol. **6**, 1279–1289
167. Zhang, Q., Magnusson, M.K. and Mosher, D.F. (1997) Mol. Biol. Cell **8**, 1415–1425
168. Pelham, R.J. and Wang, Y.l. (1997) Proc. Natl. Acad. Sci. U.S.A. **94**, 13661–13665
169. Hamasaki, K., Mimura, T., Furuya, H., Morino, N., Yamazaki, T., Komuro, I., Yazaki, Y. and Nojima, Y. (1995) Biochem. Biophys. Res. Commun. **212**, 544–549
170. Cramer, L.P. (1997) Frontiers Biosci. **2**, D260–D270
171. Choquet, D., Felsenfeld, D.P. and Sheetz, M.P. (1997) Cell **88**, 39–48
172. Li, X. and Nicklas, R.B. (1995) Nature (London) **373**, 630–632
173. Nicklas, R.B., Ward, S.C. and Gorbsky, G.J. (1995) J. Cell Biol. **130**, 929–939
174. Zhai, Y., Kronebusch, P.J., Simon, P.M. and Borisy, G.G. (1996) J. Cell Biol. **135**, 201–214
175. Liu, S.-H., Lee, H.-H., Chen, J.-J., Chuang, C.-F. and Ng, S.-Y. (1994) Cell Growth Diff. **5**, 447–455
176. Galbraith, C.G. and Sheetz, M.P., (1997) Proc. Natl. Acad. Sci. U.S.A. **94**, 9114–9118
177. Welch, M.D., DePace, A.H., Verma, S., Iwamatsu, A. and Mitchison, T.J. (1997) J. Cell Biol. **138**, 375–384
178. Rosenblatt, J. and Mitchison, T.J. (1998) Nature (London) **393**, 739–740
179. Downing, K.H. and Nogales, E. (1998) Curr. Opin. Cell Biol. **10**, 16–22
180. Jordan, M.A. and Wilson, L. (1998) Curr. Opin. Cell Biol. **10**, 123–130
181. Pierre, P., Pepperkok, R. and Kreis, T.E. (1994) J. Cell Sci. **107**, 1909–1920
182. Häckl, W., Bärmann, M. and Sackmann, E. (1998) Phys. Rev. Lett. **80**, 1786–1789
183. Welch, M.D., Mallavarapu, A., Rosenblatt, J. and Mitchison, T.J. (1997) Curr. Opin. Cell Biol. **9**, 54–61
184. Bretscher, M.S. and Aguado-Velasco, C. (1998) Curr. Biol. **8**, 721–724
185. Mullins, R.D., Heuser, J.A. and Pollard, T.D. (1998) Proc. Natl. Acad. Sci. U.S.A. **95**, 6181–6186
186. Svitkina,T.M., Verkhovsky, A.B., McQuade, K.M. and Borisy, G.G. (1997) J.Cell Biol. **139**, 397–415
187. Kelleher, J.F. and Titus, M.A. (1998) Curr. Biol. **8**, R394-R397
188. Canman, J.C. and Bement, W.M. (1997) J. Cell Sci. **110**, 1907–1917
189. Felgner, H., Frank, R., Biernat, J., Mandelkow, E.M., Mandelkow, E., Ludin, B., Matus, A. and Schliwa, M. (1997) J. Cell Biol. **138**, 1067–1075
190. Waterman-Storer, C.M. and Salmon, E.D. (1997) J. Cell Biol. **139**, 417–434
191. Dunn, G.A., Zicha, D. and Fraylich, P.E. (1997) J. Cell Sci. **110**, 3091–3098
192. Jockusch, B.M., Bubeck, P., Giehl, K., Kroemker, M., Moschner, J., Rothkegel, M., Rüdiger, M., Schlüter, K., Stanke, G. and Winkler, J. (1995) Annu. Rev. Cell Dev. Biol. **11**, 379–416
193. Bottazzi, M. and Assoian, R.K. (1997) Trends Cell Biol. **7**, 348–352
194. Meredith, J. and Schwartz, M. (1997) Trends Cell Biol. **7**, 146–150
195. Enomoto, T. (1996) Cell Struct. Funct. **21**, 317–326
196. Pletjushkina, O.P., Belkin, A.M., Ivanova, O.J., Oliver, T., Vasiliev, J.M., and Jacobson, K. (1998) Cell Adhes. Commun. **5**, 121–135
197. Kaverina, I., Rottner, K. and Small, J.V. (1998) J. Cell Biol. **142**, 181–190

Biochem. Soc. Symp. **65**, 173–205
Printed in Great Britain

11

Organization and polarity of actin filament networks in cells: implications for the mechanism of myosin-based cell motility

Louise P. Cramer

MRC Laboratory for Molecular Cell Biology, University College London, Gower Street, London WC1E 6BT, U.K.

Abstract

Force arising from myosin activity drives a number of different types of motility in eukaryotic cells. Outside of muscle tissue, the precise mechanism of myosin-based cell motility is for the most part theoretical. A large part of the problem is that, aside from cell surface features such as lamellipodia and microvilli, relatively little is known about the structural organization of potential actin substrates for myosin in non-muscle motile cells. Several groups [Cramer, Siebert and Mitchison (1997) J. Cell Biol. **136**, 1287–1305; Guild, Connelly, Shaw and Tilney (1997) J. Cell Biol. **138**, 783–797; Svitkina, Verkhovsky, McQuade and Borisy (1997) J. Cell Biol. **139**, 397–415] have begun to address this issue by determining actin organization throughout entire non-muscle motile cells. These studies reveal that a single motile cell comprises up to four distinct structural groups of actin organization, distinguished by differences in actin filament polarity: alternating, uniform, mixed or graded. The relative abundance and spatial location in cells of a particular actin organization varies with cell type. The existence in non-muscle motile cells of alternating-polarity actin filament bundles, the organization of muscle sarcomeres, provides direct structural evidence that some forms of motility in non-muscle cells are based on sarcomeric contraction, a recurring theory in the literature since the early days of muscle research. In this scenario, as in muscle sarcomeres, myosin generates isometric force, which is ideally suited to driving symmetrical types of motility, e.g. healing of circular wounds in coherent groups of cells. In contrast, uniform-polarity actin filament bundles and oriented meshworks in cells allow oriented movement of myosin, potentially over relatively long distances. In this simple 'transport-based' scenario, the direction in which myosin generates force is inherently polarized, and is well placed for driving asymmetrical or polarized types of motility, e.g. as expected for long-range transport of membrane organelles. In the more complex situation of cell locomotion, the predominant actin organization detected in locomoting fish keratocytes and locomoting primary heart fibroblasts excludes sarcomeric contraction force from having a major role in pulling these cell types forward during locomotion. Instead Svitkina et al. propose that 'dynamic network con-

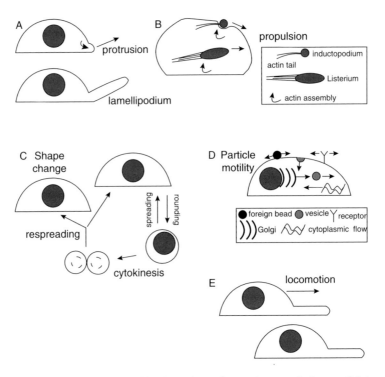

Fig. 1. Types of cell motility based on the actin cytoskeleton. Of the types of motility shown, all occur in eukaryotic cell types adherent to solid substrata. In non-adherent cell types of those types of motility shown, only cytokinesis (C) and particle motility (D) occur. However, also occurring in non-adherent cell types, but not shown here, is motility related to lamellipodium protrusion (e.g. tip growth in yeast and filamentous fungi) and certain other shape changes (e.g. deformation of red blood cells). For adherent cell types, motility shown in (A)–(E) is relative to the substratum. (A) New actin assembly (curly arrow) at the tip of the cell margin is coupled to protrusion of a lamellipodium. (B) New actin assembly (curly arrow) behind a *Listerium* bacterium and under a cell-surface-attached bead (termed an inductopodium) is coupled to propulsion of the bacterium and the bead. Panels (C) and (D) show cell motility based on previously assembled actin networks. Panel (C) shows shape changes that occur at different stages of the cell cycle. Cells round-up at the onset of mitosis. After nuclear division, cytokinesis cleaves the mother cell into two daughter cells. After mitosis, and starting towards the end of cytokinesis, ☞

traction' of a weakly adherent uniform-polarity actin filament meshwork is the basis of keratocyte locomotion. For fibroblast locomotion, however, Cramer et al. prefer a transport mechanism based on graded-polarity actin filament bundles.

Introduction

Cell motility is key to cell function and survival. In the eukaryotic cell kingdom, there are a large number of distinct types of cell motility. These are diverse, and examples vary from migration of neuronal cells in an embryo, which is essential for the normal development of an organism, to protein transport inside cells, which is important for vital biochemistry in an organism. Most types of cell motility in eukaryotic cells are based on tubulin or actin cytoskeleton proteins. One exception is in motile sperm cells of some organisms, where actin is replaced by major sperm protein. For the most part tubulin and actin systems function independently in an individual cell, but intercommunication is important for overall co-ordination during motility. One clear case of this is during cell division, where tubulin-based activity first drives the separation of sister chromatids to opposite ends of the mother cell, then actin-based activity is required to cleave the mother cell into two daughter cells. Temporal co-ordination between these two processes ensures that each daughter cell inherits the correct complement of genetic information.

Motility based on the actin cytoskeleton (bundles and meshworks of actin filaments and actin-binding proteins, e.g. myosin) is wide ranging in cells (Fig. 1), although not all cell types exhibit all forms of motility. Diverse types of actin-based cell motility can be categorized according to the source of actin used during motion. Either the motility is coupled to new actin assembly (Figs. 1A and 1B), or it is based on previously assembled actin filament bundles and meshworks (Figs. 1C and 1D). In this review, I will use 'actin network' as a general term for actin filament bundles and meshworks. Motility coupled to new actin assembly includes types of protrusion of the cell margin which result

☞ **Fig. 1 (contd.)** each daughter cell respreads and interphase morphology is regained. Additionally, independently of entering normal mitosis, cells in response to certain stimuli may round-up (e.g. amoeba cells in the presence of chemoattractant; and during cell apoptosis and tumorigenesis) or spread (e.g. fibroblasts in response to certain GTPase signalling pathways, and during replating of tissue culture cells). Panel (D) shows a number of different types of particle motility that occur in cells. Moving on the cell surface are receptors (black fork) and surface-attached foreign beads (black sphere). Inside cells, endocytosis and exocytosis are in themselves forms of motility; in addition, both endocytic (light grey half-circle) and exocytic (light grey full circle) vesicles are transported through the cell. Certain components in the cytoplasm(e.g. actin filaments in lamellipodia, and inhomogeneous material elsewhere in the cell) undergo cytoplasmic flow (wavy lines). (E) Cell locomotion: locomoting cells exhibit motility coupled to new actin assembly (protrusion) and motility based on previously assembled actin networks (cell body translocation and tail retraction; see Fig. 2 for details).

in the formation of either lamellipodia (Fig. 1A) or filopodia. A lamellipodium comprises a thin, cellular, actin-rich band, typically less than 0.5 μm thick, which is situated at the periphery of many motile cell types. Long, thin, cylindrical extensions of the lamellipodium are termed filopodia, or microspikes, which are shorter. Motility related to filopodial protrusion occurs during the acrosome reaction of thyone sperm [1]. Another type of cell motility coupled to new actin assembly is propulsion (Fig. 1B). This is a form of 'rocketing' motion associated with some intracellular pathogens (e.g. *Listeria monocytogenes*), and under certain cell conditions with vesicles inside cells and foreign beads attached to the surface of cells. Motility based on previously assembled actin networks is more varied, ranging from large changes in cell shape (Fig. 1C) to regular trafficking of small vesicles inside cells (Fig. 1D).

The most complex, and accordingly least understood, type of cell motility is translocation of the whole cell (termed cell locomotion or cell crawling) (Fig. 1E). Cell locomotion is the combination of both protrusion and types of cell motility based on previously assembled actin networks (see details below). A locomoting cell as observed in tissue culture is morphologically polarized, i.e. it is spatially asymmetrical. The front of the cell is typically broader than the cell rear, and the whole cell is composed of morphologically distinct cell regions. These cell regions are the lamellipodium, lamella (also termed the transition zone in some cells), cell body and rear of the cell (Fig. 2A). The lamellipodium in a locomoting cell is situated at the extreme front-end of the cell. The lamella is located immediately behind the lamellipodium, and is thicker. Behind the lamella is the cell body, which is the largest and bulkiest region of the cell, comprising the nucleus and most of the organelles. The rear of the cell is a rounded or drawn-out tail. During cell locomotion, as observed by time-lapse microscopy, the lamellipodium, cell body and tail all move forward approximately in the same direction. The lamella appears to stay in place as the cell body advances. Lamellipodial protrusion and forward movement of these other distinct cell regions can be tracked separately in the same locomoting cell and have characteristic properties [2,3]. For these reasons cell locomotion is considered to be the combination of distinct types of cell motility: lamellipodial protrusion, cell body translocation (also termed cell traction or cell body motility), and tail retraction/de-adhesion (also termed rear release) (Fig. 2B). Lamellipodial protrusion in a locomoting cell, as in non-locomoting motile cell types, is coupled to new actin assembly. In contrast, it is widely accepted that cell body translocation and tail retraction are based on previously assembled actin networks [3–6].

For any type of motility to occur in a cell (other than passive diffusion), the cell must generate force. The whole process of cell motility requires integration of motile force with cell adhesion, membrane and signalling systems in the cell. For example, during cell locomotion motile force must be transmitted to the substratum through sites of cell adhesion to facilitate a crawling or gliding motion of the cell. While cell adhesion, membranes and cell signalling undoubtedly are key players (for reviews, see [5,7]), determining the mechanism of force generation at a molecular level is the most basic requirement for elucidating how the whole process works. Types of cell motility coupled to

A
LOCOMOTING FIBROBLAST

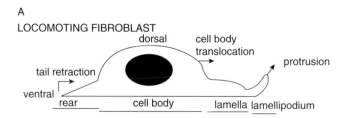

LOCOMOTING KERATOCYTE

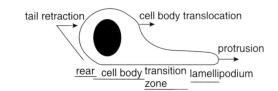

B

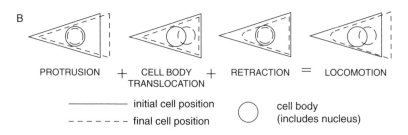

PROTRUSION $+$ CELL BODY $+$ RETRACTION $=$ LOCOMOTION
TRANSLOCATION

———————— initial cell position   ◯   cell body
– – – – – – final cell position   (includes nucleus)

Fig. 2. Cell locomotion. (A) Different cell regions and cell locations compared in a locomoting fibroblast (upper panel) and a locomoting fish keratocyte (lower panel). All locomoting cell types are composed of the same basic cell regions, but detailed morphology varies from cell type to cell type. In a fibroblast, the lamellipodium often appears ruffled, due to lifting up off the substratum. In keratocytes and several other motile cell systems, e.g. aplysia bag cell neuronal growth cones, the lamellipodium is in constant contact with the substratum and is morphologically less distinguishable from the lamella. For this reason, in keratocytes sometimes the term 'transition zone' replaces the term 'lamella' in the literature. The rear of a fibroblast and other strongly adhesive locomoting cell types is a drawn-out tail. The rear of weakly adhesive locomoting cell types, such as keratocytes, amoeba cells and neutrophils, is rounded. Fibroblasts locomote relatively slowly, the fastest movers reaching top speeds of around 1 μm/min. Keratocytes, neutrophils and amoeba cells locomote relatively quickly, up to around 15 μm/min. (B) Cell locomotion is the combination of distinct types of cell motility. In cells where these processes occur simultaneously, morphology is constant during locomotion (e.g. keratocytes and other fast-locomoting cell types). In non-locomoting motile cell types, lamellipodial protrusion may occur, but typically not cell body translocation or tail retraction.

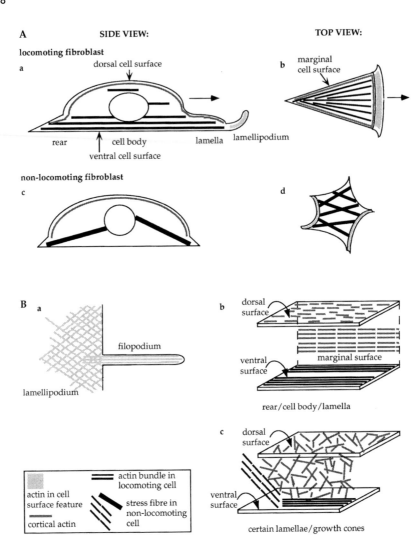

Fig. 3. Organization of actin networks in motile cells. (A) Distribution
and spatial orientation of actin networks. In all motile cell types (e.g. see panel
a), actin is located in three general cell areas: in cell surface features [shown
here is a lamellipodium (light grey patch)], within a fraction of a micron of the
plasma membrane [cortical actin (dark grey, curved line)] and throughout the
remainder of the cell (black lines). In each of these cell locations, the relative
abundance, spatial orientation and structural organization of the actin network
can vary with cell type. For example, compared here is a typical locomoting
fibroblast (panels a and b) [16] and a typical non-locomoting fibroblast (panels c
and d). (B) Arrangement of actin filaments in actin networks. Panel (a) shows
protrusive structures typical of many motile cell types. In the lamellipodium,
actin filaments (light grey lines) form a dense oriented meshwork arranged at 🖙

new actin assembly, particularly lamellipodial protrusion, have been studied extensively over recent years. In lamellipodia the relative roles of actin assembly and myosin activity in generating motile force for protrusion are being tested in different systems [6,8].

In contrast, the generation of force elsewhere in the cell to drive motility based on previously assembled actin networks is poorly understood. While it is known or suspected that force to drive this type of motility comes from the activity of myosins, the mechanism is not well understood at a molecular level. Part of the problem in understanding cell motility driven by myosin activity outside of lamellipodia is that, unlike actin in lamellipodia, the structural organization and filament polarity of actin elsewhere in motile cells is for the most part unknown. Considering filament polarity, an actin filament is polar beause each of its two ends behaves differently. One end is termed the barbed (or plus) end, and the other is the pointed (minus) end. The barbed end is the preferred end for actin assembly, and the end towards which the activity of all known myosins is directed. Thus not knowing the filament polarity of an actin network makes it difficult to predict the direction in which myosin would generate force, for example. In lamellipodia, data on the structural organization of actin filaments and their polarity have important implications for the generation of motile force for protrusion [9–13]. Similarly, determining the structural organization of actin filaments in an entire motile cell would have important implications for the generation of force for motility occurring elsewhere in the cell. In particular, it would provide essential information on the organization of potential actin substrates for myosin in a cell. Recent progress on this problem has been made in several motile cell types. Incorporating this new information, in this review I will first describe the structural organization and filament polarity of known actin networks in motile cells. Then, focusing outside of lamellipodia, I will discuss the implications this has for the mechanism of myosin-based motility in cells. I will not discuss the mechanism of lamellipodial protrusion and related motility, nor several types of motility that occur in lamellipodia, as these have been covered extensively in other reviews (e.g. see [2–6,8,14]).

☞ **Fig. 3 (contd.)** roughly 45° [13], but in the filopodium they form a tight, parallel bundle. Panel (b) represents a cut-away view of a locomoting primary heart fibroblast, showing actin networks associated with cell surfaces [16]. On the ventral cell surface, longitudinal actin bundles are composed of long, oriented, densely packed actin filaments (black lines). Cortical actin meshworks and bundles under the dorsal and marginal cell surfaces respectively are composed of short, oriented, loosely packed actin filaments (dark grey lines). Panel (c) represents lamellae of certain tissue culture cell types [9] and certain neuronal growth cones [11], where cortical actin under the dorsal cell surface is composed of short filaments (dark grey lines) randomly arranged in a loose meshwork. The same meshwork fills the space between stress fibres (diagonal black lines) in the tissue culture cells [9], and between the dorsal cell surface and actin bundles (straight black lines) on the ventral cell surface in the growth cones [11].

Structural organization of actin networks in motile cells

Variation in actin-based cell motility must ultimately arise from differences in underlying actin organization and filament polarity. In a motile cell, distinct actin networks comprise filament bundles and filament meshworks. Actin is a highly abundant protein, and actin filament bundles and meshworks are distributed throughout the cell. For simplicity, actin networks can be considered to be located in three general cell areas (e.g. Fig. 3A, panel a). One cell area is cell surface features, such as microvilli, filopodia and lamellipodia (Fig. 3A, panel a, light grey area). Another cell area is within a fraction of a micron of the dorsal cell surface (the 'top' or 'roof' of the cell) and marginal cell surfaces (the 'walls' or 'sides' of the cell), also referred to as the cell cortex in the literature (Fig. 3A, panels a and b, dark grey line). The third cell area comprises the remaining bulk volume of the cell (Fig. 3A, panel a, black lines), which, for reasons explained later, also includes the ventral cell surface (the 'bottom' or 'floor' of the cell).

It has long been recognized [10,15] that cell surface features are composed of dense, oriented actin networks in almost all motile cell types tested (see [6,14]). In microvilli and filopodia, actin filaments are tightly bundled (see filopodium in Fig. 3B, panel a). In lamellipodia, filaments are arranged in a dense orthogonal meshwork (Fig. 3B, panel a). One exception is in a type of thicker lamellipodium, termed a pseudopodium, in amoeba cells, where filaments appear to be more randomly arranged [12]. However, from the images presented, some oriented filaments are evident at the extreme margin of these pseudopodia.

Moving to dorsal and marginal cell surfaces, electron microscopy of cultured cell types shows that cortical actin coating the inside of the plasma membrane is composed of short actin filaments arranged in loose meshworks and bundles. While such meshworks and bundles may also coat the inside of the ventral cell surface, so far actin detected in this location has a distinct organization (and is considered separately below). Loose cortical actin meshworks underlie the dorsal cell surface. The precise organization of the meshwork varies with cell type. In locomoting heart fibroblasts in primary culture, actin filaments are oriented in the meshwork (Fig. 3B, panel b, short dark grey lines) [16]. This organization is most clear under the dorsal cell surface of the cell body of these fibroblasts, but may also exist under the dorsal cell surface of the lamella and cell rear. In several longer-term cultured motile cell types, organization of cortical actin has been experimentally sought under the dorsal cell surface of the lamella. In these cell types, in contrast to locomoting heart fibroblasts, cortical actin filaments appear to be more randomly arranged (Fig. 3B, panel c, short dark grey lines) [9]. It is not known if the acquisition of oriented or more randomly arranged cortical actin meshworks reflects cell type, cell behaviour (e.g. locomoting versus non-locomoting) or differential preservation of actin organization during experimental manipulation. One possibility is that actin filaments underlying the dorsal cell surface are arranged more randomly where they also perform a 'space-filling' role. This is observed between the dorsal cell surface and more ventral cell locations in lamellae of certain motile cells [9] and in growth cones of certain neurons [11] (Fig. 3B, panel c). Whether cortical actin exists under the dorsal cell surface of lamellipodia, in addition to the oriented dense filament bundles and meshworks that comprise the predominant actin organization in this cell location, has been difficult to assess, because lamellipodia are sensitive to experimental manipulation. In fish keratocytes it is estimated that most actin filaments associated with the dorsal cell surface of lamellipodia are lost during chemical fixation procedures [13]. This issue is relevant for determining filament length distribution in lamellipodia [13] and for a certain type of motility that occurs in lamellipodia [14]. Cortical actin filaments underlying marginal cell surfaces in the lamella, cell body and rear of the cell are loosely bundled (Fig. 3B, panel b), at least in locomoting heart fibroblasts (the sub-plasma-membrane bundles of [16]). Whether the apparent differential location of cortical actin meshworks and bundles observed in locomoting heart fibroblasts reflects a general pattern in all motile cell types is not known.

In the remainder of the cell, detailed information on the structural organization of actin meshworks and bundles varies with cell type. In locomoting cell types, the structural organization of actin outside of cell surface features and the cell cortex is generally not well documented. However, two recent careful electron-microscopic studies clearly reveal actin organization throughout locomoting heart fibroblasts [16] and fish keratocytes [17]. In the lamella, cell body and rear of locomoting primary heart fibroblasts (cultured for 12–20 h), actin mostly comprises thin filament bundles (termed longitudinal actin bundles). These bundles are predominantly located on the ventral cell surface and ventral region of the cell (Fig. 3A, panel a, black lines). In contrast with locomoting heart fibroblasts, actin bundles in locomoting fish keratocytes are not abundant and are mostly only observed at the boundary between the cell body and transition zone in these cells (termed boundary actin bundles). Instead, in locomoting keratocytes much of the actin outside of the lamellipodium comprises an oriented dense filament meshwork in the transition zone of these cells [17], similar to the organization of actin in the lamellipodium (see lamellipodium in Fig. 3B, panel a) [13,17]. In contrast, in non- or relatively very slowly locomoting motile cell types, such as established fibroblast cell lines, many other cell lines (e.g. PtK2 epithelial cells) and primary cells (mostly fibroblasts) grown for longer than a few days or so in culture, actin has been extensively documented. Here, actin filaments are predominantly bundled in the cell body and in these cell types are termed stress fibres (reviewed in [18]) (e.g. Fig. 3A, panel c, diagonal black lines). It is clear that in these cell types stress fibres have a distinct organization compared with longitudinal actin bundles observed in locomoting primary heart fibroblasts (cultured for 12–20 h) [16] and with actin bundles in the cell body of other motile cell types, such as nurse cells in *Drosophila* embryos [19]. In both these cases, this is notably due to differences in actin filament polarity (see below). This is not surprising with respect to the longitudinal actin bundles in locomoting heart fibroblasts, since activity of stress fibres is generally thought to prevent cells from locomoting [18]. Distinguishing stress fibres from other types of actin bundle is discussed below.

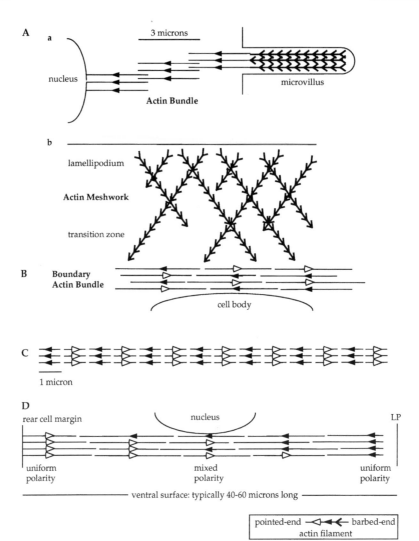

Fig. 4. Polarity of actin bundles and meshworks in motile cells. (A) Uniform polarity. Panel (a) shows a uniform-polarity actin filament bundle in a nurse cell in a *Drosophila* embryo [19]. The bundle radiates back from the base of a microvillus on the cell surface to the nucleus. The whole actin bundle is composed of shorter sub-actin bundles. Sub-bundles are about 3 μm ('microns') long and are aligned in a staggered arrangement. Panel (b) shows a uniform-polarity actin filament meshwork in the transition zone of a locomoting fish keratocyte [17]. The filament meshwork is organized similarly to, and appears to be continuous with, the filament meshwork in the lamellipodium. Individual actin filaments have not been traced the entire distance covered by the lamellipodium and the transition zone. (B) Mixed polarity. In the same keratoctye, actin filament polarity is mixed in the boundary actin bundle located between the transition zone and the cell body. (C) Alternating polarity. So far, alternating polarity has been reported for stress fibres in PtK2 epithelial cells ([16]; but see [25]), cortical

When actin bundles are the predominant type of actin organization in a motile cell type, the spatial orientation of the bundles correlates well with overall cell shape and behaviour. For example, in locomoting heart fibroblasts which are morphologically polarized, longitudinal actin filament bundles are largely oriented parallel to the substratum (Fig. 3A, panel a, black lines) and in the direction of locomotion, approx. 30° (Fig. 3A, panel b, black lines). In contrast, in non-locomoting cell types, stress fibres typically pass obliquely through the cell towards the nucleus (Fig. 3A, panel c, diagonal black lines) and adopt various spatial orientations in the cell (Fig. 3A, panel d, criss-cross black lines) which reflects the orthogonal morphology of these cell types.

Polarity of actin networks in motile cells

The polarity of any given actin network in the cell depends on the precise arrangement of individual actin filaments in that network. An individual actin filament is arranged with either it's barbed or pointed end facing a given direction. The arrangement adopted, or 'filament polarity', can change depending on the position of a given filament in the actin network. This allows a number of distinct types of actin organization to exist in the cell.

Actin filament polarity is determined by decorating the filament with fragments of myosin: either S1 myosin (myosin head domains; see Fig. 6A) or heavy meromyosin (head domain plus part of the tail domain). As viewed by electron microscopy, decorated actin filaments appear as a line of arrowheads or chevrons (e.g. see Fig. 4 key, and the microvillus in Fig. 4A, panel a). The 'barbed' and 'pointed' appearance of the arrowheads coined the names given to the ends of an actin filament (see Fig. 4 key). Filament decoration was originally used as a tool to identify actin filaments in cells outside of muscle systems, in the days when muscle proteins were just beginning to be identified in non-muscle motile cells. Today, filament decoration and electron microscopy is the only established method for identifying actin filament polarity.

☞   **Fig. 4 (contd.)** actin bundles in locomoting heart fibroblasts [16] and muscle sarcomeres. Shown is a whole bundle which is composed of very short sub-actin bundles and short filaments, each about 1 μm long. (D) Graded polarity. Filament polarity is graded in longitudinal actin bundles in locomoting heart fibroblasts [16]. Shown is a simplified version of filament polarity observed across the length of the ventral cell surface. For simplicity only a single bundle is drawn. In cells, bundles of this length have not been observed (on average bundles are 13 μm long in ventral cell locations, but can reach up to 30 μm long). In cells, graded-polarity actin bundles overlap or are co-linear with each other so as to span the length of the cell almost continuously [they do not appear to be located in the lamellipodium (LP)]. In typical locomoting fibroblasts that have drawn-out tails (see Fig. 2A), mixed polarity is observed roughly mid-cell, in the region of the nucleus (as shown in the diagram).

Four types of actin filament organization

Information on actin filament polarity in motile cells has largely been restricted to actin in cell surface features, notably lamellipodia, filopodia, microspikes, microvilli, cilia, hair bristles and retraction fibres (e.g. [9,11,13, 15]). Almost without exception (see pseudopodia above), filament polarity is uniform in these features, with filament barbed ends facing outward, away from the nucleus (e.g. see the microvillus in Fig. 4A, panel a, and the lamellipodium in Fig. 4A, panel b). Recently, the polarity of actin networks located elsewhere in motile cells has been determined in a number of cell types. Four types of actin organization have been observed (uniform, mixed, alternating and graded polarity actin meshworks and bundles; Fig. 4). Since actin filament polarity is uniform in cell surface features, and mixed and alternating polarity have previously been identified (outlined below), the identification of these three types of actin organization in motile cells is not surprising. Graded polarity, on the other hand, is newly identified and is an initially surprising type of actin organization. Outside of cell surface features, uniform-polarity actin filament networks have been identified in two motile cell types. In *Drosophila* nurse cells, actin filament bundles of uniform polarity radiate back from the base of microvilli towards the nucleus (Fig. 4A, panel a) [19]. In locomoting fish keratocytes, the polarity of the oriented dense actin meshwork in the transition zone of these cells is uniform (Fig. 4A, panel b) [17]. In both these cases, as with the organization of actin in cell surface features, barbed ends of filaments face outwards, away from the nucleus (Fig. 4A). In contrast, both stress fibres (e.g. Fig. 3A, panel c) in PtK2 epithelial cells and cortical actin bundles in locomoting heart fibroblasts (Fig. 3B, panel b, dark grey lines) are actin filament bundles with alternating polarity (Fig. 4C) [16] (see also PtK2 study in [25] and discussion below). Alternating actin filament polarity is the organization found in muscle sarcomeres, and for many years has been suspected to exist in non-muscle motile cells. This idea traces its origin to the identification by many laboratories of muscle proteins in non-muscle motile cells [20]. Graded polarity (Fig. 4D) is the organization of longitudinal actin bundles (Fig. 3A, panel a, black lines) in the lamella, cell body and rear of locomoting heart fibroblasts [16]. Graded-polarity actin bundles account for 85% of the total actin bundles observed in these cells and 100% of the bundles on the ventral cell surface. Lastly, mixed polarity (Fig. 4B) is observed in boundary actin bundles in locomoting fish keratocytes, at least towards the centre of the bundle [17]. Mixed polarity is also observed roughly mid-cell in graded-polarity actin filament bundles (Fig. 4D).

These four types of actin organization are readily distinguished. Filament polarity in uniform-polarity actin networks is constant as a function of network length (e.g. in Fig. 4A, panel a, the barbed ends of all filaments face towards the right, regardless of position in the bundle). In contrast, filament polarity in alternating-polarity actin bundles alternates over short bundle lengths, roughly every 1 μm. Thus over the first 1 μm in the alternating-polarity actin bundle in Fig. 4C, the barbed ends of all filaments face towards the right, then over the next 1 μm the pointed ends of all filaments face towards the right, and so on. This is quite different from graded-polarity actin bundles, where filament polarity changes gradually along the length of the bundle (changing roughly by 1–2% per μm). Actual filament polarity at any given point in a graded-polarity actin bundle is determined by its position in the cell; the closer to the front of the cell, the more barbed ends of filaments face forward (see Fig. 4D for a simplified version). A characteristic of a mixed-polarity actin bundle is that, regardless of position in the bundle, roughly 50% of filament barbed ends and 50% of filament pointed ends always face the same direction (Fig. 4B).

How widespread are these four types of actin organization in other motile cell types? In *Drosophila* embryos, the staggered actin filament bundles observed in nurse cells (Fig. 4A, panel a) are also found in hair bristles [21] and ring canals [22], and are perhaps a feature of embryos in general. This organization is thought to form a rapidly collapsible network, an elegant mechanism for coping with constant shape change in a developing organism. Uniform-polarity actin filament meshworks in keratocytes are extensive, comprising both the lamellipodium and the transition zone. This may be a feature of other rapidly locomoting cell types. Graded polarity, on the other hand, may be more common in more slowly locomoting cell types. In PtK2 epithelial cells, a type of actin filament bundle (termed the circumferential ring or actin belt) circumscribes the cell (L.P. Cramer, P. Siebert and T. Mitchison, unpublished work). This may represent a more general phenomenon, such that alternating polarity is the organization of actin belts that circumscribe epithelia in whole tissue.

Turning to stress fibres, when is an actin filament bundle a stress fibre and in which cell types? The conclusion from several studies in established cell lines and primary cell types grown for longer than a few days or so in tissue culture is that the organization of actin filament bundles in the bulk cytoplasm of the cell body of these cells is similar to that of muscle sarcomeres [16,23–27]. Since in these studies the actin bundles are referred to as stress fibres, the term stress fibre should be reserved for only those actin filament bundles where organization is similar to muscle sarcomeres. This definition should probably be further restricted to actin bundles located in the bulk cell cytoplasm, since other actin bundles located elsewhere in the cell, such as cortical actin bundles, are also related to muscle sarcomeres. There are several tests that determine if organization of an actin bundle in a non-muscle motile cell is similar to that of muscle sarcomeres. A conclusive test is that, as in muscle sarcomeres, actin filament polarity should alternate as a function of bundle length over short distances (Fig. 4C). Also conclusive is the ultrastructural organization of myosin II in the actin bundle, which should form bipolar filaments and be situated where actin filament polarity alternates in the bundle (see the single sarcomere shown in Fig. 6B, panels a and b). Also indicative, but less conclusive, are tests done at the light-microscope level, where the relative staining patterns of certain muscle proteins (e.g. myosin II, α-actinin, tropomyosin) in the actin bundle should have a periodicity similar to that observed in sarcomeric muscle.

The importance of these criteria is emphasized with two examples. First is the case of fibroblasts, in which the term 'stress fibre' has been used extensively to refer to actin bundles in cell biology research, but in most studies the identity of the actin bundle has not been determined. When identity has been determined, it is clear that there are differences, and this apparently is a function of locomotion capacity. For example, in locomoting primary heart fibroblasts (cultured for 12–20 h) the ultrastructural organization of longitudinal actin bundles in the lamella, cell body and rear of the cell is distinct from that of muscle sarcomeres; actin filament polarity in the fibroblast bundles is graded rather than alternating (compare Figs. 4C and 4D) [16]. In contrast, when fibroblasts have been cultured under conditions that caused or were in the process of causing loss of locomotion capacity (e.g. cells kept in longer-term culture), actin bundles in the cell body are similar to muscle sarcomeres, as determined by the ultrastructural organization of myosin II [27] and relative staining patterns of muscle proteins [26]. The second example is that, in some cases, stress fibres have been identified by the relative staining patterns of muscle proteins but, instead of having the expected alternating polarity, they are reported to have mixed polarity (although, importantly, filament polarity was not determined as a function of fibre length) [23,25]. One simple explanation for this is that the overlap between actin filaments of opposite polarity (where mixed filament polarity would occur in an alternating-polarity actin filament bundle; e.g. see Fig. 6B, panel b) is greater in some cultured cell types, or reflects the cell environment. Alternatively, I have observed in several cases that thicker stress fibres in PtK2 epithelial cells are composed of several narrower bundles, rather like fibrils in muscle. Each narrower bundle is an alternating-polarity actin filament bundle, but switches in polarity are not aligned between adjacent narrow bundles. Mixed polarity is thus expected across the whole bundle. If either of these explanations turns out to be true for stress fibres in general, then a logical question is: do observations of mixed polarity in a particular actin bundle in a non-muscle motile cell indicate that the bundle is related to a stress fibre? This is a formal possibility in cases where mixed polarity is the predominant composition of an individual actin bundle. This is perhaps the situation for boundary actin bundles in locomoting keratocytes (Fig. 4B). In contrast, this is not the case in individual graded-polarity actin filament bundles in locomoting heart fibroblasts, where the mixed polarity is only a small part of a gradient in polarity and not the composition of the entire bundle (Fig. 4D).

Filament polarity reversal over the whole cell

There is a striking feature of uniform- and graded-polarity actin filament networks when considered over the whole cell. Filament polarity relative to a given region of the cell surface or cell margin reverses in the region of the nucleus, with filament barbed ends preferentially facing outwards (Fig. 5). Because these actin networks are the most abundant in their respective cell types, filament polarity reversal occurs in most of the actin cytoskeleton, at least on the ventral cell surface. That this reversal is observed in at least five distinct motile cell types in fish, flies and mammals suggests conservation of an important function through evolution. Although the full significance of this simple arrangement is unclear, it has several key implications for motility. Biasing the number of barbed-end sites facing a discrete region of cell surface/cell margin provides a simple mechanism for converting a morpho-

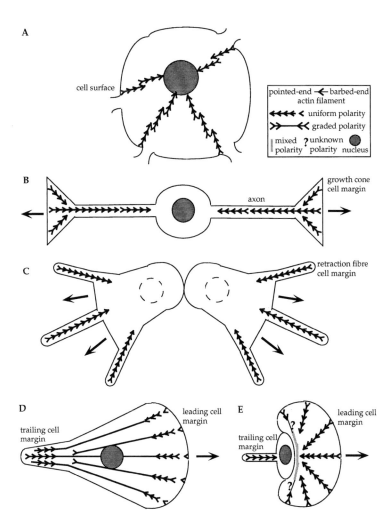

Fig. 5. Filament polarity reversal in several motile cell types. Actin filament polarity over the whole cell reverses in the region of the nucleus. Filament barbed ends preferentially face outwards, towards the cell surface (A) or towards a given cell margin (B–E). (A) Uniform-polarity actin filament bundles in a *Drosophila* nurse cell [19]. (B) Uniform-polarity actin filament networks in a bipolar neuron. Actin filament polarity has not been determined over a whole individual neuron. It is known that, in certain neuronal growth cones, filament barbed ends on the ventral cell surface face the growth cone cell margin [11], and actin filaments in axons are assembled preferentially in growth cones [97]. (C) Uniform-polarity actin filament bundles in a pair of daughter cells respreading after mitosis in PtK2 cell lines [65]. (D) Graded-polarity actin filament bundles in a locomoting heart fibroblast [16]. (E) Uniform-polarity actin filament networks in a locomoting fish keratocyte [17].

logically non-polarized, primitive cell into one with the capacity for polarized motility in the direction of bias. Polarized motility is achieved regardless of the precise mechanism of producing motile force, since the barbed end of an actin filament is both the preferred end for assembly and the end towards which the activity of known myosins is directed.

This theory explains in a most fundamental way why motility in some motile cell types is unidirectional, while in others it is bi- or multi-directional. For example, in cases of extreme bias where filament barbed ends predominantly face one spatial location, e.g. as observed in locomoting cell types, motility is polarized in a single direction (Figs. 5D and 5E). However, where filament barbed ends are distributed equally between two spatial locations, such as in bipolar neurons (Fig. 5B) and in two daughter cells respreading after mitosis (Fig. 5C), motility is polarized simultaneously in two directions. Also explained is why keratocyte locomotion is more persistent in one direction than is fibroblast locomotion. Keratocytes are very highly polarized, with almost all barbed ends facing the leading cell margin (Fig. 5E). Fibroblasts are less highly polarized, with a more equal distribution of barbed ends facing both the leading and trailing cell margins (Fig. 5D). A fibroblast, then, is more likely to convert its rear, trailing margin into a leading margin and start moving off in the opposite direction. Consistent with this, I have frequently observed that primary heart fibroblasts turn by reversing back on themselves. In contrast, keratocytes rarely reverse, as predicted by filament polarity, but instead steer towards one side of their leading cell margin. While polarity reversal over the whole cell, by definition, does not occur in alternating-polarity actin filament bundles, interestingly, at the ends of stress fibres (in focal adhesions), filament barbed ends face the ventral cell surface.

Mechanism of formation

What is the mechanism of formation of actin networks with a specific actin organization? This must be tightly spatially controlled where distinct types of actin organization are located in discrete sites in an individual cell. The issue for any uniform-polarity actin filament network is whether new filaments form by elongation from existing filament ends, or by nucleation of new ends (as discussed in [6] for actin in cell surface features). Uniform-polarity actin filament bundles in the cell body of nurse cells probably elongate from the base of uniform-polarity actin filament bundles in microvilli (Fig. 4A, panel a) [19], since this actin is very stable, nucleating infrequently. In keratocytes, actin filaments in the lamellipodium and transition zone are long, also indicative of a preferential elongation mechanism [13,17]. Stress fibres and cortical actin bundles are in the same structural group as muscle sarcomeres and are likely to share elements of sarcomeric muscle formation. In cultured cell types, stress fibres form at the back of the lamellipodium. A careful analysis of this formation has shown that there is progressive zipping together of actin filaments mediated by myosin II [28]. This fits observations that pure myosin II and actin filaments sort into alternating-polarity actin filament bundles [29]. For graded-polarity actin filament bundles, the issue is the source of polarity. While several mechanisms are conceivable, e.g. graded distribution of parallel and anti-parallel actin bundling proteins, or graded distribution of actin nucleation sites, there are insufficient data to favour one over the other.

Myosin force generation: contraction and transport mechanisms

How precisely myosin generates force is determined by actin organization and the properties of individual myosins. Myosin proteins comprise an ever growing superfamily [30,31]. To date, myosins implicated in cell motility primarily fall into classes I, II, V and VI. Each myosin has a similar overall structure, composed of a head, neck and tail domain, and different myosins exist in either monomeric, dimeric or filamentous form (Fig. 6A). Based on sequence identity, all myosins have the capacity to bind actin filaments and hydrolyse ATP. Using energy from ATP hydrolysis to generate mechanical force, the activity of all known myosins, where tested, is directed towards the barbed end of actin filaments. No pointed-end-directed actin motor has been identified but, by analogy with microtubule-based motors, one may exist. Considering, first, known actin organization in motile cells, conceptually myosin can generate force by two extreme mechanisms, based on contraction or transport. A well-known form of contraction is shortening of muscle sarcomeres driven by myosin II filaments (Fig. 6B, compare panels a and b). A simple transport mechanism is exemplified by *in vitro* (cell-free) motility assays with pure proteins (Fig. 6C) [32–37]. Muscle shortening is a form of whole-tissue motility and has for many years influenced the way we think about force generation and motility in non-muscle motile cells. Transport mechanisms, although less discussed in cells, are likely to be as important as contraction for some forms of motility. Contraction and transport mechanisms are distinguished by the type of interaction between actin and myosin (Fig. 6D), and the polarity of the actin substrate for myosin (compare Fig. 6B, panel a, with Fig. 6C). In the exemplary sarcomere, myosin II filaments and actin filaments form a large molecular assembly. The interaction between actin and myosin is internal, within the assembly (Fig. 6D, left panel). In this assembly a myosin II filament lies between individual actin filaments of opposite polarity from an alternating-polarity actin bundle (Fig. 6B, panel a). During sarcomeric contraction, myosin II generates equal force in two opposite directions by pulling actin filaments of opposite polarity towards each other (Fig. 6B, compare panels a and b). In contrast, to allow unhindered transport, myosin and actin do not form a large molecular assembly. Instead, myosin interacts with the surface of the actin substrate (Fig. 6D, right panel). For persistent transport in one direction, the actin substrate surface has uniform polarity (Fig. 6C).

The known properties of myosin family members restrict contraction-based mechanisms in motile cells, as illustrated for muscle sarcomeres, to myosin II. Myosin II is the only known myosin that forms filaments (Fig. 6A). Other as yet unidentified contractile assemblies of distinct myosin family members and actin may exist in motile cells. For example, certain myosin I isoforms have a second, ATP-insensitive actin-binding site in the tail, allowing the myosin to cross-link actin filaments [30]. In cells, such myosins have the poten-

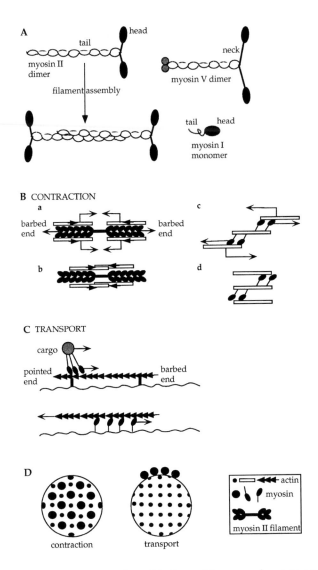

Fig. 6. Myosin force generation. (A) General features of myosin protein family members. In all myosin classes, the head is globular and contains an ATP-binding domain and an ATP-sensitive actin-binding domain. Neck regions vary in length depending on the myosin class and typically bind regulatory light chains. Tail regions may comprise membrane-binding or ATP-insensitive actin-binding domains. Myosin classes most frequently implicated in cell motility are myosin I, which is monomeric, and myosins II and V, which form homodimers through coil–coil interactions in the tail. Only myosin II is known to assemble into filaments; dimers assemble tail-to-tail, which forms a bipolar filament. Myosin II filaments in sarcomeric muscle are large, comprising around 300 myosin molecules per sarcomere. In non-muscle motile cells, myosin filaments are around 10-fold smaller or less (referred to as minifilaments; e.g. [17,98]). (B) Contraction-based force. Panels (a) and (b) show a muscle sarcomere. ☞

tial to generate force by sliding actin filaments within the cross-linked assembly [38], perhaps as illustrated (Fig. 6B, compare panels c and d). In contrast, several myosin classes have the potential to function as transport motors [39,40]. For these myosins, known properties of myosin tail domains either specify the cargo to be transported or localization in the cell [30]. For example, membrane-binding domains in the tails of class I and V myosins could specify either binding to vesicle cargo (e.g. Fig. 6C, upper panel) or anchoring to a cell membrane (e.g. Fig. 6C, lower panel). Similarly, a second actin-binding site in a myosin I tail, or in myosin II filaments, could specify transport of actin cargo over a second actin surface or anchoring to the actin cytoskeleton in cells. During transport, mechanisms must prevent myosin diffusing away from the actin surface, before the next actin-binding site is reached. Theoretical considerations [41], based on known structural and biophysical characteristics of several myosin classes, predict that this is possible for a single myosin V dimer. Myosin I monomers and myosin II dimers, on the other hand, must work in teams, e.g. by multiple myosin I molecules binding to a membrane surface (Fig. 6C, upper panel) or as a myosin II minifilament. In contractile muscle tissue, diffusion is not an issue, since the large molecular assembly of actin and myosin II is supported by other muscle proteins. For contractile assemblies in non-muscle motile cells, similar structural support is expected (see [24,27,42–44]).

Simpler types of myosin-based cell motility

What is the evidence that contraction or transport mechanisms exist in non-muscle motile cells? Contraction force is inherently non-polarized, since myosin generates equal but opposite force in two directions (Fig. 6B, compare panels a and b). Thus contraction force is better suited to driving morphologically non-polarized or spatially symmetrical types of myosin-based cell motility. Conversely, transport-based force is inherently polarized, since myosin generates force in one constant direction (Fig. 6C). Thus transport-based force is better suited to driving polarized types of myosin-based cell motility. Although variations on these mechanisms are likely, for simpler types of myosin-based motility this is exactly what is seen in cells. For more complex types of motility, such as cell locomotion, both mechanisms are plausible (dis-

☞ **Fig. 6 (contd.)** In panel (a), myosin II heads at each end of a myosin II bipolar filament move towards the barbed ends of opposing actin filaments in an alternating-polarity actin filament bundle. In panel (b), contraction force in the sarcomere is generated as myosin pulls actin filaments of opposite polarity towards each other. Panels (c) and (d) show the hypothetical generation of contraction force in a myosin I/actin filament assembly. (C) Transport-based force. Myosin moves continuously towards the barbed end of a uniform-polarity actin filament substrate. Depending on the geometry, either the cargo is transported (upper panel) or the actin substrate is transported (lower panel). (D) End-on view of actin filament substrates for myosin. In a contractile assembly (left panel), the interaction between actin and myosin is internal, within one large molecular assembly. For transport (right panel), myosin heads interact with the surface of the actin substrate.

cussed in a later section). That both contraction and transport mechanisms have evolved in cells probably reflects diversification of myosin function through evolution. This appears to be a general feature of the cytoskeleton, since parallels are observed for microtubules and microtubule-based motor proteins. For example, the microtubule motor ncd forms a large molecular assembly with tubulin in the mitotic spindle, whereas conventional kinesin interacts with the surface of microtubules to transport vesicles and organelles.

Contraction-driven cell motility

In motile cells, contraction force is generated in the form of cell tension. Maintenance of cell shape requires tension over the whole cell, evenly distributed in the cell surface or cell cytoplasm. Cortical actin filament bundles have

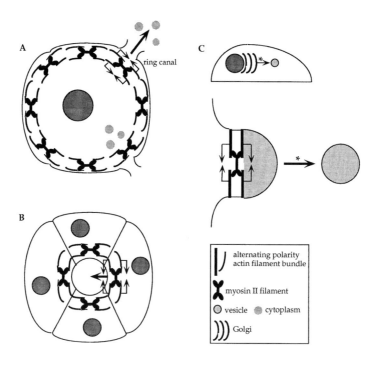

Fig. 7. Contraction-driven cell motility. (A) Nurse cell in a *Drosophila* embryo. Cortical actin networks under the cell surface are thought to contract to expel cytoplasm through ring canals during female oogenesis [19]. The filament polarity of cortical actin has not been reported in these cells. (B) Group of coherent cells in a wounded epithelial monolayer. A cable of actin bundles at the edge of the wound undergoes purse-string closure to heal the wound (discussed in [99]). It has not been determined whether cables are composed of alternating-polarity actin filament bundles. In (C) a secretory cell is shown (upper panel), in which a hypothetical actin bundle contracts (lower panel) around a vesicle to facilitate budding of the vesicle from the *trans*-Golgi network. Myosin II has been localized to Golgi buds [61,62] and a low-affinity myosin inhibitor (BDM) inhibits budding [62].

the correct spatial location (Fig. 3A, panel b) and structural organization (alternating polarity; Fig. 4C) to generate tension under the cell surface. Indeed, genetic studies in amoeba cells demonstrate that myosin II is required for cortical tension [45]. Similarly, in permanently cultured cells where myosin II is localized to stress fibres, the spatial orientation (Fig. 3A, panel d) and structural organization (Fig. 4C) of stress fibres is consistent with generation of tension throughout the cell cytoplasm. This may provide more robust maintenance of cell shape and cell integrity for cells that are permanently cultured. Although maintenance of cell shape is not on-going motility, precisely timed changes in whole-cell tension are the likely basis for several types of spatially symmetrical cell motility. For example, in amoeba cells a form of cell rounding (termed cringing) occurs in response to chemoattractant. Since cringing is comparatively fast (within 30–60 s), and myosin II function is essential for this response [46], it could easily be driven by a rapid increase in whole-cell tension that pulls the cell inward/away from sites of cell adhesion. This may be related to cell rounding that occurs in permeabilized tissue-culture cell models in response to exogenous ATP [47,48]. Among other responses, exogenous ATP probably causes hypercontraction of both stress fibres and cortical actin. Similar cell rounding situations occur during oogenesis in *Drosophila* female embryos. *Drosophila* nurse cells undergo an essential, precisely timed, reduction in cell size. This squeezes out cytoplasm through ring canals into the developing egg chamber, an event termed dumping. Nurse cell dumping is dependent on myosin activity [49]. Dumping is thought to be driven by contraction of cortical actin networks (Fig. 7A), since actin networks elsewhere in these cells have filament polarity inappropriate for contraction (uniform-polarity actin bundles; Fig. 4A, panel a) [19]. Cell rounding and other symmetrical shape changes based on whole-cell tension are likely to be important in multicellular organisms in general. However, some cell rounding situations in organisms must have evolved to be driven by distinct mechanisms. In particular, cell rounding at mitosis in amoeba [50] and certain tissue culture cells [51] is independent of functional myosin II, and much of the actin cytoskeleton in the cell disassembles. This may have evolved because myosin II needs to be switched off during mitosis, so that cytokinesis, the process that divides a cell into two daughter cells at the end of mitosis (see below), is timed appropriately [52,53]. Whether rounding at mitosis is driven instead primarily by loss of cell adhesion is not known. In this case, existing cell tension, in the absence of increased myosin II activity, is sufficient to round-up cells only after they have become less adherent. Another theory implicates a contribution to force generation from a putative pointed-end-directed actin motor by a transport-based mechanism [51].

Next I will consider more localized, but still spatially symmetrical, forms of motility in cells. A contraction-based mechanism seems ideal for healing small circular wounds in groups of coherent cells, by roughly equal generation of force around the edge of the wound. For this, a cable of actin filaments and myosin II of the correct spatial orientation needs to line the edge of the wound. Such cables have been observed in groups of wound-healing epithelial cells [54] and whole organisms [55]. The dynamic behaviour of actin and myosin II during wound healing [54] predicts that the cables are alternating-polarity actin filament bundles. This type of motility has been termed 'purse-string' closure (Fig. 7B) and is also thought to drive myosin II-dependent closure of symmetrical openings during development [56]. Another local contraction is cytokinesis. Cytokinesis has been studied for about 100 years and in marine organisms has long been thought to derive from increased cell tension in the cell equator (reviewed in [57]). Recently this has also been shown in mammalian cells [58]. Myosin II is the sole source of force that drives cytokinesis in cells growing in suspension [59,60]. This is not the case for cells growing on solid substrata [50]. Adherent amoeba cells that are null for myosin II go though cytokinesis at an overall slower rate than normal cells, most notably (from the values presented) due to a slower component at the end of cytokinesis. A simple interpretation of this is that, in adherent amoeba, there is functional redundancy in the myosin family of proteins. An alternative explanation is that, in adherent amoeba, additional proteins, perhaps one of the myosin I class, generate force early in cytokinesis, leaving myosin II a role in strong, rapid force generation to completely pinch off daughter cells at the end of cytokinesis.

The tiniest forms of spatially symmetrical cell motility, based on highly localized contraction, may exist within cells. Myosin II has been localized to vesicles budding from the *trans*-Golgi network [61,62], and myosin IC to the contractile vacuole in *Acanthamoeba* [63]. Vacuole contraction is dependent on functional myosin IC [64], and both budding [62] and vacuole contraction are sensitive to a low-affinity, but specific, inhibitor of myosin activity (butanedione monoxime; BDM) [65]. The detailed organization of the actin cytoskeleton has not been determined either on buds emerging from the *trans*-Golgi network or on vacuoles. Two types of contractile assembly are plausible: a contractile cage around the bud/vacuole surface, or contractile actin bundles around the bud neck (Fig. 7C, lower panel). Myosin IC has the potential to form a contractile assembly, through a second actin-binding site in the tail [66] (e.g. Fig. 6B, panels c and d). Similar forms of 'micro-contraction' may function at the cell surface, e.g. to help pinch off invaginating vesicles or assist deformation of the membrane during endocytosis. In yeast, deletion of a class I myosin, which has a predicted second actin-binding site in the tail, causes defects in internalization from the cell surface [67].

Transport-driven cell motility

When polarized motility occurs in cells and is oriented with the long axis of actin filaments in uniform-polarity actin networks, it is natural to suppose that motion is driven by a transport mechanism. One such situation is the directed transport of chloroplasts, and probably other organelles, on oriented actin tracks in plants [68]. The myosin responsible has not been identified, although several plant myosins have been purified and are known to translocate actin filaments *in vitro*, at the same rate as organelles stream *in vivo* [69,70]. In yeast and higher eukaryotes there is also accumulating evidence that vesicles and particles are transported through the cell on oriented actin tracks [39,71]. Unlike in plants, the filament polarity of these actin tracks has not been determined in these eukaryotes. It is known that uniform-polarity actin filament networks (Fig. 4A) and graded-polarity actin filament bundles (Fig. 4D), which would bias transport in a persistent direction, exist in higher organisms. As with plants, positive identification of actin motors that drive vesicle and particle transport on actin tracks is largely speculative. Although it is known from genetic and localization studies that several classes of myosins associate with various membranes in the cell, and are required for membrane trafficking steps [39,71], this in itself does not implicate long-range transport on actin tracks. For example, defects in membrane trafficking resulting from genetic disruption of myosin activity could equally be explained by defects in contraction mechanisms, e.g. during budding or endocytosis (as in Fig. 7C), or loss of organized actin tracks. One exception is myosin V (Fig. 8A) [40]. In yeast and mammalian cells, lack of function of certain myosin V isotypes causes no obvious defects in biogenesis of vesicles and particles or in organization of the actin cytoskeleton, but targeted delivery of the vesicles/particles does not occur. Related, shorter-range transport on uniform-polarity actin networks is also thought to occur in lamellipodia of motile cells (see [14]).

Considering motility of cell margins, two possibly related types of polarized motility occur over previously assembled, oriented, actin filament bundles. One type occurs when certain tissue culture cells respread after mitosis (Fig. 8B). The two daughter cells move outwards preferentially along retraction fibres, which provide a uniform-polarity actin filament transport track [65,72]. Inhibition of spreading by the myosin inhibitor BDM suggests that spreading may be driven by myosin activity. This is supported by immunolocalization of myosin II [65], and forward transport of myosin II-containing filaments [73], at the spreading edge. Similarly, in certain types of neuronal growth cones, the body of the growth cone moves forward over long actin bundles on the ventral cell surface. These actin bundles are oriented with their barbed ends mostly in the direction of forward movement, and are continuous with uniform-polarity actin filament bundles in filopodia at the leading edge of the growth cone [11]. Since filopodia produce pulling force (the 'contractile force' of [74]), this may be generated by myosin pulling the growth cone body forward on the bundles. Myosin II is present in the correct location to generate such force [75], and in *Aplysia* neurons BDM inhibits forward movement of the growth cone body (the 'central domain' of [76]). In this view growth cones (and other filopodia-rich cells, such as secondary mesenchyme cells) could first use protrusive force (see Fig. 1A) to send out filopodia, which then act as transport tracks for myosin to pull the growth cone forward (Fig. 8C). This model may be more generally applicable. For example, the uniform-polarity actin meshwork deposited by lamellipodia of several locomoting cell types may provide a transport mat for myosin to pull the cell body forward (see below).

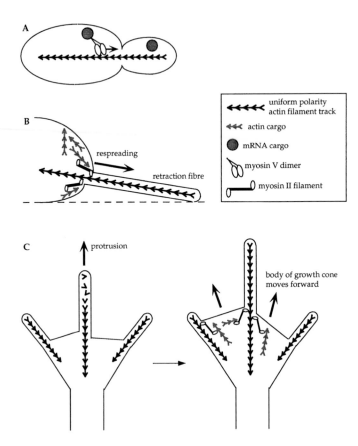

Fig. 8. Transport-driven cell motility. (A) Myosin V is shown transporting a particle of mRNA into the developing bud of *Saccharomyces cerevisiae* (discussed in [40]). The polarity of the actin track has not been determined in yeast. (B) A side view of a PtK2 cell shown respreading after mitosis. An oriented transport track is provided by uniform-polarity actin filament bundles in retraction fibres. Experimental evidence supports a model in which myosin II drives respreading by pulling the cell over these actin bundles [65,72]. (C) The model in (B) is extended to growth cone motility. Protrusion of filopodia is driven by a mechanism coupled to actin filament assembly (see [6]), and is followed by adhesive contact of the filopodium with the substratum. The growth cone is then pulled outwards over this oriented filamentous actin track provided by myosin II activity. There is no direct experimental support for such a model in growth cones; however, myosin II is present at the base of filopodia [75] and uniform-polarity actin filament bundles have been observed in growth cones [11].

Contraction- and transport-based motility during cell locomotion

Cell locomotion research has a distinguished history, going back centuries in the study of amoeba motility. Long before actin and myosin were identified in motile cells in the 1950s and 1960s, many theories on mechanism were debated (e.g. [77]). Over the last decade, myosin II activity has emerged as the basic engine that drives cell body translocation and tail retraction/de-adhesion during cell locomotion [78–80] (see Fig. 2B for a diagram of these types of cell motility).

Cell body translocation

While other classes of myosin are likely to have important roles, myosin II-generated force, based on contraction or transport, can conceivably pull the cell body forward during cell locomotion (discussed in [6]). Although contraction force is not inherently polarized, net forward motility occurs by superimposing a source of polarity, e.g. polarized adhesion, which is stronger at the front relative to the rear of the cell. Are contraction or transport mechanisms supported, now that the organization of actin has been determined in locomoting fish keratocytes (Fig. 4A, panel b, and Fig. 4B) [17] and locomoting heart fibroblasts (Fig. 4D) [16]? The absence of alternating-polarity actin filament bundles as a major type of actin organization in both cell types discounts sarcomeric contraction as the most plausible mechanism. Neither, however, is a simple transport mechanism supported. In keratocytes, myosin II does not move forward long distances, as predicted for a transport motor (see Fig. 6C). In fibroblasts, uniform-polarity actin filament bundles or meshworks, the predicted actin organization for simple transport tracks or mats (see Fig. 6C), are not located in front of the cell body. Instead, in keratocytes a mechanism is proposed based on dynamic contraction of the actin cytoskeleton in the transition zone of these cells (Fig. 9A). In fibroblasts, a transport mechanism is still favoured (Fig. 9B). The only difference is that uniform-polarity actin tracks are replaced by graded-polarity actin tracks, in which filament barbed ends preferentially face the direction of motility (Fig. 4D).

What are the distinguishing features of these models? An essential feature of the keratocyte model is that dynamic contraction of the filament network causes re-orientation of actin filaments and myosin II to form a boundary actin bundle (oriented perpendicular to the direction of locomotion) in front of the cell body (Fig. 9A, compare left and right panels). This model may be more generally applicable. For example, cytokinesis may be driven by local dynamic contraction of cortical actin and myosin II. In this scenario, the 'contractile ring' (a bundle of actin and myosin II separating daughter nuclei) does not drive cytokinesis, but instead is a consequence of it. Another feature of the keratocyte model is that depolymerization of the actin filament network in the transition zone, forward of the cell body, is coupled with the likelihood that myosin II generates force (Fig. 9A, panels a and b). This fits mathematical modelling in these cells, where depolymerization in the transition zone (the 'rear of the lamellipodium' in [81]) is coupled to the development of tension. In this sense, 'dynamic contraction' in keratocytes is similar to 'solation contraction' models proposed for force generation during amoeba locomotion (another rapidly locomoting cell type) [82]. In the amoeba model, 'contraction' is concomitant with local disassembly/severing ('solation') of the actin meshwork. Also in keratocytes, the cell body rolls forward [17,83]. This is estimated to contribute 46% to the total distance that the cell body translocates during loco-

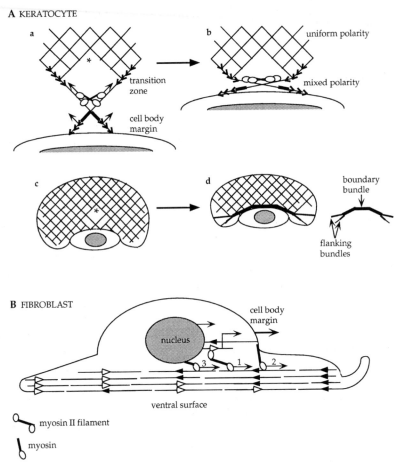

Fig. 9. Models for translocation of the cell body during cell locomotion. (A) Dynamic network contraction model in fish keratocytes [17]. Panels (a) and (b) are high-magnification views of panels (c) and (d) respectively. In panel (a), myosin II minifilaments in the transition zone pull forward uniform-polarity actin filaments connected to the cell body (compare panels a and b). This drives the cell body forward. A predicted consequence of force generation is the observed re-orientation of actin filaments and myosin II to form a mixed-polarity actin filament bundle (boundary bundle) in front of the cell body (compare panels a and b). Accordingly, panel (c) shows the predicted organization of actin filaments in a keratocyte in which the cell body has never previously translocated. Panel (d) shows concomitant formation of a boundary bundle (black bars) during the first cell body translocation event. Dynamic ☞

motion, the remainder coming from forward sliding of the cell body [17]. Whether dynamic contraction generates sufficient force to drive both forward rolling and forward sliding is not known. Myosin II force, as predicted by dynamic contraction in keratocytes, is generated in the transition zone roughly parallel/orthogonal to the direction of locomotion (see Fig. 9A, panel a). In the cell body proper, there is a strong, additonal force generated perpendicular to the direction of locomotion [84,85]. This distinct source of force may drive cell body rolling in keratocytes [83].

In contrast with keratocytes, in the fibroblast model neither concomitant actin filament bundle formation nor depolymerization is featured. Also, the cell body does not roll during fibroblast locomotion. Although a type of boundary actin bundle, termed an arc, can form in the lamella of fibroblasts [86], these are not directly associated with the ventral cell surface against which motile force needs to be generated for locomotion. Also, arcs have mostly been noted in fibroblasts likely to be losing or to have lost locomotion capacity. Instead, a feature of the fibroblast model is that myosin moves on the surface of preformed graded-polarity actin filament bundles located on the ventral cell surface (Fig. 9B). Due to filament polarity reversal in these bundles (see Fig. 5D), roughly mid-cell in the region of the nucleus (see Fig. 4D), myosin has the capacity to move towards both the leading and trailing cell margins. This notion is supported by direct measurements of force in fibroblasts, where motile force is directed towards both leading and trailing cell margins [87,88] and reverses in the region of the nucleus during cell locomotion [88]. What, then, determines net forward movement of the cell body in locomoting fibroblasts? One possibility is that net actin filament polarity in graded-polarity actin bundles over the whole ventral surface is tipped in favour of filament barbed ends forward. Indeed, measurements of such net actin filament polarity reveal a small bias in barbed ends forward, ranging between 5 and 13% in individual cells. Interestingly, in these cells only a small bias in integrated force forward of the cell body need occur for net forward translocation [7]. While filament polarity reversal also occurs in keratocytes, it is not an issue for net forward cell body translocation, because, unlike in fibroblasts, almost all filament barbed ends face forward (Fig. 5E).

Although the keratocyte and fibroblast models need to be tested, are they in effect variations of the same basic mechanism? In both cell types, the actin

☞ **Fig. 9 (contd.)** contraction can also explain formation of flanking actin bundles (panel d; splayed black lines). Equally, however, these may be a consequence of lateral flow of filament barbed ends, a predicted event coupled with protrusion of the lamellipodium [100]. (B) Transport model in heart fibroblasts [16]. In locomoting heart fibroblasts, graded-polarity actin filament bundles, fixed on the ventral cell surface, provide an oriented transport track. Experimental evidence supports a model whereby myosin (e.g myosin II; mechanism 1) pulls a second dynamic population of actin filaments over these tracks, which drives the cell body forward. Data do not exclude other myosins acting as transport motors to pull the cell body forward, e.g. by pulling the margin of the cell body forward directly (mechanism 2) or by pulling directly on the nucleus (mechanism 3).

substrate for myosin, forward of the nucleus, is composed of filament barbed ends predominantly facing forward (compare Figs. 9A and 9B). This alone is more suggestive of a transport mechanism. In locomoting keratocytes, newly synthesized myosin II clusters (composed of myosin II minifilaments) move forward short distances in the transition zone [17], where dynamic contraction is proposed to occur. One interpretation of these data is that 'dynamic contraction' is simply the same as 'transport', but over short distances. Transport mechanisms are not favoured by the authors [17], because detected myosin II does not move forward in the lamellipodium. However, for several reasons, forward movement of myosin II is not expected in this region of the cell. Myosin II clusters in the lamellipodium are probably too small to act as transport motors without diffusing away from the actin substrate. Also, there is simply no cell body cargo to transport in the lamellipodium. My alternative view of the keratocyte data is that an individual myosin II minifilament (which is bigger in size in the transition zone) only actively moves forward short distances in the transition zone, because it and its actin substrate re-orient sideways during force generation (Fig. 9A, compare panels a and b). In contrast, in fibroblasts major re-orientation of filaments is not a consequence of myosin force generation during cell body translocation, presumably because these cells are more strongly adherent, and so potentially myosin travels longer distances. In the keratocyte model, then, persistent translocation of the cell body is mediated by a succession of newly formed myosin II minifilaments, in a process of 'dynamic contraction–release', or 'transport–release'. Arguably, in keratocytes the detected organization of actin filaments and myosin II minifilaments in the transition zone does not distinguish between 'contraction' and 'transport' mechanisms (as outlined in Fig. 6D). In the transition zone, myosin II minifilaments consist of a few myosin dimers which interact with a few actin filaments: are these small contractile elements or the interaction of a transport motor with the surface of its actin substrate? In a similar vein, in fibroblasts it is theoretically possible that graded-polarity actin filament bundles generate a form of contractile force. While this is neither sarcomeric nor dynamic (as discussed above), other variations may exist. For example, force from filament sliding generated within an individual graded-polarity actin bundle is theoretically possible where filament polarity is mixed. This model predicts that the region of highest mixed polarity in a fibroblast (typically, roughly mid-cell in the region of the nucleus; see Fig. 4D) will generate the greatest force in the cell body during locomotion. Testing this requires the measurement of force in small zones, every few microns from the front to the back of the ventral surface of an individual cell as it locomotes. This should be possible using a new device, which so far has made measurements in small zones in restricted cell locations in different locomoting fibroblasts [88].

Tail retraction

Less well understood is the force that drives retraction of the cell rear during cell locomotion. While myosin II [80] and regulation of adhesion [5,7] are key, it is not known if retraction is based on contraction or transport mechanisms. Equally problematic is spatial location of retraction force, is it generated in the cell rear, or is the cell rear simply pulled forward by the force that drives translocation of the cell body, for example? Where myosin II is highly concentrated in the cell rear, such as in amoeba cells [89], sufficient force for tail retraction is likely to be generated in the cell rear. In theory, in these cells a contraction mechanism may primarily drive tail retraction. Locomoting amoeba cells are highly rounded, and much of the filamentous actin in this cell type appears to comprise a relatively thick layer of cortical actin, which in fibroblasts has the correct polarity for sarcomeric contraction (Fig. 4C). Consistent with a contraction theory is the observation that, in the absence of myosin II, surface-attached particles and receptors are unable to form a tight spot on the rear of amoeba cells; instead, they disperse along the entire rear cell margin [90,91]. In contrast, in other locomoting cell types, such as keratocytes and fibroblasts, myosin II is less highly biased in the tail, if at all [16,17,92]. In keratocytes, which are weakly adhesive, myosin forces generated in the transition zone and cell body (as discussed above) may be sufficient to pull both the cell body and the rear of the cell forward. This idea is supported by the simultaneous occurrence of cell body translocation and tail retraction in these cells, and the high correlation between the rates of these two motile events. In individual locomoting fibroblasts the correlation is less good, and additional force generation in the cell rear seems likely. The pattern of traction forces generated in the cell rear and cell body in locomoting fibroblasts is highly consistent with this idea [88]. One possible additional source of force for tail retraction in fibroblasts is sarcomeric contraction in cortical actin filament bundles. Contraction force alone is unlikely to be sufficient for tail retraction in locomoting fibroblasts due to the relatively low abundance of cortical actin bundles in these cells. Another possibility is that a hypothetical pointed-end-directed transport motor pulls the cell rear over uniform-polarity actin filament tracks that reside in tips of fibroblast tails (Figs. 4D and 5D).

Contribution of motile force to net forward cell locomotion

In order to locomote, a cell integrates several distinct types of motile force. What is the relative contribution of force for cell body translocation and tail retraction to the net forward locomotion of the cell? Are these forces more important for net locomotion than is protrusive force at the front of the cell? I think that, for individual locomoting cell types, this depends on cell speed, adhesion and the spatial location of myosin activity. It will probably turn out that, for all locomoting cell types, protrusive force contributes the least to overall locomotory force. Although in weakly adhesive, fast locomoting cell types, such as amoeba cells and keratocytes, it is theoretically possible that protrusive force alone is sufficient to drag the rest of the cell forward [93], in practice this does not occur. For example, in keratocytes, blocking protrusion in the short term with cytochalasin does not change the rate of cell body translocation [83]. Also, protrusive activity is unable to overcome a severe or complete deficiency in locomotion capacity in amoeba cells null for myosin II [78–80]. Similar observations have also been made in relatively slow, strongly adherent locomoting cell types, such as fibroblasts [16,94]. What, then, is the major role of protrusion? One possibility is that a protruding structure provides 'feelers' in front of the cell that respond to extracellular guidance cues and thus steer the cell. This has long been thought to be the role of filopodia in neuronal growth cones and perhaps is generally the case for locomoting cell types. Consistent with this idea, data support a model whereby protrusion is required for directed, but not random, fibroblast locomotion [94]. An alternative, possibly related, view is that assembly of actin during protrusion provides a substrate which myosin uses later to generate motile force (as proposed in Fig. 8C). In keratocytes this is feasible, as actin assembled in lamellipodia appears continuous with actin in the transition zone, which myosin II may then use to drive the cell body forward (as discussed above) [17]. This would provide a simple mechanism for the tight overall co-ordination observed between protrusion and cell body translocation in locomoting keratocytes.

Conversely, data infer that force generated in the lamella/transition zone, cell body and tail is used solely or almost solely to drive the cell body and tail forward, i.e. it is not utilized to drive protrusion. For example, fragments of cells do not need a cell body to protrude; they protrude almost as rapidly as intact cells [95]. Also, in keratocytes, applying external force to the cell body has no effect on the rate of protrusion [83]. In weakly adhesive locomoting cell types, it may turn out that only one of the two forces for cell body translocation and tail retraction is required to combine with protrusion to drive a cell forward. Which one is determined simply by the spatial location of myosin II activity during cell locomotion. In keratocytes, myosin II is most probably active in front of [17] and perhaps within [83] the cell body; thus force for cell body translocation is the more important in this cell type. In contrast, in amoeba cells, the evidence is that myosin II is highly active in the cell rear (see above), indicating that force for tail retraction is the more important in this cell type. In strongly adhesive locomoting cell types such as fibroblasts, it is more likely that generation of force is needed in both the cell body and the cell rear to overcome attachment to the substratum.

Future directions

In this review I have described several distinct types of myosin-based cell motility. It may turn out that some of these will have overlapping functions in the cell. For example, during fibroblast locomotion, organelle transport may contribute to the force that drives cell body translocation (perhaps as illustrated in Fig. 9B, mechanism 3). A challenge for the future will be to distinguish different types of myosin force generation. This is especially important as multitudes of myosin family members are identified and attributed to various functions in motile cells. For example, in the same neuronal growth cone, myosin V is implicated in vesicle transport both out to the cell surface and back again [96]. A prediction of this is that two subpopulations of actin transport tracks or mats exist in these growth cones: one with filament barbed ends facing outwards, and the other with filament barbed ends facing inwards.

Significant progress has been made in locomotion research, and it is now clear that generation of myosin II force to drive cell body translocation is distinct from shortening of muscle sarcomeres. Testing alternative proposed mechanisms is a necessary step towards solving the whole problem of force generation during cell locomotion. One key approach will be the development of techniques that measure simultaneously the dynamic behaviour of the actin cytoskeleton and the cell–substratum adhesion force in the same discrete cell zone as a cell locomotes. For example, is differential adhesion the only true distinguishing feature of the keratocyte and fibroblast models discussed in this review (Fig. 9)?

Also described here is the identification of multiple types of actin organization with distinct filament polarity in individual motile cell types. The prevalent view that actin bundles in the cell body of all types of motile cells are the same type of bundle, i.e. 'stress fibres', is clearly not the case. Determining the organization and polarity of a particular actin network has key implications for regulation research. For example, one approach to drug discovery is to identify regulatory molecules that are specific for distinct types of actin organization.

Many thanks to Paul Martin, Gareth Jones, Bob Simmons and Barbara Daniel for critically reading the manuscript and for helpful suggestions. I am currently funded by a Wellcome Trust (U.K.) career development award, a Royal Society grant and a University research fellowship.

References

1. Tilney, L.G. and Inoue, S. (1982) J. Cell Biol. **93**, 820–827
2. Heath, J.P. and Holifield, B.F. (1991) Cell Motil. Cytoskeleton **18**, 245–257
3. Sheetz, M.P. (1994) Semin. Cell Biol. **5**, 149–155
4. Grebecki, A. (1994) Int. Rev. Cytol. **148**, 37–79
5. Lauffenburger, D.A. and Horwitz, A.F. (1996) Cell **84**, 359–369
6. Mitchison, T.J. and Cramer, L.P. (1996) Cell **84**, 371–379
7. Sheetz, M.P., Felsenfeld, D.P. and Galbraith, C.G. (1998) Trends Cell Biol. **8**, 51–54
8. Welch, M.D., Mallavarapu, A., Rosenblatt, J. and Mitchison, T.J. (1997) Curr. Opin. Cell Biol. **9**, 54–61
9. Small, J.V. (1988) Electron Microsc. Rev. **1**, 155–174
10. Small, J.V., Isneberg, G. and Celis, J.E. (1978) Nature (London) **272**, 638–639
11. Lewis, A.K. and Bridgman, P.C. (1992) J. Cell Biol. **119**, 1219–1243
12. Cox, D., Ridsdale, J.A., Condeelis, J. and Hartwig, J. (1995) J. Cell Biol. **128**, 819–835
13. Small, J.V., Herzog, M. and Anderson, K. (1995) J. Cell Biol. **129**, 1275–1286
14. Cramer, L.P. (1997) Front. Biosci. **2**, 260–270; also on websites http://www.bioscience.org/current/vol2.htm and http://www.bioscience.org/1997/v2/d/cramer/list.htm
15. Mooseker, M.S. and Tilney, L.G. (1975) J. Cell Biol. **67**, 725–743
16. Cramer, L.P., Siebert, M. and Mitchison, T.J. (1997) J. Cell Biol. **136**, 1287–1305
17. Svitkina, T.M., Verkhovsky, A.B., McQuade, K.M. and Borisy, G.G. (1997) J. Cell Biol. **139**, 397–415
18. Byers, H.R., White, G.E. and Fujiwara, K. (1984) in Cell and Motility (Shay, J.W., ed.), pp. 83–137, Plenum Press, New York and London
19. Guild, G.M., Connelly, P.S., Shaw, M.K. and Tilney, L.G. (1997) J. Cell Biol. **138**, 783–797
20. Huxley, H.E. (1973) Nature (London) **243**, 445–449
21. Tilney, L.G., Connelly, P., Smith, S. and Guild, G.M. (1996) J. Cell Biol. **135**, 1291–1308
22. Tilney, L.G., Tilney, M.S. and Guild, G.M. (1996) J. Cell Biol. **133**, 61–74

23. Begg, D.A., Rodewald, R. and Rebhun, L.I. (1978) J. Cell Biol. **79**, 846–852
24. Gordon, W.E. (1978) Exp. Cell Res. **117**, 253–260
25. Sanger, J.M. and Sanger, J.W. (1980) J. Cell Biol. **86**, 568–575
26. Sanger, J.W., Sanger, J.M. and Jockusch, B.M. (1983) J. Cell Biol. **96**, 961–969
27. Langanger, G., Moeremans, M., Daneels, G., Sobieszek, A., De Brabander, M. and Mey, J.D. (1986) J. Cell Biol. **102**, 200–209
28. Verkhovsky, A.B., Svitkina, T.M. and Borisy, G.G. (1995) J. Cell Biol. **131**, 989–1002
29. Nakazawa, H. and Sekimoto, K. (1996) J. Phys. Soc. Jpn. **65**, 2404–2407
30. Mooseker, M.S. and Cheney, R.E. (1995) Annu. Rev. Cell Dev. Biol. **11**, 633–675
31. Cope, M.J., Whisstock, J., Rayment, I. and Kendrick-Jones, J. (1996) Structure **4**, 969–987
32. Sheetz, M.P. and Spudich, J.A. (1983) Nature (London) **303**, 35–39
33. Adams, R.J. and Pollard, T.D. (1986) Nature (London) **322**, 754–756
34. Kron, S.J. and Spudich, J.A. (1986) Proc. Natl. Acad. Sci. U.S.A. **83**, 6272–6276
35. Mooseker, M.S. and Coleman, T.R. (1989) J. Cell Biol. **108**, 2395–2400
36. Zot, H.G., Doberstein, S.K. and Pollard, T.D. (1992) J. Cell Biol. **116**, 367–376
37. Wolenski, J.S., Cheney, R.E., Mooseker, M.S. and Forscher, P. (1995) J. Cell Sci. **108**, 1489–1496
38. Pollard, T.D., Doberstein, S.K. and Zot, H.G. (1991) Annu. Rev. Physiol. **53**, 653–681
39. Goodson, H.V., Valetti, C. and Kreis, T.E. (1997) Curr. Opin. Cell Biol. **9**, 18–28
40. Titus, M.A. (1997) Curr. Biol. **7**, R301–R304
41. Howard, J. (1997) Nature (London) **389**, 561–567
42. Sanger, J.M., Mittal, B., Pochapin, M. and Sanger, J.W. (1986) J. Cell. Sci. Suppl. **5**, 17–44
43. Hitt, A.L. and Luna, E.J. (1994) Curr. Opin. Cell Biol. **6**, 120–130
44. Cowin, P. and Burke, B. (1996) Curr. Opin. Cell Biol. **8**, 56–65
45. Pasternak, C., Spudich, J.A. and Elson, E.L. (1989) Nature (London) **341**, 549–551
46. Fukui, Y., De Lozanne, A. and Spudich, J.A. (1990) J. Cell Biol. **110**, 367–378
47. Crowley, E. and Horwitz, A.F. (1995) J. Cell Biol. **131**, 525–537
48. Sims, J.R., Karp, S. and Ingber, D.E. (1992) J. Cell Sci. **103**, 1215–1222
49. Wheatley, S., Kulkarni, S. and Karess, R. (1995) Development **121**, 1937–1946
50. Neujahr, R., Heizer, C. and Gerisch, G. (1997) J. Cell Sci. **110**, 123–137
51. Cramer, L.P. and Mitchison, T.J. (1997) Mol. Biol. Cell **8**, 109–119
52. Satterwhite, L.L., Lohka, M.J., Wilson, K.L., Scherson, T.Y., Cisek, L.J., Corden, J.L. and Pollard, T.D. (1992) J. Cell Biol **118**, 595–605
53. Yamakita, Y., Yamashiro, S. and Matsumura, F. (1994) J. Cell Biol. **124**, 129–137
54. Bement, W.M., Forscher, P. and Mooseker, M.S. (1993) J. Cell Biol. **121**, 565–578
55. Martin, P. and Lewis, J. (1992) Nature (London) **360**, 179–183
56. Young, P.E., Richman, A.M., Ketchum, A.S. and Kiehart, D.P. (1993) Genes Dev. **7**, 29–41
57. Rappaport, R. (1996) Cytokinesis in Animal Cells, Cambridge University Press, New York
58. Burton, K. and Taylor, D.L. (1997) Nature (London) **385**, 450–454
59. DeLozanne, A. and Spudich, J.A. (1987) Science **236**, 1086–1091
60. Knecht, D. and Loomis, W.F. (1987) Science **236**, 1081–1086
61. Ikonen, E., Bruno de Almeid, J., Fath, K.R., Burgess, D.R., Ashman, K., Simons, K. and Stow, J.L. (1997) J. Cell Sci. **110**, 2155–2164
62. Musch, A., Cohen, D. and Rodriguez-Boulan, E. (1997) J. Cell Biol. **138**, 291–306
63. Baines, I.C., Brzeska, H. and Korn, E.D. (1992) J. Cell Biol. **119**, 1193–1203
64. Doberstein, S.K., Baines, I.C., Wiegand, G., Korn, E.D. and Pollard, T.D. (1993) Nature (London) **365**, 841–843
65. Cramer, L.P. and Mitchison, T.J. (1995) J. Cell Biol. **131**, 179–189
66. Doberstein, S.K. and Pollard, T.D. (1992) J. Cell Biol. **117**, 1241–1249
67. Geli, M.I. and Riezman, H. (1997) Science **272**, 533–535
68. Kersey, Y.M., Hepler, P.K., Palevitz, B.A. and Wessels, N.K. (1976) Proc. Natl. Acad. Sci. U.S.A. **73**, 165–167

69. Higashi-Fujime, S., Ishikawa, R., Iwasawa, H., Kagami, O., Kurimoto, E., Kohama, K. and Hozumi, T. (1995) FEBS Lett. **375**, 151–154

70. Yamamoto, K., Kikuyama, M., Sutoh-Yamamoto, N., Kamitsubo, E. and Katayama, E. (1995) J. Mol. Biol. **254**, 109–112

71. Fath, K.R. and Burgess, D.R. (1994) Curr. Opin. Cell Biol. **6**, 131–135

72. Cramer, L.P. and Mitchison, T.J. (1993) J. Cell Biol. **122**, 833–843

73. DeBiasio, R.L., LaRocca, G.M., Post, P.L. and Taylor, D.L. (1996) Mol. Biol. Cell **7**, 1259–1282

74. Heidemann, S.R., Lamoureux, P. and Buxbaum, R.E. (1991) J. Cell Sci. Suppl. **15**, 35–44

75. Bridgman, P.C. and Dailey, M.E. (1989) J. Cell Biol. **108**, 95–109

76. Lin, C.-H., Espreafico, E.M., Mooseker, M.S. and Forscher, P. (1996) Neuron **16**, 769–782

77. De Bruyn, P.P.H. (1947) Q. Rev. Biol. **22**, 1–24

78. Wessels, D., Soll, D.R., Knecht, D., Loomis, W.F., De Lozanne, A. and Spudich, J. (1988) Dev. Biol. **128**, 164–177

79. Doolittle, K.W., Reddy, I. and McNally, J.G. (1995) Dev. Biol. **167**, 118–129

80. Jay, P.Y., Pharm, P.A., Wong, S.A. and Elson, E.L. (1995) J. Cell Sci. **108**, 387–393

81. Mogilner, A. and Oster, G. (1996) Eur. J. Biophys. **25**, 47–53

82. Taylor, D.L. and Fechheimer, M. (1982) Philos. Trans. R. Soc. London B **299**, 183–197

83. Anderson, K., Wang, Y.-L. and Small, J.V. (1996) J. Cell Biol. **134**, 1209–1218

84. Lee, J., Leonard, M., Oliver, T., Ishihara, A. and Jacobson, K. (1994) J. Cell Biol. **127**, 1957–1964

85. Oliver, T., Dembo, M. and Jacobson, K. (1995) Cell Motil. Cytoskeleton **31**, 225–240

86. Heath, J.P. (1983) J. Cell Sci. **60**, 331–354

87. Harris, A.K., Wild, P. and Stopak, D. (1980) Science **208**, 177–179

88. Galbraith, C.G. and Sheetz, M.P. (1997) Proc. Natl. Acad. Sci. U.S.A. **94**, 9114–9118

89. Fukui, Y., Lynch, T.J., Brzeska, H. and Korn, E.D. (1989) Nature (London) **341**, 328–331

90. Jay, P.Y. and Elson, E.L. (1992) Nature (London) **356**, 438–440

91. Aguado-Velasco, C. and Bretscher, M.S. (1997) Proc. Natl. Acad. Sci. U.S.A. **94**, 9684–9686

92. Conrad, P.A., Giuliano, K.A., Fisher, G., Collins, K., Matsudaira, P.T. and Taylor, D.L. (1993) J. Cell Biol. **120**, 1381–1391

93. Oliver, T., Lee, J. and Jacobson, K. (1994) Semin. Cell Biol. **5**, 139–147

94. Arnand-Apte, B., Zetter, B.R., Viswanathan, A., Qiu, R.-g., Chen, J., Ruggieri, R. and Symons, M. (1997) J. Biol. Chem. **272**, 30688–30692

95. Euteneuer, U. and Schliwa, M. (1984) Nature (London) **310**, 58–61

96. Prekeris, R. and Terrian, D.M. (1997) J. Cell Biol. **137**, 1589–1601

97. Okabe, S. and Hirokawa, N. (1991) J. Neurosci. **11**, 1918–1929

98. Sinard, J.H., Stafford, W.F. and Pollard, T.D. (1989) J. Cell Biol. **109**, 1537–1547

99. Cramer, L.P., Mitchison, T.J. and Theriot, J.A. (1994) Curr. Opin. Cell Biol. **6**, 82–86

100. Small, J.V., Anderson, K. and Rottner, K. (1996) Biosci. Rep. **16**, 351–368

Biochem. Soc. Symp. **65**, 207–222
Printed in Great Britain

12

Network contraction model for cell translocation and retrograde flow

A.B.Verkhovsky,T.M. Svitkina and G.G. Borisy

Laboratory of Molecular Biology, University of Wisconsin, Madison, WI 53706, U.S.A.

Abstract

Kinetic and structural analysis of the actin–myosin II system in mammalian fibroblasts and fish epidermal keratocytes suggests that the cell's motility machinery arises behind the leading edge in the form of myosin filament clusters immersed in an actin filament network. We discuss how the contraction of this actin–myosin II network is related to the formation of actin–myosin filament bundles, cell translocation and retrograde flow.

Introduction

The crawling motion of animal cells involves three basic steps: formation of a lamellipodial protrusion at the front of the cell, adhesion of the lamellipodium to the substratum, and translocation forward of the cell body. Numerous studies indicate that protrusion is driven by polymerization of actin at the leading edge of the lamellipodium (for reviews, see [1–4]). Actin filaments grow at their barbed, forward-facing ends, thus providing the force for protrusion, but polymerized domains do not themselves move forward relative to the substratum. In contrast, major components of the cell body, such as the nucleus and other organelles, actually move forward. Consequently, translocation of the cell body requires elements in addition to lamellar protrusion.

A minimalistic scheme for cell body translocation could rely on a passive means of maintaining cell integrity, such as the mechanical continuity and elasticity of the plasma membrane and/or cortical cytoskeleton. In such a scheme, filament polymerization inside a closed container propels the container forward, with its contents following passively. Translocation of the cell body by this mechanism would be fully dependent on lamellipodial extension. Such a 'treadmilling' mechanism has been proposed for *Ascaris* sperm translocation [5], although in this novel example of cell motility the filament-forming protein is unrelated to actin, and both actin and myosin are absent from the sperm.

In general, however, cell motility is based on actin and myosin, and it is reasonable to consider that actin-dependent motor proteins may generate forces contributing to the translocation of the cell body relative to the lamellipodium. If such forces exist, translocation of the cell body may occur even in the absence of lamellipodial protrusion. This was actually demonstrated for fish epidermal keratocytes, where the cell body continued to move forward after protrusion was blocked with cytochalasin [6]. The same forces that drive the cell body forward relative to the lamellipodium and the substratum would be expected to produce a retrograde flow of lamellipodial components relative to the substratum if the cell body, for any reason, could not move forward (Fig. 1). Thus the phenomenon of retrograde flow can, in principle, be explained in the same terms as cell locomotion [7]. A variety of other mechanisms have been proposed for retrograde flow [8], but the idea that flow is a mirror reflection of cell translocation seems simple and attractive. Consistent with this idea, retrograde flow in *Aplysia* growth cones was shown to be independent of actin assembly at the edge [9], to be inversely related to the rate of neurite translocation [10] and to require the activity of a myosin motor [11]. The specific myosin protein involved in growth cone motility has not been identified. In general, among the members of the myosin superfamily, myosin II is a likely candidate because it is the only myosin species with the ability to form polymeric supramolecular assemblies [12,13] which could generate and transmit forces over significant distances within the cell. Knockout of myosin II in *Dictyostelium* resulted in a dramatic decrease in the rate of cell locomotion [14] or in a block of locomotion in an environment of increased resistance [15,16], thus clearly demonstrating the role of this protein in motility.

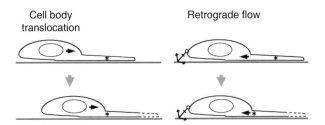

Fig. 1. Forces acting between the lamellipodium and the cell body may drive both cell body translocation and retrograde flow. If the lamellipodium is attached more strongly than the cell body (left), the lamellipodium remains stationary relative to the substratum, but the cell body moves forward. If the cell body is attached more strongly than the lamellipodium (right), the cell body remains stationary, but the lamellipodium moves back. The dotted boundary shows the newly assembled region of the lamellipodium, and the asterisk indicates a reference mark that remains stationary with respect to cytoskeletal components of the lamellipodium.

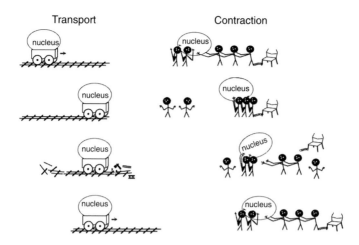

Fig. 2. Schematic representation of mechanisms of cell body translocation. In the transport mechanism (left), the cell body (represented by the nucleus) moves forward along the railroad track. After the carriage reaches the end of the track, more track has to be built in front of the carriage. To obtain construction material, track at the back should be taken apart. In the contraction model (right), the nucleus is pulled forward by a series of contractile elements (cartoon figures). In order to achieve forward directional movement, the figure at the front is immobilized with respect to the substratum (chained to a chair). After the contraction is complete, the nucleus should be passed on to a newly assembled series of contractile elements.

All myosins investigated to date have a force polarity which drives them towards the barbed ends of actin filaments. However, it is not clear how the molecular event of myosin translocation on an actin filament is translated at the cellular level into directional translocation of the cell body. As discussed in the literature [2,17], two basic mechanisms can be envisioned. The simplest possibility (Fig. 2; transport hypothesis) is that directional movement at the cellular level results from polar organization of actin filaments, which function like railroad tracks. In this hypothesis, the actin filaments have to be oriented with their barbed ends in the direction of translocation. Myosin II would travel forward along these uniformly polarized tracks, providing the motive force for the translocation of the cargo (cell body) to which it is mechanically connected. An alternative model (Fig. 2; contraction hypothesis) holds that myosin II interacts with actin filaments to produce contraction in the cell. In this model, directional translocation is explained not by the polarity of actin itself but by a non-uniform distribution of contractile force and/or strength of cell adhesion to the substratum (contractility increases and/or adhesion decreases towards the rear of the cell). Unlike the transport model, the extent of translocation of individual myosin molecules in the contraction model could be small compared with the size of the cell, but, being multiplied by the number of contractile elements, it could result in cell body movement of sufficient magnitude. It is not clear what specific organization of actin and myosin II is required to support contraction. By analogy with striated muscle, a sarcomeric-like organization of alternately polarized arrays of actin filaments anchored at their barbed ends is often deemed necessary [18,19]. However, such organization is not readily apparent in most non-muscle cells, and *in vitro* studies clearly demonstrated that actin–myosin II gels [20–22] and actin–heavy-meromyosin bundles [23] are capable of contraction, despite the lack of sarcomeric organization. Unidentified contractile elements in the cortical cytoskeleton have been postulated to generate and participate in cortical flow and cell translocation [7,24]. How can the transport and contraction models be tested? While inhibitory and genetic knockout approaches are crucial to identify functional components at the molecular level, the only way to penetrate the design of a machine is to determine how the elements of its mechanism are connected and how they move during action. This calls for a study of the supramolecular organization and dynamics of the actin–myosin II system.

An additional reason why cytoskeletal dynamics are relevant to the mechanism of translocation is that the motile machinery needs to be dynamic to support continuous translocation of the cell. As the cell advances, what was once a new protrusion becomes a part of the cell body. Whether the mechanism consists of transport tracks or contractile elements, these components of the transport machinery need to arise continuously in newly extended regions of the cell to support the translocation of the cell body beyond the limits of the cell at an earlier time (Fig. 2). Conversely, components need to be taken up at the rear of the cell. It seems plausible that the cytoskeletal elements need to be re-organized through continuous assembly and degradation in order to support cell body translocation. Thus a comprehensive model of translocation should consider not only the mechanism of translocation itself, but also the mechanism of formation of the transport machinery (contractile elements or transport tracks) and the recycling of the parts.

While the organization and dynamics of actin filaments have been the subject of numerous studies, the lack of information on the supramolecular organization of myosin II represented a significant gap in our knowledge. Here we provide a model for both the pathway of formation and the function of the actin–myosin II machinery in cell translocation and retrograde flow. We will discuss our recent data obtained in two model systems: mammalian fibroblasts and fish epidermal keratocytes. Fibroblasts have long been a favourite model for studies of cytoskeletal organization and dynamics. They possess a complex actin–myosin cytoskeletal system complete with lamellipodial and cortical actin networks and longitudinal, transverse and circumferential filament bundles often displaying a periodic arrangement of myosin II [25]. Thus fibroblasts represent a model system for the morphogenesis of complex cytoskeletal structure. However, fibroblast locomotion is slow and usually lacks persistent polarization. The lack of regularity makes it difficult to relate cytoskeletal organization and dynamics to the translocation process. Fish keratocytes possess a simpler actin–myosin organization, consisting mostly of a filament network in lamellipodia and filament bundles parallel to the leading edge at the lamellipodial/cell body transition zone. These cells exhibit remarkably fast and persistent locomotion, which makes them an ideal model system for the study of the basic mechanisms of motility [26].

Clusters of myosin filaments as tension-generating elements

What are the units of myosin organization in the cytoskeleton? Traditionally, myosin organization was described as punctate, periodic or continuous along bundles of actin filaments (stress fibres) [27] or within a cortical sheath of actin filaments [28]. In this description, it was implicit that the organization of myosin is dependent on and secondary to actin organization. However, careful examination of light-microscopic images of cells double stained for actin and myosin (Fig. 3) shows that myosin at the cell periphery is present in the form of small discrete spots that are not aligned along actin filament bundles, but seem to represent structural entities on their own [29]. Dynamic studies utilizing fluorescently labelled microinjected myosin II showed that it is in the form of these spots that myosin II first appears at the active edge of the cell [29–31]. Spots appear spontaneously in both fibroblast lamella and keratocyte lamellipodia, and subsequently exhibit dramatic growth (Fig. 4) suggestive of seeded macromolecular assembly.

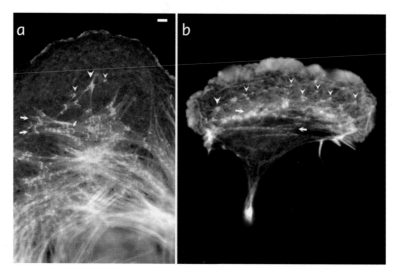

Fig. 3. Overall distribution of myosin II (red) and actin (cyan) in a REF-52 fibroblast (a) and a fish epidermal keratocyte (b). Close to the active cell edge (top), myosin is found in the form of discrete spots (arrowheads) that are not associated with well-defined actin structures. Some spots (large arrowheads) appear as centres of convergence of several small actin bundles. Further from the edge, myosin spots align along actin filament bundles (arrows). In fibroblasts, myosin spots appear as ribbons forming a periodic banding pattern. Some ribbons cross several actin bundles and exhibit sharp bending at crossover points indicative of pulling forces (small arrow). Bar = 2 μm.

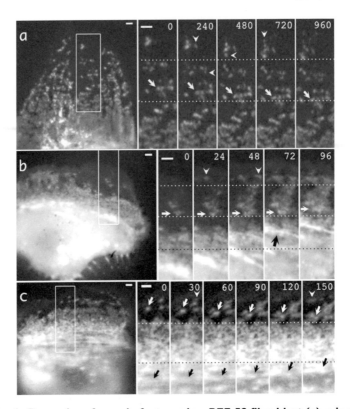

Fig. 4. Dynamics of myosin features in a REF-52 fibroblast (a), a loco-moting fish keratocyte (b) and a fish keratocyte tethered at the edge of an epithelioid colony (c). Cell overviews are shown on the left, and myosin dynamics in enlarged boxed areas are shown on the right, with time indicated in seconds. Broken lines represent reference marks fixed with respect to the substratum. New myosin spots (arrowheads) arise continuously close to the edge. Myosin spots move away from the edge with respect to the substratum in the fibroblast and the tethered keratocyte (white arrows). In the locomoting keratocyte, the myosin spot indicated by the arrow is stationary with respect to the substratum until it reaches the vicinity of the cell body (at 48 s), where it flattens and starts to move forward. A new myosin fibre (large black arrow) condenses from myosin spots in the vicinity of the cell body and moves forward. Myosin fibres (black arrows) also move forward in the teth-ered keratocyte, reflecting the advance of the cell as part of an epithelial sheet. Bars = 2 μm.

What is the molecular nature of these spots? The ultrastructural organiza-tion of myosin, although long suggested to be in the form of bipolar filaments [32–34], until recently had escaped electron-microscopic visualization. The rea-son for this is that myosin structures are obscured by other, more abundant elements of the cytoskeleton, primarily actin filaments. We have accomplished morphological identification of myosin structures by the removal of actin fila-ments from permeabilized cells with the actin-severing protein, gelsolin [35]. Correlative light and electron microscopy showed that the permeabilization

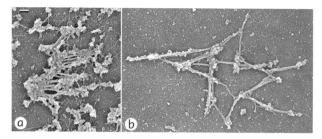

Fig. 5. Clusters of myosin filaments in a REF-52 fibroblast (a) and a fish epidermal keratocyte (b). Clusters of a similar size, each corresponding to a discrete myosin spot, are shown. In both types of cells myosin filaments in clusters contact each other mostly at their myosin head-containing terminal regions. In fibroblasts, filaments are mostly oriented in parallel stacks, while in keratocytes they form non-aligned clusters. Bar = 100 nm.

procedure and the removal of actin did not induce significant redistribution of myosin II [29,36]. Thus the organization of myosin II revealed in actin-depleted cytoskeletons faithfully reflected its organization in living cells. Numerous dumb-bell-shaped structures 0.3–0.5 μm in length were uncovered. These were identified as myosin bipolar minifilaments, with a rod-like central bare zone and globular termini containing myosin head domains. Myosin filaments in the cytoskeleton were rarely found as singles; mostly they were organized in aggregates of various sizes ranging from clusters of several filaments to large stack-like and network-like assemblies (Fig. 5). In these aggregates, myosin filaments apparently contacted each other at their head-containing terminal regions and were oriented with respect to each other at a variety of angles, suggesting the possibility that the sites of contact could serve as flexible joints around which the filaments may rotate [36]. Filament clusters, but not the single filaments, corresponded to individual myosin spots [29]. Thus formation of a new myosin spot was equivalent in molecular terms to formation of a cluster of bipolar filaments. It is not yet known whether growth of a filament cluster occurs by addition of bipolar filaments or individual myosin molecules, or both.

Dynamic studies show that, in both fibroblasts and keratocytes, nascent myosin clusters grow and move away from the active edge and eventually give rise to all the myosin structures of the cell [29–31,37]. The ways in which the clusters are delivered from the periphery to the central cell regions may vary. In fibroblasts and in keratocytes tethered at the edge of an epithelioid colony, myosin clusters move away from the edge with respect to the substratum (Figs. 4a and 4c). In contrast, in freely locomoting keratocytes, myosin clusters in the lamellipodium are stationary with respect to the substratum, and thus their apparent movement relative to the leading edge is due solely to the advance of the edge (Fig. 4b). On the way from the edge, fibroblast myosin clusters elongate to form ribbon-like stacks of parallel filaments which eventually align into a periodic pattern of transverse bands along actin filament bundles in the lamella and the cell body. In keratocytes, clusters approach the cell body and contribute to actin–myosin bundles at the lamellipodial/cell body transition zone. Thus clusters of bipolar filaments might be considered as building blocks of the myosin II machinery of the cell.

Are myosin clusters also functional units of this machinery and, if so, what are their functional properties? One might expect that myosin function is related to its ability to translocate actin filaments or to be translocated relative to them, and that enzymically active myosin would move towards the barbed ends of actin filaments and eventually co-localize with them. Here we encounter an apparent paradox. Co-localization of myosin II with the barbed ends of actin filaments is clearly not the case in either the fibroblast or the keratocyte model. We have performed simultaneous immunogold localization of myosin filament clusters and determination of actin filament polarity using myosin S1 decoration in platinum replicas of the fibroblast cytoskeleton. Myosin clusters were found mostly in association with actin filaments of mixed polarity, with approximately equal numbers of filaments oriented with their barbed or pointed ends towards myosin [38]. The situation is even more paradoxical in the keratocyte lamellipodium [31]. Here we demonstrated that actin filaments throughout the lamellipodium are directed uniformly with their barbed ends towards the edge, although they may be oriented over a range of angles. Barbed ends of filaments were clearly seen only within a narrow zone adjacent to the edge. However, there was no enrichment of myosin II in this zone. Moreover, myosin clusters did not move towards the leading edge but, instead, were stationary relative to the substratum in locomoting cells or moved away from the edge in tethered cells (Fig. 4). Since actin in locomoting keratocytes has been shown to be stationary with respect to the substratum [39], there is in fact no relative movement of actin and myosin in these cells.

There are two possible explanations for this paradox. First, myosin clusters may somehow not be active in actin translocation. Instead, they may have another function, e.g. cross-linking of actin filaments. The second explanation is that myosin is fully capable of translocating filaments, but that translocation is constrained due to rigidity of the actin network [40]. The second explanation was tested experimentally by reducing actin rigidity. Consistent with this idea, cell contraction was shown to be promoted by disruption of the actin network by cytochalasin without the additional activation of myosin II [41]. Recently we have shown directly that myosin clusters translocate actin filaments and eventually co-localize with actin barbed ends if actin filaments are rendered free by cytochalasin treatment [38]. As a result, actin asters form, with actin filament barbed ends oriented uniformly towards myosin at the centre of an aster (Fig. 6). New myosin clusters continue to form after the application of cytochalasin and become centres of actin asters. Thus myosin clusters are capable of translocating actin but, normally, significant relative translocation does not occur because of mechanical constraints.

What would be the result of activity of myosin clusters when translocation of actin filaments is limited by network rigidity? Obviously, it would be the generation of tension in the surrounding actin network (Fig. 6). The size of the cluster might be an important parameter for the efficiency of tension generation. A single myosin filament or a small cluster is likely to interact with a single actin filament or a group of filaments of similar orientation.

Formation of an aster   Tension generation

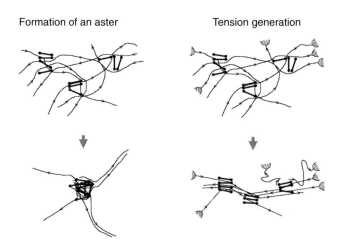

Fig. 6. Diagram showing how myosin filament clusters can transform an actin network. Myosin filaments are shown as dumb-bell figures, and actin filaments are shown as lines with polarity indicated by arrowheads. Myosin clusters are immersed in an actin network of mixed polarity. If actin filaments are not tethered, e.g. in the presence of cytochalasin (left), they are translocated with their pointed ends forward and form an aster-like structure surrounding aggregated myosin clusters. Interaction of myosin clusters with tethered actin filaments (right) results in tension generation. Tethered actin filaments are predicted to behave differently depending on the polarity of their attachment to substratum adhesion sites or to other filaments in the network. Filaments tethered in the network at their barbed ends experience tension that would be transmitted throughout the network, inducing co-alignment of filaments. In contrast, filaments tethered at their pointed ends would be 'polarity sorted' and expelled from the forming bundle.

Consequently, a small myosin cluster would travel along actin filament(s) without constraint and, therefore, would impose no tension on the network. The bigger the myosin cluster, the greater the likelihood that it would interact simultaneously with actin filaments of different orientation. A large cluster would not be able to follow any of the numerous divergent actin tracks (such as filaments in keratocyte lamellipodia that radiate at various angles even though they are directed with barbed ends to the edge), but would instead 'pull' on actin filaments and impose tension on the network. Thus aggregation of myosin filaments into clusters might be functionally significant for tension generation. We propose that myosin clusters embedded in the actin network are the long-sought tension-generating units of non-muscle cells. The tension that they develop, although not yet measured directly, is manifested in the formation of actin filament bundles, cell translocation and retrograde flow.

Formation of actin–myosin filament bundles

The idea that tension in an actin network produces alignment of actin filaments into bundles was suggested long ago for rearrangements observed in the
Physarum cytoskeleton [42]. A more recent concept, the tensegrity model [43],
considers tension as a factor defining cytoskeletal architecture. Our structural
and dynamic observations are consistent with these ideas and demonstrate a
kinetic pathway for actin bundle formation [29]. Since myosin clusters arise in
the vicinity of the active edge and then move away from it, the age of a cluster is
roughly proportional to its distance from the edge. Thus structural arrangements seen at different distances from the edge can be interpreted as sequential
phases of one morphogenetic process. Examination of both fibroblasts and keratocytes at the light- and electron-microscopic levels demonstrates that nascent
myosin clusters at the cell periphery are associated with a network of non-
aligned actin filaments, while the growth of myosin clusters correlates with
progressive alignment of actin filaments around them and myosin filaments
themselves. Several additional findings suggest the active role of myosin clusters in actin bundle formation. In particular, myosin features often appear to be
the centres of organization of other proteins: several small actin bundles frequently converge to a single myosin feature (Fig. 3; see also [29]), and the stress
fibre protein α-actinin exhibits paired accumulations flanking single myosin
clusters [29]. Elongated ribbon-like myosin clusters are often seen to bend
sharply at the intersections with actin filament bundles, suggestive of a force
exerted on actin at this site and of a reactive force causing deformation of the
myosin structure. All these observations suggest that myosin clusters precede
actin filament bundles and induce their formation by exerting tension on the
actin network, and that both actin and myosin filaments are aligned in the
process of their interaction.

Fibroblasts and keratocytes show similar sequential phases of mutual
organization of actin and myosin filaments, with two minor differences. First,
myosin clusters that have not yet aligned actin filaments around them are only
rarely observed in fibroblasts, but are abundant in keratocyte lamellipodia (Fig.
3). Secondly, myosin organization in the form of periodic ribbons along actin
filament bundles is common in fibroblasts, but virtually absent in keratocytes
(Fig. 3). These differences may reflect different regulation of myosin assembly
and activity by some additional factors in these two types of cells. Another
possibility is that these differences result from a difference in the distribution of
tension.

The alignment of actin filaments into a bundle reflects either an
anisotropic tension force or an anisotropic distribution of network resistance,
or both. Consequently, the direction of alignment depends on the distribution
of myosin clusters and on the local rigidity of the actin network which, in turn,
is dependent on filament density, attachment and cross-linking. Analysis of
these parameters in the general case is complicated and has not yet been accomplished. However, in the case of locomoting keratocytes, several simple
considerations permit an explanation of why actin–myosin bundles always
form parallel to the leading edge. Actin and myosin exhibit inversely related,

graded distributions in lamellipodia: while actin density and, therefore, network rigidity decrease towards the cell body, myosin clusters increase in number and size. These considerations suggest that small myosin clusters at the front of a lamellipodium cannot change the geometry of the rigid actin network, but at some critical distance from the edge the myosin clusters acquire enough strength to overcome actin rigidity and start contraction of the network. Tension developing due to loss of actin filaments from the network, as predicted by Mogilner and Oster [44], may also contribute to the contraction. Since both actin rigidity and myosin strength depend on the distance from the edge, the boundary between contracting and rigid regions would be parallel to the leading edge. Given that the contracting region of the network is physically continuous with the rigid network at the leading edge, the contracting portion would compress to the border of the rigid area, forming a bundle parallel to the leading edge. At the supramolecular level, bending and aligning of actin filaments into a transverse bundle would allow individual myosin clusters to travel simultaneously towards the barbed ends of many divergent actin filaments (Fig. 7). Similar considerations might apply to the formation of circumferential bundles in radially spreading fibroblasts and transverse bundles in polarized fibroblasts [45]. The bundles formed parallel to the leading edge would be expected to have a mixed polarity of actin filaments, and this expectation has

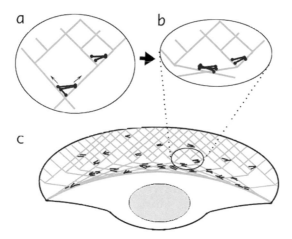

Fig. 7. Compression of an actin–myosin network into a bundle at the lamellipodial/cell body transition zone of a fish keratocyte is coupled to forward translocation. (a) In the lamellipodium, the network of divergent actin filaments interacts with clusters of myosin bipolar filaments. Whereas small myosin clusters situated in the dense network close to the cell front cannot move, larger clusters in the sparser network in the transition zone are capable of approaching barbed ends of diverging filaments and moving forward. (b) As myosin clusters move forward, they align actin filaments parallel to the leading edge. (c) Overall, network contraction at the lamellipodial/cell body transition zone results in formation of actin–myosin bundles and forward translocation of the cell body.

born true for keratocytes [31]. Such a mechanism might offer a general explanation of how actin bundles of mixed polarity found in the cytoplasm could be derived from a lamellipodial actin network of uniform polarity. Abundant longitudinal actin bundles in polarized fibroblasts may arise as the result of strong substrate attachment at the cell centre and/or rear preventing compression of the network parallel to the edge and favouring longitudinal compression. Longitudinal bundles might include portions of the uniformly polarized network at the opposite cell edges and thus might be expected to have a polarity that changes gradually from uniform with barbed ends towards one edge to uniform with barbed ends towards the opposite edge, as described by Cramer et al. [46].

Translocation of the cell body

On the assumption that formation of an actin bundle is equivalent to compression of the network perpendicular to the axis of the forming bundle, the presence and orientation of bundles reflects the pattern of network contraction. Polarized fibroblasts are characterized by a complex pattern of filament bundles, making it difficult to analyse the consequences of contraction in terms of cell body translocation. In contrast, contraction in keratocytes has an obvious relationship with motility. In keratocytes, bundles are oriented parallel to the leading edge, suggesting that compression of the network has occurred along the anterior/posterior axis. This compression is actually visualized in video sequences of locomoting cells as the shortening of distances between neighbouring myosin features and changes in the shape of individual features in the lamellipodial/cell body transition zone. More specifically, myosin features in front of the transition zone remain stationary with respect to the substratum, while those adjacent to the cell body move forward, flatten and align into fibres (Fig. 4). We propose that the cell body is attached to the actin–myosin network in the transition zone either directly or through other cytoskeletal structures, e.g. intermediate filaments. Consequently, compression of the actin–myosin network in the forward direction (Fig. 7) pulls the cell body along. A region of the network compressed to a transverse bundle can no longer produce force along the anterior/posterior axis. However, the region forward of the transverse bundle would start to compress and take on the pulling of the cell body. Thus the cell body rides on a wave of contraction that is continuously taking place at the lamellipodial/cell body boundary. The by-products of network compression, i.e. actin–myosin bundles, are observed to be overrun by the cell body and then contract, fragment and disappear [31]. These findings may reflect the recycling of the actin–myosin machinery in the cell body which is necessary to provide a pool of subunits for new assembly in lamellipodia.

Our model of cell body movement is essentially a contraction model, because relatively small translocations of myosin clusters along actin filaments in the transition zone are proposed to result in filament re-alignment and compression of the network at the cellular level (Fig. 7). An alternative possibility, i.e. directional transport of myosin along uniform actin arrays, does not seem to fit the experimental data. No transport of myosin clusters is observed in lamellipodia, in which actin filaments are oriented almost uniformly with barbed ends forward and could serve as transport tracks. In the transition zone and the cell body, where myosin features actually move forward, actin filaments concomitantly align into bundles perpendicular to the direction of locomotion, inconsistent with the idea of actin being a stationary track for a transport vehicle. Thus we conclude that contraction, not transport, drives cell body translocation.

The mechanism of contraction involved in cell body translocation could be of the network type that we have discussed, or of a sarcomeric type. Our results permit us to exclude, at least for keratocytes, a sarcomeric model based on contraction of alternately polarized actin arrays. No alternately polarized actin arrays have been found in keratocytes. Even if such arrays exist in the transverse actin bundles, these bundles are oriented perpendicular to the direction of cell locomotion and, therefore, cannot provide the driving force. Moreover, since these bundles are arc-shaped with their convex side forward and presumably attached most strongly at their trailing lateral edges (as suggested by the wrinkling of elastic substrata [47]), their contraction would create a force component in a direction opposite to the direction of locomotion. It is necessary to note that the contraction of actin–myosin bundles could contribute to cell body translocation if the bundles are oriented along the direction of locomotion. In fibroblasts, such contraction seems to be involved in the detachment of the tail [48]. However, other evidence shows an inverse relationship between the density of stress fibres and the speed of fibroblast locomotion [49]. This inverse relationship may be a consequence of an increased density of focal adhesions which invariably are found at the termini of stress fibres. A more detailed study of myosin dynamics during the locomotion of fibroblasts seems to be required in order to evaluate the relative contributions of network and bundle contraction in cell motility.

Our findings also seem to be inconsistent with another recent model of keratocyte translocation, a model of rotation driven by actin–myosin axles flanking the cell body [6]. This model was apparently motivated by the finding that the cell body rotates concomitantly with forward motion. The authors reasoned that, because of the rotation, connection between the cell body and the lamellipodium must be transient and, consequently, cannot provide the driving force. They proposed that the actin–myosin bundles flanking the cell body serve as axles, providing a permanent connection to the lamellipodium and driving the rotation. We confirmed the existence of rotation, although this did not occur at a speed sufficient to account for the observed translocation. Consequently we arrived at a different interpretation for the significance of rotation and the flanking bundles. Although flanking bundles sometimes splay forward into the lamellipodium, the forward splaying is slight, and the same cells show as much or more backward splaying as well. In general, in our observations, actin–myosin bundles in keratocytes were arc-shaped with the convex side forward and thus, as discussed earlier, poorly positioned to provide forward force. We propose that the forward force is produced by contraction of the actin–myosin network along the entire lamellipodial/cell body transition zone, and not just at the flanks. Actin–myosin bundles in our model are considered to be the result of network contraction, and not the driving force. We also do not find that the cell body needs or could have permanent attachment to the lamellipodium. The attachment is necessarily transient (irrespective of rotation) due to the fact that the lamellipodial cytoskeleton is stationary relative to the substratum, but the cell body moves forward and thus continuously encounters new regions of the lamellipodium. Contracting elements of the lamellipodial network must continuously replace each other and together produce a permanent pulling action on the cell body. Rotation of the body may be explained by assuming that the top surface of the body experiences less resistance than the substrate-facing bottom surface and thus moves faster. Cell body rotation might be a specific feature of keratocyte motility, but our model of network contraction is applicable to other cells that do not rotate during locomotion.

Retrograde flow

Retrograde flow is a necessary consequence of contractile force between the lamellipodium and the cell body (although contributions from other mechanisms cannot be excluded). In locomoting keratocytes, the lamellipodium is attached to the substratum more strongly than is the cell body, thus contraction of the network in the transition zone moves the cell body forward relative to the stationary lamellipodium. In this case, there is no retrograde flow of cytoskeletal components with respect to the substratum, although there is relative motion of the lamellipodium with respect to the cell body. In keratocytes tethered at the edge of an epithelial island, cell body translocation is constrained. Consequently, contraction in the transition zone drives a retrograde flow of myosin features in the lamellipodium relative to the substratum. In keratocytes, the retrograde flow with respect to the substratum in tethered cells is slower than the motion with respect to the cell body in locomoting cells. This difference could be explained by substrate resistance experienced by lamellipodia. In fibroblasts, contraction of network and/or filament bundles could also participate in retrograde flow. In particular, in radially spreading fibroblasts retrograde flow could be at least partially accounted for by contraction of the network occurring either in the central region of the cell or uniformly along most of the cell area. In either case, flow would be expected to be directed towards the centre.

In both locomoting and stationary cells, retrograde flow delivers newly assembled contractile elements from the cell periphery to their site of action. Contraction of these elements, in turn, drives the flow which brings new contractile elements to fuel contraction. Thus contraction becomes a self-supporting process and flow provides the logical link between the assembly and function of the contractile machinery.

Conclusion

We have described a simple mechanism for how contraction of a system composed of myosin II filament clusters and an actin network could provide the force for cell body translocation and at the same time lead to the formation of actin–myosin bundles and retrograde flow. By describing how the actin–myosin network is assembled at the cell periphery and then delivered to the site of action by retrograde flow, our model provides a framework for understanding the function of the motility machinery in the context of its assembly and structure. The mechanism described is intentionally a minimal one and many additional elements, both structural and regulatory, are likely to contribute to the overall process. For example, what is the mechanism of association of myosin filaments into clusters? How are the assembly and disassembly of actin and myosin at the cell periphery and in the cell body regulated? How are local contraction events integrated spatially in large and complex cells such as fibroblasts? These are some avenues of future research.

This work was supported by grants ACS CB-95 and NIH GM 25062.

References

1. Condeelis, J. (1993) Annu. Rev. Cell Biol. **9**, 411–444
2. Mitchison, T.J. and Cramer, L.P. (1996) Cell **84**, 371–379
3. Small, J.V., Rohlfs, A. and Herzog, M. (1993) in Cell Behaviour: Adhesion and Motility (Jones, G., Wigley, C. and Warn, R., eds.), pp. 57–71, The Company of Biologists, Cambridge
4. Mogilner, A. and Oster, G. (1996) Biophys. J. **71**, 3030–3045
5. Roberts, T.M. and King, K.L. (1991) Cell Motil. Cytoskeleton **20**, 228–241
6. Anderson, K.I., Wang, Y.-L. and Small, J.V. (1996) J. Cell Biol. **134**, 1209–1218
7. Bray, D. and White, J.G. (1988) Science **239**, 883–888
8. Cramer, L.P. (1997) Front. Biosci. **2**, 260–270
9. Forscher, P. and Smith, S.J. (1988) J. Cell Biol. **107**, 1505–1516
10. Lin, C.-H. and Forscher, P. (1995) Neuron **14**, 763–771
11. Lin, C.-H., Espreafico, E.M., Mooseker, M.S. and Forscher, P. (1996) Neuron **16**, 769–782
12. Cheney, R.E., Riley, M.A. and Mooseker, M.S. (1993) Cell Motil. Cytoskeleton **24**, 215–223
13. Goodson, H.V. (1994) Soc. Gen. Physiol. Ser. **49**, 141–157
14. Wessels, D., Soll, D.R., Knecht, D., Loomis, W.F., De Lozanne, A. and Spudich, J. (1988) Dev. Biol. **128**, 164–177
15. Doolittle, K.W., Reddy, I. and McNally, J.G. (1995) Dev. Biol. **167**, 118–129
16. Jay, P.Y., Pham, P.A., Wong, S.A. and Elson, E.L. (1995) J. Cell Sci. **108**, 387–393
17. Maciver, S.K. (1996) BioEssays **18**, 179–182
18. Huxley, H.E. (1973) Nature (London) **243**, 445–449
19. Sanger, J.M. and Sanger, J.W. (1980) J. Cell Biol. **86**, 568–575
20. Pollard, T.P. (1976) J. Cell Biol. **68**, 579–601
21. Stossel, T.P. and Hartwig, J.H. (1976) J. Cell Biol. **68**, 602–619
22. Janson, L.W., Kolega, J. and Taylor, D.L. (1991) J. Cell Biol. **114**, 1005–1015
23. Takiguchi, K. (1991) J. Biochem. (Tokyo) **109**, 520–527
24. White, J.G. and Borisy, G.G. (1983) J. Theor. Biol. **101**, 289–316
25. Heath, J.P. and Holifield, B.F. (1993) in Cell Behaviour: Adhesion and Motility (Jones, G., Wigley, C. and Warn, R., eds.) pp. 35–56, The Company of Biologists, Cambridge

26. Lee, J., Ishihara, A. and Jacobson, K. (1993) in Cell Behaviour: Adhesion and Motility (Jones, G., Wigley, C. and Warn, R., eds.), pp. 73–89, The Company of Biologists, Cambridge

27. Byers, H.R., White, G.E. and Fujiwara, K. (1984) Cell Muscle Motil. **5**, 83–137

28. Zigmond, S.H., Otto, J.J. and Bryan, J. (1979) Exp. Cell Res. **119**, 205–219

29. Verkhovsky, A.B., Svitkina, T.M. and Borisy, G.G. (1995) J. Cell Biol., **131**, 989–1002

30. McKenna, N.M., Wang, Y.-L. and Konkel, M.E. (1989) J. Cell Biol. **109**, 1163–1172

31. Svitkina, T.M., Verkhovsky, A.B., McQuade, K.M. and Borisy, G.G. (1997) J. Cell Biol. **139**, 397–415

32. Burridge, K. and Bray, D. (1975) J. Mol. Biol. **99**, 1–14

33. Karlsson, R. and Lindberg, U. (1985) Exp. Cell Res. **157**, 95–115

34. Langanger, G., Moeremans, M., Daneels, G., Sobieszek, A., De Brabander, M. and De Mey, J. (1986) J. Cell Biol. **102**, 200–209

35. Svitkina, T.M., Surgucheva, I.G., Verkhovsky, A.B., Gelfand, V.I., Moeremans, M. and DeMay, J. (1989) Cell Motil. Cytoskeleton **12**, 150–156

36. Verkhovsky, A.B. and Borisy, G.G. (1993) J. Cell Biol. **123**, 637–652

37. Giuliano, K.A. and Taylor, D.L. (1990) Cell Motil. Cytoskeleton **16**, 14–21

38. Verkhovsky, A.B., Svitkina,T.M. and Borisy, G.G. (1997) J. Cell Sci. **110**, 1693–1704

39. Theriot, J.A. and Mitchison, T.J. (1991) Nature (London) **352**, 126–131

40. Taylor, D.L. and Fechheimer, M. (1982) Philos. Trans. R. Soc. London B: Biol. Sci. **299**, 185–197

41. Kolega, J., Janson, L.W. and Taylor, D.L. (1991) J. Cell Biol. **114**, 993–1003

42. Fleischer, M. and Wohlfarth-Bottermann, K.E. (1975) Cytobiologie **10**, 339–365

43. Ingber, D.E. (1993) J. Cell Sci. **104**, 613–627

44. Mogilner, A. and Oster, G. (1996) Eur. J. Biophys. **25**, 47–53

45. Heath, J. (1983) J. Cell Sci. **60**, 331–354

46. Cramer, L.P., Siebert, M. and Mitchison, T.J. (1997) J. Cell Biol. **136**, 1287–1305

47. Lee, J., Leonard, M., Oliver, T., Ishihara, A. and Jacobson, K. (1994) J. Cell Biol. **127**, 1957–1964

48. Chen, W.T. (1981) J. Cell Biol. **90**, 187–200

49. Herman, I.M., Crisona, N.J. and Pollard, T.D. (1981) J. Cell Biol. **90**, 84–91

Biochem. Soc. Symp. **65**, 223–231
Printed in Great Britain

13

Centrosomes, microtubules and cell migration

Manfred Schliwa[1], Ursula Euteneuer, Ralph Gräf and Masahiro Ueda

Adolf Butenandt Institut, Zellbiologie, Universität München, Schillerstrasse 42, 80336 München, Germany

Abstract

Directed cell movement is an immensely complex process that depends on the co-operative interaction of numerous cellular components. Work over the past three decades has suggested that microtubules play an important role in the establishment and maintenance of the direction of cell migration. This chapter summarizes recent work from our laboratory designed to determine the roles of the microtubules and centrosome position relative to the direction of cell migration in a variety of cell types, and discusses these observations in the context of work from other laboratories. The results suggest that microtubules are required for stabilization of the direction of migration in many, but not all, cell types. For the centrosome to act as a stabilizer of cell migration requires that it is repositioned behind the leading edge. However, the process of repositioning does not precede the extension of a leading edge and the establishment of a new direction of cell migration. Rather, the centrosome follows the repositioning of the leading edge in response to other stimuli and, in doing so, stabilizes cell movement.

Introduction

Cell locomotion is the most easily visible and yet one of the most complex processes exhibited by a living cell: complex because numerous organelles and macromolecular components are implicated and the entire cell is involved, and obvious because a brief look into a microscope discloses the phenomenon to even a casual observer and, at the same time, gives an impression of the complexity of the process. It is still difficult to state how many different types of molecule are required for the various aspects of cell locomotion, although cer-

[1]To whom correspondence should be addressed.

tainly a few major players have been identified. Actin is probably the single most important protein, along with a number of actin-binding proteins and factors that regulate their activity. Next, plasma membrane components are without doubt required, if only to provide new membrane for extension of a pseudopod or to receive chemical signals for direction from the environment. And then there is the system of microtubules and its organizing centre, which in most cell types is the centrosome, an as yet poorly understood organelle that determines the nucleation, number, protofilament composition and initial direction of polymerization of a cell's microtubules. It is the possible role of microtubules and the centrosome in cell locomotion that are dealt with in this brief review, which summarizes work from our laboratory on this subject.

An actively migrating cell has a distinct polarity that is based on the fact that protrusive activity is restricted to one cell pole, the lamellipod or pseudopod, while retraction takes place at the opposite end. An involvement of microtubules in the directed movement of cells was first suggested by Vasiliev et al. [1]. Polarity is lost and cell movement is impaired when fibroblasts in culture are treated with a microtubule-depolymerizing agent such as colchicine [2–4]. These findings provided strong evidence for the notion that microtubules influence, or even govern, overall cell shape and cell polarity. Moreover, in several cell types an intriguing correlation exists between the position of the microtubule-organizing centre, the centrosome, and the direction of extension of a lamellipod, since the centrosome is placed behind the leading edge and in front of the nucleus ([5,6]; for a review, see [7]). These observations have led to the suggestion that the position of the centrosome determines the direction of movement of a cell and serves as a steering device whose repositioning presages changes in the direction of migration.

Centrosome repositioning

To study the role of microtubules and centrosome position in the establishment and maintenance of cell polarity and cell movement, wounded monolayers of cells in culture have long served as a good experimental system. In the *in vitro* wound assay, a strip of cells several hundred μm wide is removed by scratching a monolayer of confluent cells in a culture dish or on a coverslip. Cells at the wound edge will reposition their centrosome (as well as the Golgi apparatus) towards the open wound within about 1 h [6–9]. However, even though the phenomenon of centrosome repositioning in cells at the wound edge is well documented, neither the mechanism of repositioning nor, indeed, the cause and effect of the phenomenon are well understood.

Our studies of the mechanism of repositioning of the centrosome in cells at the wound edge have made use of African green monkey kidney (BSC-1) cells. Wounding of the monolayer causes the cells at the wound edge to extend a lamellipod towards the open substrate (Fig. 1) and to re-orient the centrosome towards the lamellipod and ahead of the nucleus. The process is gradual, starting after about 15 min. More than 50% of the centrosomes are still located at the back or on the sides of the nucleus after 30 min, whereas after 3 h more than 80% are located clearly in front of the nucleus [10]. This process is accom-

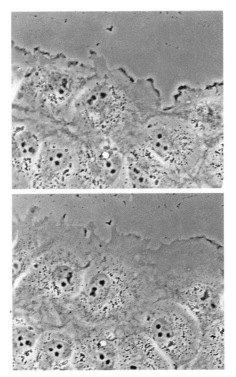

Fig. 1. Behaviour of BSC-1 cells at the edge of an *in vitro* wound.
Upper panel, 20 min after wounding; lower panel, same field 1.5 h after wounding. Note the extension of broad lamellae into the wound.

panied by a decrease in the number of detyrosinated (presumably more stable) microtubules extending in the direction of the lamellipod near the wound edge (Fig. 2), which is opposite to what was observed in 3T3 fibroblasts [9]. Does this imply that repositioning of the centrosome has caused an extension of the lamellipod in the direction of the wound? To test this, cells were treated with nocodazole, a microtubule-depolymerizing compound, prior to wounding. Surprisingly, extension of a lamellipod was not at all inhibited, even though centrosome repositioning was effectively blocked: only about 30% of the cells had their centrosome in front of the nucleus 2 h after wounding. However, the centrosome will reposition effectively and rapidly if the drug is removed at this time, even though the wound has already existed for 2 h and a lamellipod has formed. Thus repositioning still occurs after completion of the initial wound response, and the centrosome still ends up in a position ahead of the nucleus *after* the lamellipod has been fully extended. So what does that tell us about the cause and effect of repositioning? Why does the centrosome still come to lie ahead of the nucleus even though cell asymmetry has already been established?

The answers to these questions lie in a closer consideration of cell geometry after wounding, and the explanation turns out to be quite trivial. BSC-1 cells, either when solitary or when embedded in a confluent monolayer, are polygonal, whereas at a wound edge they are rectangular due to the extension

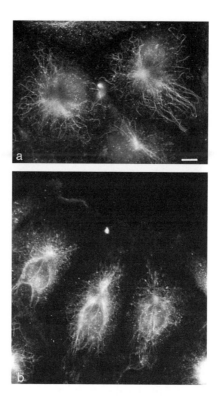

Fig. 2. Arrangement of detyrosinated microtubules in BSC-1 cells.
(a) Within a confluent monolayer, detyrosinated microtubules extend from the
cell centre in a radial arrangement. (b) At the wound edge (top), the microtubules are virtually absent from the side facing the wound. Bar = 10 μm.

of a broad lamellipod at one side of the cell. The nucleus is near the cell centre
in polygonal cells, whereas in cells at the wound edge it ends up in the cell posterior due to the asymmetrical extension of a lamellipod. Analysis of both
intact and enucleated cells [10] shows that the majority of centrosomes in cells
both in a confluent monolayer and at the edge of a wound are in a centroid
position, or not very far from it. So the rather trivial explanation for the re-
orientation process in cells at the edge of the wound is that the centrosome
attempts to maintain a position near the cell's centroid, which, in the asymmetrical cells at the wound edge, puts it ahead of the nucleus due to the extension
of a broad lamellipod. The centrosome does not 'move ahead of the nucleus' or
're-orient in the new direction of movement'; it simply tries to stay in the cell
centre. 'Self-centring' of the microtubule system is probably a fitting expression for the phenomenon. Indeed, the process depends on the presence of intact
microtubules extending to the cell cortex [10], suggesting that microtubule
dynamic instability is the driving force for the process of centring of the microtubule aster. There are many examples of centrosome centring in cells due to
length changes of the microtubules extending from the centrosome [11–14]. An
elegant experimental verification of the self-centring capacity of a microtubule

aster comes from the work of Holy et al. [15], who fabricated microscopic chambers that mimic the geometry of cells. Amazingly, centrosome-based microtubule asters move towards the central position solely on the basis of microtubule dynamic instability. While this is a simple experimental system, it nevertheless offers a possible explanation of the processes observed to take place in intact cells.

One might argue, and justifiably so, that the behaviour of cells at the edge of an *in vitro* wound is not representative of the behaviour of a freely moving cell. How important might repositioning of the centrosome be for the initiation of directional changes during long-range movements of single cells on a featureless substrate? If, in fact, the centrosome determines the direction of movement, it should re-orient in the intended direction of migration prior to the extension of a lamellipod in that direction. A test of this hypothesis requires a reliable assay for analysis of the precise temporal and spatial relationship between centrosome position and pseudopodial activity, which is difficult to accomplish using cells fixed for immunofluorescence or electron microscopy. We have developed a real-time assay that takes advantage of the green fluorescent protein (GFP) as a reliable, stable and non-toxic marker of cellular constituents [16]. A fusion protein of GFP and the universal centrosomal constituent γ-tubulin, a protein involved in microtubule nucleation [17–19], was generated in *Dictyostelium discoideum*. In all cell types examined so far, including *Dictyostelium*, γ-tubulin is localized at the centrosome [20,21]. In *Dictyostelium* amoebae transfected with the fusion construct (Fig. 3), the centrosome is brightly labelled, allowing continuous observation of living, locomoting cells for at least 10 min without deleterious effects. During this time period, single amoebae may move randomly on a glass surface for many cell diameters, making numerous turns in the course of their movement. Using appropriate procedures for observation in the light microscope, we followed both the position of the centrosome and pseudopodial activities at video rates [22]. We found that the centrosome never repositions prior to a change in the direction of migration, i.e. the extension of a new pseudopod in a new direction of movement. This is true both under conditions of non-stimulated random migration and during chemotaxis towards cAMP. Thus the centrosome does not direct a cell's direction of migration by re-orienting prior to the initiation of a directional change. Centrosomes do re-orient in the new direction, but always after a lag period.

A second important finding has emerged from this analysis. Cells often extend two or three pseudopods in different directions in rapid succession. We have never found that a cell would move in a direction (i.e. continue to extend a pseudopod) that was not eventually followed by the centrosome (i.e. the microtubule system). Those pseudopods into which the centrosome does not re-orient are retracted within about 30 s. These observations are consistent with an important role for the microtubule aster extending from the centrosome in the stabilization of an extending pseudopod, a finding also confirmed for cell types other than *Dictyostelium*. For example, whereas the initiation of cell polarization at the wound edge is independent of microtubules, as described here, subsequent migration into the wound depends, at least in part,

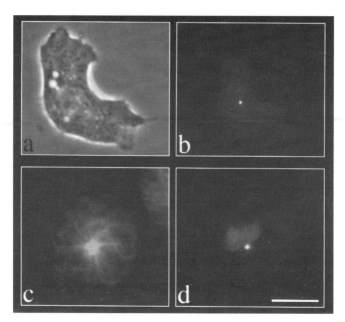

Fig. 3. Distribution of GFP–γ-tubulin in live (a and b) and fixed (c and d) *Dictyostelium* **amoebae.** (a) Phase-contrast micrograph of living amoeba. (b) Same cell viewed by fluorescence microscopy, showing the location of the centrosome. (c) Distribution of microtubules as seen by immunofluorescence microscopy with an antibody against tubulin. (d) Same cell viewed in the fluorescein channel, demonstrating that GFP–γ-tubulin is localized in the centre of the microtubule aster. The nucleus is weakly stained with 4′,6-diamidino-2-phenylindole (DAPI). Bar = 10 μm.

on a system of intact microtubules [23]. Fibroblasts likewise require microtubules for persistent cell movements [2,24,25]. On the other hand, directional movement of certain other cell types, such as keratocytes, leucocytes and macrophages, does not require microtubules [25–28], although they are essential for stabilizing cell polarization and enhancing the persistence of locomotion [29–32]. Thus the microtubule apparatus reinforces persistence of locomotion and contributes to the stabilization of directional movement, but it does not determine the direction in which a cell chooses to move. These two aspects should be clearly distinguished when considering the role of microtubules or the centrosome in directional movements of the cell.

Effect of substrate

Cell locomotion not only requires numerous intracellular components; it is also a process that depends on the fine-tuned interaction between the cell and its environment. Surprisingly, different substrates have a strong influence not only on the shape of locomoting cells, but also on the deployment of microtubules and the position of the centrosome. Whereas fibroblasts migrating

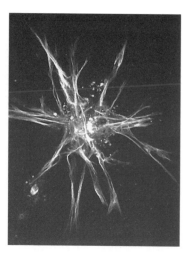

**Fig. 4. Chick embryo fibroblasts migrating away from a cell aggre-
gate in a three-dimensional collagen gel.** Confocal microscopy image of
cells stained with an antibody against tubulin. A total of 14 optical sections
spaced 1 μm apart are stacked in this composite.

from a cell aggregate on a flat glass surface are generally well spread, with a
broad anterior lamellipod, the same cells on a two-dimensional collagen surface, and even more so within a three-dimensional collagen gel (Fig. 4), adopt a

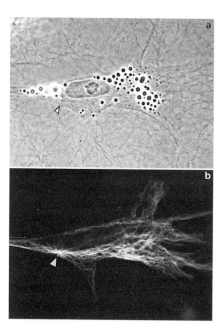

**Fig. 5. Chick embryo fibroblast moving on the surface of a two-
dimensional collagen gel.** Note the extreme posterior position of the
centrosome (arrowhead).

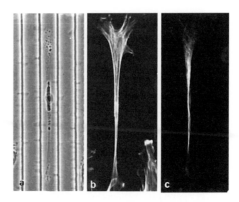

Fig. 6. Chick embryo fibroblast moving on the surface of a grooved glass slide. (a) Phase-contrast image; (b) rhodamine–phalloidin staining for actin; (c) tubulin immunofluorescence. The distance between the grooves is 10 μm.

highly elongated, spindle-shaped morphology [33]. As in epithelial cells migrating into an *in vitro* wound, the centrosome of fibroblasts migrating from a cell aggregate on a glass surface have their centrosome ahead of the nucleus (but still near the cell's centroid!). Surprisingly and unexpectedly, the centrosome in fibroblasts migrating directionally and persistently on either a two- or a three-dimensional collagen substrate is located highly asymmetrically in the cell posterior (Fig. 5). Not even a preferred position relative to the nucleus, which also is found in an extremely posterior location, could be confirmed. These observations caution against generalizations and suggest that, in an environment that mimics the *in vivo* situation more closely than a flat glass surface, centrosome position plays a less prominent role than on a two-dimensional substrate. Patterning the glass substrate by photolithographic etching of parallel grooves (Fig. 6) removes the correlation between a predominantly anterior position of the centrosome and the direction of cell movement [33].

Conclusions

Clearly, centrosomes and microtubules are important components in the complex structural and regulatory network that generates and controls cell locomotion. Centrosome repositioning accompanies directional changes during cell movement in most cell types, but this correlation does not imply a causal relationship in the sense that centrosome repositioning governs cell movement. Clearly, however, the microtubule system is required to support motile activities of the cell.

The authors' work is supported by the Deutsche Forschungsgemeinschaft (SFB 184), a grant from the European Union, a JSPS postdoctoral fellowship and a NATO travel grant.

References

1. Vasiliev, J.M., Gelfand, I.M., Domnina, L.V., Ivanova, O.Y., Komm, S.G. and Olshevkaya, L.V. (1970) J. Embryol. Exp. Morphol. **24**, 625–640
2. Goldman, R.D. (1971) J. Cell Biol. **51**, 752–762
3. Vasiliev, J.M. and Gelfand, I.M. (1976) Cold Spring Harbor Conf. Cell Prolif. **3**, 279–304
4. Gotlieb, A.I., Subrahmanyan, L. and Kalnins, V.I. (1983) J. Cell Biol. **96**, 1266–1272
5. Albrecht-Buehler, G. and Bushnell, A. (1979) Exp. Cell Res. **120**, 111–118
6. Gotlieb, A.I., May, L.M., Subrahmanyan, L. and Kalnins, V.I. (1981) J. Cell Biol. **91**, 589–594
7. Singer, S.J. and Kupfer, A. (1986) Annu. Rev. Cell Biol. **2**, 337–365
8. Bergmann, J.E., Kupfer, A. and Singer, S.J. (1983) Proc. Natl. Acad. Sci. U.S.A. **80**, 1367–1371
9. Gundersen, G.G. and Bulinski, J.C. (1988) Proc. Natl. Acad. Sci. U.S.A. **85**, 5946–5950
10. Euteneuer, U. and Schliwa, M. (1992) J. Cell Biol. **116**, 1157–1166
11. Rappaport, R. (1981) J. Exp. Zool. **217**, 365–375
12. McNiven, M.A. and Porter, K.R. (1988) J. Cell Biol. **106**, 1593–1605
13. Hamaguchi, M.S. and Hiramoto, Y. (1986) Dev. Growth Differ. **28**, 143–156
14. Euteneuer, U. and Schliwa, M. (1985) J. Cell Biol. **101**, 96–103
15. Holy, T.E., Dogterom, M., Yurke, B. and Leibler, S. (1997) Proc. Natl. Acad. Sci. U.S.A. **94**, 6228–6231
16. Chalfie, M., Tu, Y., Euskirchen, G., Ward, W.W. and Prasher, D. (1994) Science **263**, 802–805
17. Oakley, C.E. and Oakley, B.R. (1989) Nature (London) **338**, 662–664
18. Joshi, H.C., Palacios, M.J., McNamara, L. and Cleveland, D.W. (1992) Nature (London) **356**, 80–83
19. Marshall, L.G., Jeng, R.L., Mulholland, J. and Stearns, T. (1996) J. Cell Biol. **134**, 443–454
20. Kalt, A. and Schliwa, M. (1993) Trends Cell Biol. **3**, 118–128
21. Kellog, D.R., Moritz, M. and Alberts, B.M. (1994) Annu. Rev. Biochem. **63**, 639–674
22. Ueda, M. Euteneuer, U., Gräf, R., MacWilliams, H.K. and Schliwa, M. (1997) Proc. Natl. Acad. Sci. U.S.A. **94**, 9674–9678
23. Liao, G., Nagasaki, T. and Gundersen, G.G. (1995) J. Cell Sci. **108**, 3473–3483
24. Gail, M.H. and Boone, C.W. (1971) Exp. Cell Res. **65**, 221–227
25. Rodionow, V.I., Gyoeva, T.K., Tanaka, E., Bershadsky, A.D., Vasiliev, J.M. and Gelfand, V.I. (1993) J. Cell Biol. **123**, 1811–1829
26. Euteneuer, U. and Schliwa, M. (1984) Nature (London) **310**, 58–61
27. Keller, H.U., Naef, A. and Zimmermann, A. (1984) Exp. Cell Res. **153**, 173–184
28. Zigmond, S.H. and Sullivan, S.J. (1979) J. Cell Biol. **82**, 517–527
29. Rich, A.M. and Hoffstein, S.T. (1981) J. Cell Sci. **48**, 181–191
30. Devreotes, P.N. and Zigmond, S.H. (1988) Annu. Rev. Cell Biol. **4**, 649–686
31. Glasgow, J.E. and Daniele, R.P. (1994) Cell. Motil. Cytoskeleton **27**, 88–96
32. Ueda, M. and Ogihara, S. (1994) J. Cell Sci. **107**, 2071–2079
33. Schütze, K., Maniotis, A. and Schliwa, M. (1991) Proc. Natl. Acad. Sci. U.S.A. **88**, 8367–8371

Biochem. Soc. Symp. **65**, 233–243
Printed in Great Britain

14

Cell migration as a five-step cycle

Michael P. Sheetz[1], Dan Felsenfeld, Catherine G. Galbraith and Daniel Choquet[2]

Department of Cell Biology, Duke University Medical Center, Durham, NC 27710, U.S.A.

Abstract

The migration of cells over substrata is a fundamental and critical function that requires the co-ordination of several cellular processes which operate in a cycle. At the level of the light microscope, the cycle can be divided into five steps: (1) extension of the leading edge; (2) adhesion to matrix contacts; (3) contraction of the cytoplasm; (4) release from contact sites; and (5) recycling of membrane receptors from the rear to the front of the cell. Each step is dependent upon one or more cyclical biochemical processes. The development of many *in vitro* and subcellular assays for the fundamental biochemical processes involved has increased our understanding of each cycle dramatically in the last several years to include a definition of many of the protein and enzymic components, the role of the position of extracellular-matrix receptors on the cell, and the contribution of physical force. The next generation of questions are directed at resolving the roles of the many individual proteins in each step of the cell migration process. In this chapter we will examine each of the migration steps and discuss the biochemical mechanisms that may underlie them.

Introduction

A number of molecular events need to be co-ordinated to allow a cell to move across a substrate (for recent reviews, see [1–7]). We have chosen to break the process of migration into five discrete steps (Fig. 1). Although the cycle does not necessarily have a beginning or an end point, the cell must cycle through each process in order for significant displacement to occur; e.g. release at the rear enables extension at the front, recycling enables attachment, etc.

---

[1]To whom correspondence should be addressed.
[2]Present address: UMR CNRS 5541, Universite de Bordeaux 2, 146 rue Leo Saignat, 33076 Bordeaux Cedex, France.

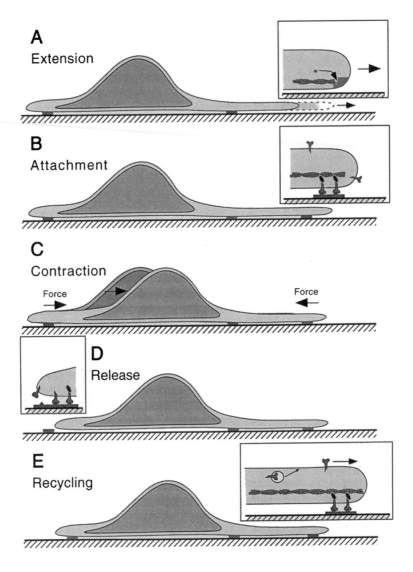

Fig. 1. The five steps of cell migration. Each step is illustrated, with insets showing the favoured subcellular mechanism involved in each case. Detailed descriptions are given in the text. (A) Extension: a membrane-localized protein complex catalyses the assembly of actin filaments, driving the protrusion of the membrane [4]. (B) Attachment: integrins (Y-shaped molecules in the membrane), which are normally free of cytoskeletal attachments, attach to retrograde-moving actin filaments after binding to ligand (dark bar) on the substrate. (C) Contraction: the cell contracts from both ends in towards the nucleus. When accompanied by release of adhesion receptors from the cytoskeleton and substrate at the back of the cell (D), cell displacement in a forward direction occurs. (E) Recycling of adhesion receptors to the front of the cell is accomplished by endocytosis and vesicular transport and/or forward-directed movement on the cell surface.

After extending into new regions, the cell attaches and pulls rearward on an adhesion site, pulling the cell body forward. This rearward pulling was reported by Abercrombie nearly 30 years ago as the rearward transport of carbon particles attached to the dorsal surface of motile cells [8]. Once the nucleus nears the attachment site, which remains stationary on the substrate, the cell tries to release from the site to enable components to be recycled to the front of the cell. This has many similarities to the process of rock climbing, where the climber reaches out to a new handhold, which he then uses to pull himself forward. In the case of the cell in tissue culture, however, the predominant resistance to forward movement comes from the attachment sites at the rear of the cell. Many different technologies make it possible to examine the molecular basis of the steps in the migration process.

Cell migration: a stochastic cycle of movement and re-orientation

The trajectories of directed and randomly migrating cells have been analysed extensively in order to understand fundamental aspects of the basic motile process. Over long periods of time, directed migration (as in chemotaxis) can be described as a linear path, and undirected migration can be described as a random walk [9]. However, over short periods of time, both cases show a periodic behaviour that involves cycling between movement and re-orientation. We have used a mathematical analysis of the movements of the cell centroid to measure the cycle time and define specific features of the motile process [10]. In directed movement, there is a partial re-orientation within each cycle that allows continual rechecking of the correct direction, prevents long periods of misdirected movement and allows for temporal reinforcement of directionality. Although many aspects of the motile process are explained by simple stochastic treatments such as described previously [9,10], there are several phenomena in chemotaxis and macrophage migration that require a non-linear control theory approach in order to reasonably approximate the biological system (A.D. Shenderov and M.P. Sheetz, unpublished work).

Extension of the leading edge

Multiple areas of extension are observed as suspension cells attach to a surface. Directed movement and polarization occur when one area of extension predominates. The critical element of the extension process is directed actin assembly [11]. The process of actin assembly must generate a protrusive force sufficient to extend the plasma membrane against the compressive forces occasionally imposed by the environment and always exerted by tension within the plasma membrane [12]. The mechanism responsible for assembly is still under debate, but several possible models have been proposed: a polymerization ratchet [13], a motor-dependent assembly, or a novel assembly protein complex that would add subunits sequentially [14,15]. The first model does not explain the localization of actin filament assembly at the leading edge of the cell. The second model is in disfavour because the general myosin ATPase inhibitor, butanedione monoxime, does not initially block filipodia or leading-edge extension [16] or the actin-polymerization-driven movement of *Listeria monocytogenes* [11], but it blocks the rearward flow of material in the ecto-plasm of neuronal growth cones [16]. There is reason to favour the third model, in which a regionally activated multiprotein complex catalyses actin assembly and force generation. Recent evidence suggests that the Arp2/3 complex has such a role in *L. monocytogenes* [17]. Assembly could involve significant addition of subunits to the barbed end of the filament, which would propel a membrane-anchored Arp2/3 complex forward.

Adhesion to the matrix

Extension of the leading edge to new extracellular-matrix (ECM) molecules will enable receptors to bind and to initiate the adhesion process. The adhesion-receptor–ECM complexes stabilize newly extended cellular domains and permit the cell to exert forces on the substrate (Fig. 2). ECM binding to adhesion receptors can initiate binding of specific cytoskeletal proteins to the cytoplasmic tail of the receptor, in addition to a cascade of other signals. At the molecular level, integrins are the best characterized receptors for ECM molecules that stimulate cell migration. Integrins are a family of heterodimeric transmembrane receptors that link ECM outside the cell with force-generating components of the cytoskeleton [18]. Ligand binding and integrin cross–linking regulate integrin–cytoskeleton interactions. Although receptor redistribution to focal adhesions occurs with ligand binding, the accumulation of some cytoskeletal focal adhesion proteins depends on cross-linking as well as ligand binding [19]. Typically, unliganded integrins diffuse freely in the membrane and bind to the cytoskeleton upon ligand activation in a rapidly reversible complex [20,21]. When integrin receptors attach to the cytoskeleton in newly extended lamellipodial regions, they are drawn rearward [21], as is commonly observed for particles attached to lamellipodia [8]. If the substrate is rigid, local integrin adhesion sites will form on the lower surface of the cell near the leading edge, remaining stationary as the cell migrates forward (Fig. 2B) [22].

Whereas ECM binding initiates integrin interactions with force-generating components of the cytoskeleton, several lines of evidence suggest that the mechanical properties of the ECM can modulate the strength of this interaction. The application of force to the cell surface affects cytomechanics [23,24], and the use of integrin ligand-coated beads shows that cytoskeleton stiffness increases in proportion to the applied force [25,26]. Furthermore, cytoskeleton assembly and ECM protein organization on the cell surface are influenced by ECM stiffness; fibroblasts cultured on stressed substrates [27,28] develop bundles of actin filaments and fibronectin fibrils that are not observed on relaxed substrates. Although there may be a role for force on the whole-cell cytoskeleton in such activation, we have found that individual submicron contacts respond to force in a tyrosine-phosphatase-dependent manner, without altering the strength of adjacent contacts [29]. Thus the cell is able to sense the rigidity of individual integrin–ECM contacts.

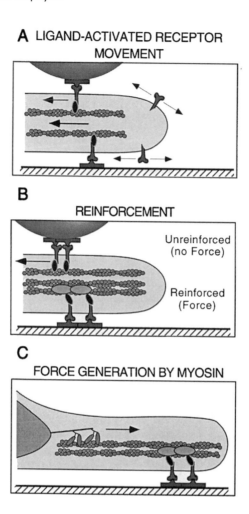

A LIGAND-ACTIVATED RECEPTOR MOVEMENT

B

REINFORCEMENT

Unreinforced (no Force)

Reinforced (Force)

C

FORCE GENERATION BY MYOSIN

Fig. 2. Fibronectin contacts and force generation. The formation of two different types of fibronectin adhesive contacts is described: one with a free latex bead and the other with substrate-bound fibronectin. Unliganded integrins are free to diffuse in the plane of the membrane, even when concentrated by edge curvature [40]. Once an $\alpha_5\beta_1$ integrin binds to fibronectin, the integrin is linked to the actin cytoskeleton that is moving rearward [21]. Fibronectin beads on the upper surface will move rearward without generating a force, whereas fibronectin attached to the glass surface will remain stationary and significant forces will be generated on them that lead to strengthened (reinforced) contacts [29].

Force-dependent strengthening of cytoskeletal linkages could be a critical signal in many cell functions (Fig. 2B). Cells may use force or rigidity of the surrounding matrix to orient or start signalling cascades. Pathfinding by growing neurons and other cells relies on the recognition of specific ECM or adherence molecules in the environment [30], and modification of the mechanical properties of adhesion molecules on a neighbouring cell or the ECM could orient the path of movement. In support of this argument, neutrophils moving in three-dimensional matrices probe the environment and choose to move along the most rigid fibrils [31]. The rigidity of the ECM or adhesion molecules and forces applied to them, therefore, provide mechanisms for signalling and orienting cell migration that have not been explored.

Contraction of the cytoplasm (cortical or radial?)

The force required to pull the body of the cell forward is generated within the cell by the cytoskeleton, and is expressed outside the cell as a traction force on the substratum. Significant traction forces are needed for wound healing and tissue penetration; fibroblasts can generate 1–5 nN/μm^2 (the equivalent of the force of 1000 myosin molecules) on substrata [32]. These forces are much larger than the forces needed to move the cell against fluid drag (for a cell moving at 40 μm/min, the drag in medium alone, the force required would be less than that generated by a single myosin head). In leading lamellipodia, traction forces are oriented rearward; this could result from attachment of ECM-bound integrins to cytoskeletal material that is normally moving centripetally rearward. Since steady rearward movement of the filaments that assemble at the leading edge of the cell is seen both in actin-based cell motility and in *Ascaris major* sperm-protein-based motility, which is actin independent [33], this implies that coupling to a rearward filament flow is a general mechanism for powering cell migration.

Large traction forces such as those generated by motile fibroblasts or keratocytes appear to depend upon myosin II (Fig. 2C), although cell migration *per se* can occur in some amoeboid cells without myosin II. Forces are likely to be generated via one of three possible mechanisms: (1) cortical tension [34], (2) muscle-like contraction, or (3) central contraction of ventral actin. The fact that surface beads only move rearward to the endoplasm–ectoplasm boundary in fibroblasts indicates that the cortex is not contracting from the front to the back of the cell as would be required in the cortical tension model, whereas in amoeboid cells and macrophages beads and cross-linking antibodies are capped at the rear of the cell. With regard to a myosin-like contraction mechanism, recent evidence [35] demonstrates that the majority of ventral actin filaments do not have a muscle-like organization with alternating polarities along their length, as would be needed for a muscle-like contraction model. Instead, these filaments have graded polarity, with barbed ends oriented outwards at the anterior and posterior regions and mixed polarity in the cell centre [35]. There is an obvious condensation of myosin (the force-generating component) behind the leading edge of fibroblastic cells [36], and there is greater modulation of both the direction and the magnitude of traction forces in the centre of cells [32]. All of these findings are consistent with contraction of the peripheral actin filaments by centrally activated myosin, as in the central contraction model. In such a model, migration is dependent upon the asymmetry of actin assembly and/or of the linkage of contracting actin to the ECM. How the central endoplasm is coupled to the ectoplasmic actin and myosin is unclear, but attachment of myosin to the endoplasm could explain many aspects of forward migration (Fig. 2C) and the rolling of the endoplasm (which is observed in some migratory keratocytes) [37]. Such a model would explain why extended actin-rich lamellae are typically drawn radially inwards even in non-motile cells, and it focuses questions of control of migration on the regulation of extension and adhesion in cells with active lamellae.

Release from contact sites: physical and enzymic

Models describing migration in simple physical terms suggest that for forward migration to occur there must be an asymmetry in the adhesion process, such that traction forces in the front are greater than those in the rear [38]. In other words, release from attachments must occur preferentially in the rear. Although much interest has focused on focal contacts, the number and size of these structures correlates inversely with migration speed. In cases where focal contacts form in migratory fibroblasts, contacts generally increase in size (measured by anti-integrin antibody fluorescence intensity) as they approach the nucleus and disperse towards the tail of the cell (see [3] for a review). Although some contacts persist for the whole time that the cell is over them, there is evidence to suggest that the integrins in the contacts are dynamically exchanging with others in the surrounding membrane [22,39]. This dynamic behaviour would facilitate motility if there was even a weak bias for breaking contacts in the rear. For example, decreased cytoskeletal attachment of integrins in the rear would favour dissociation from the ECM [40] (Fig. 3A). Alternatively, the fact that rearward-pulling forces in the front of the cell are counterbalanced by fewer matrix contacts in the rear of the cell would result in greater force per contact [32] and cause preferential release in the rear. Very high forces are generated on the tail contacts, which explains why cells move forward abruptly after rear contacts are broken [41]. The observed decrease in integrin attachment to the cytoskeleton at the rear of the cell [40] would make it possible for the integrins to be pulled out of the cell, as has been observed [22,39]. Thus there is considerable evidence for physical release occurring at the rear of fibroblasts.

However, since more rapidly moving cells exert smaller forces on matrix contacts [42], it is unlikely that the binding is reversed by physical means in all cases. The detachment of integrins from matrix contacts in highly motile cells may be enzymic (Fig. 3B). The trigonal shape of the fibroblasts indicates a loss of adhesive contacts near the nucleus of the cell, and we have evidence for a site-specific release of ECM in the front half of fibroblasts (T. Nishizaka and M.P. Sheetz, unpublished work). Reversible changes in ligand–receptor affinity provide a logical mechanism for release [43] if the affinity was decreased in the rear. As an example, the adhesion complexes between $\alpha_v\beta_3$ integrin and vitronectin on migrating neutrophils require intracellular Ca^{2+} to break them apart, and buffering of Ca^{2+} or inhibition of the Ca^{2+}-dependent phosphatase calcineurin causes the cells to become stuck on vitronectin [44]. Thus there is evidence for both biochemical and physical release mechanisms.

(a) Mechanical release

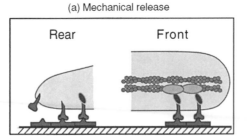

(b) Enzymic release

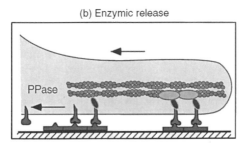

(c) Recycling

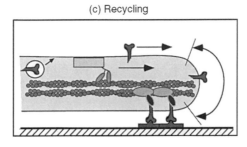

Fig. 3. Release and recycling of integrins. (a) Mechanical release. The presence of the cytoskeleton attached to the new contacts in the front of the cell spreads traction forces over all contacts whereas, at the rear, the traction forces are applied to those few integrins at the extreme rear, causing preferential release from rear contacts. (b) Enzymic release. A time-delayed enzymic modification of the integrin decreases its affinity for the substrate {the one known case involves a calcium-dependent phosphatase (PPase) [44]}. (c) Recycling of integrins to the front of the leading edge can occur by vesicle and surface transport. Because the surface transport process is rapid (1–3 μm/s), it is postulated that a myosin motor (two heads bound to a bar) associated with the membrane drives the forward movement [40,47]. Once at the curved region of the leading edge, integrins stay at the leading edge because of curvature-dependent trapping [40,47].

Receptor recycling: surface or endocytic transport?

In order to maintain a continuous retrograde flow of integrins on the cell surface, translocating cells must replenish receptor at the leading edge. Although the insertion of newly synthesized protein accounts for some of the available receptor, several lines of evidence suggest that protein must also be recycled (reviewed in [45]). Two different models have been suggested to explain the recycling of these proteins: (1) endocytosis and re-insertion into the leading edge, and (2) forward transport of the protein in the plasma membrane. These two models are not mutually exclusive: a combination of the two mechanisms could function in receptor recycling. The labelling of cell surface proteins has revealed the endocytosis of various integrin isoforms, including the fibronectin receptor ($\alpha_5\beta_1$) [22,39,44,46], suggesting that some, but not all, molecules are taken up by endocytosis. As an alternative mechanism of recycling, there is a forward surface transport process [47,48], which could account for recycling at the front of the cell (Fig. 3c). This forward movement is only seen in actively translocating cells, is actin dependent [48], and is dependent upon specific amino acids in the β-integrin cytoplasmic domain [40]. When some receptors, including integrins, reach the highly curved leading edge they diffuse along the edge for several minutes and only occasionally diffuse away from the edge (a phenomenon that we describe as curvature trapping; Fig. 3c) [40,47]. Although both endocytic vesicle and surface transport may be involved in recycling, some integrins are left behind on the substrate as a component of the membrane torn from the cell, implying that not all receptor is recycled, at least *in vitro* [22,39].

Summary

We have provided an integrated model of cell migration involving five interlocking steps. Extension of the leading edge by actin assembly relies upon activation of complexes that appear to be concentrated in curved membrane regions [49]. Adhesion through ECM–integrin interaction involves ligand binding and ligand rigidity, which may feed back on extension activity. Contraction forces are radial under the lamellipodium, and they serve to power forward movement, test matrix rigidity and stimulate release at the rear. Release at the rear appears to be a function of both decreased membrane–cytoskeleton interactions at the rear and, perhaps, modification of receptor–ligand affinity. Recycling of integrins to the leading edge can be driven by a combination of intracellular vesicle and membrane surface transport, with curvature-induced trapping to hold the integrins at the very leading edge. As we increase our understanding of these processes, the mechanisms of co-ordination of the processes in migration will be easier to define.

We thank past and present members of the Sheetz lab for helpful comments, and many members of the scientific community for discussions which were essential in forming ideas presented here. Work in this lab was supported by NIH.

References

1. Gumbiner, B.M. (1996) Cell **84**, 345–357
2. Jay, P.Y., Pasternak, C. and Elson, E.L. (1993) Blood Cells **19**, 375–386 (and discussion 386–388)
3. Lauffenburger, D.A. and Horwitz, A.F. (1996) Cell **84**, 359–369
4. Mitchison, T.J. and Cramer, L.P. (1996) Cell **84**, 371–379
5. Oliver, T., Lee, J. and Jacobson, K. (1994) Semin. Cell Biol. **5**, 139–147
6. Parsons, J.T., Schaller, M.D., Hildebrand, J., Leu, T.H., Richardson, A. and Otey, C. (1994) J. Cell Sci. Suppl. **18**, 109–113
7. Small, J.V., Anderson, K. and Rottner, K. (1996) Biosci. Rep. **16**, 351–368
8. Abercrombie, M., Heaysman, J. and Pergrum, S. (1970) Exp. Cell Res. **60**, 437–444
9. Gruler, H. (1993) Blood Cells **19**, 91–110 (and discussion 110–113)
10. Shenderov, A.D. and Sheetz, M.P. (1997) Biophys. J. **72**, 2382–2389
11. Cramer, L.P., Mitchison, T.J. and Theriot, J.A. (1994) Curr. Opin. Cell Biol. **6**, 82–86
12. Sheetz, M.P. and Dai, J. (1996) Trends Cell Biol. **6**, 85–89
13. Peskin, C., Odell, G. and Oster, G. (1993) Biophys. J. **65**, 316–324
14. Marchand, J.B., Moreau, P., Paoletti, A., Cossart, P., Carlier, M.F. and Pantaloni, D. (1995) J. Cell Biol. **130**, 331–343
15. Cossart, P. (1995) Curr. Opin. Cell Biol. **7**, 94–101
16. Lin, C., Espreafico, E., Mooseker, M. and Forscher, P. (1996) Cell **16**, 769–782
17. Welch, M., Iwamatsu, A. and Mitchison, T. (1997) Nature (London) **385**, 265–269
18. Hynes, R.O. (1992) Cell **69**, 11–25
19. Miyamoto, S., Akiyama, S.K. and Yamada, K.M. (1995) Science **267**, 883–885
20. Duband, J.L., Nuckolls, G.H., Ishihara, A., Hasegawa, T., Yamada, K.M., Thiery, J.P. and Jacobson, K. (1988) J. Cell Biol. **107**, 1385–1396
21. Felsenfeld, D.P., Choquet, D. and Sheetz, M.P. (1996) Nature (London) **383**, 438–440
22. Regen, C.M. and Horwitz, A.F. (1992) J. Cell Biol. **119**, 1347–1359
23. Zhelev, D.V. and Hochmuth, R.M. (1995) Biophys. J. **68**, 2004–2014
24. Sato, M., Levesque, M.J. and Nerem, R.M. (1987) Arteriosclerosis **7**, 276–286
25. Wang, N., Butler, J.P. and Ingber, D.E. (1993) Science **260**, 1124–1127
26. Wang, N. and Ingber, D.E. (1994) Biophys. J. **66**, 2181–2189
27. Halliday, N.L. and Tomasek, J.J. (1995) Exp. Cell Res. **217**, 109–117
28. Shirinsky, V.P., Antonov, A.S., Birukov, K.G., Sobolevsky, A.V., Romanov, Y.A., Kabaeva, N.V., Antonova, G.N. and Smirnov, V.N. (1989) J. Cell Biol. **109**, 331–339
29. Choquet, D., Felsenfeld, D.P. and Sheetz, M.P. (1997) Cell **88**, 39–48
30. Tessier-Lavigne, M. and Goodman, C.S. (1996) Science **274**, 1123–1133
31. Mandeville, J.T., Lawson, M.A. and Maxfield, F.R. (1997) J. Leukocyte Biol. **61**, 188–200
32. Galbraith, C. and Sheetz, M. (1997) Proc. Natl. Acad. Sci. U.S.A. **94**, 9114–9118
33. Italiano, Jr., J.E., Roberts, T.M., Stewart, M. and Fontana, C.A. (1996) Cell **84**, 105–114
34. Bray, D. and White, J. (1988) Science **239**, 883–888
35. Cramer, L.P., Siebert, M. and Mitchison, T.J. (1997) J. Cell Biol. **136**, 1287–1305
36. Svitkina, T.M., Verkhovsky, A.B., McQuade, K.M. and Borisy, G.G. (1997) J. Cell Biol. **139**, 397–415
37. Anderson, K.I., Wang, Y.L. and Small, J.V. (1996) J. Cell Biol. **134**, 1209–1218
38. DiMilla, P., Barbee, K. and Lauffenburger, D. (1991) Biophys. J. **60**, 15–37
39. Palecek, S.P., Schmidt, C.E., Lauffenburger, D.A. and Horwitz, A.F. (1996) J. Cell Sci. **109**, 941–952
40. Schmidt, C.E., Horwitz, A.F., Lauffenburger, D.A. and Sheetz, M.P. (1993) J. Cell Biol. **123**, 977–991
41. Chen, W.T. (1981) J. Cell Biol. **90**, 187–200
42. Harris, A., Wild, P. and Stopak, D. (1980) Science **208**, 117–118

43. Huttenlocher, A., Ginsberg, M.H. and Horwitz, A.F. (1996) J. Cell Biol. **134**, 1551–1562
44. Lawson, M. and Maxfield, F. (1995) Nature (London) **377**, 75–79
45. Bretscher, M.S. (1989) EMBO J. **8**, 1341–1348
46. Bretscher, M.S. (1992) EMBO J. **11**, 405–410
47. Sheetz, M.P., Baumrind, N.L., Wayne, D.S. and Pearlman, A.L. (1990) Cell **61**, 231–241
48. Kucik, D.F., Elson, E.L. and Sheetz, M.P. (1989) Nature (London) **340**, 315–317
49. Symons, M.H. and Mitchison, T.J. (1991) J. Cell Biol. **114**, 503–513

Biochem. Soc. Symp. **65**, 245–265
Printed in Great Britain

15

Cytoskeletal protein mutations and cell motility in *Dictyostelium*

Igor Weber[1], Jens Niewöhner and Jan Faix

Max-Planck-Institut für Biochemie, Abteilung Zellbiologie, Am Klopferspitz 18a, D-82152 Martinsried, Germany

Abstract

Dictyostelium is a suitable experimental system in which to study the effects of mutations in actin-binding proteins on cell motility. Three cytoskeletal mutants that show distinct alterations in cell shape, chemotactic movement and cytokinesis serve to illustrate the diversity of phenotypes. Cells lacking talin, a protein which in many mammalian cell types is a constituent of focal complexes that link the actin cytoskeleton to the plasma membrane, are strongly impaired in adhesion to external surfaces. Coronin is an actin-associated protein that belongs to the WD-repeat family of proteins, which are engaged in protein–protein interactions involved in signalling pathways. Cells lacking coronin build large hyaline protrusions at their leading edge, diagnostic of an imbalance in the actin polymerization/depolymerization cycle. Cells devoid of a pair of cortexillins, which are novel members of the spectrin/ α-actinin superfamily of actin-binding proteins, form an atypical cleavage furrow on a solid surface and fail to divide in suspension. Other mutants in which one or more actin-binding proteins have been knocked out have weaker phenotypes. With these mutants, cells need to be subjected to special conditions in order to reveal an effect on cell motility. For instance, only on weakly adhesive surfaces is a disturbance in the spatio–temporal co-ordination of protrusion and retraction of the cell body, and of the attachment to and detachment from a substratum, observed in a mutant that lacks three actin-binding proteins: α-actinin, 120 kDa F-actin gelation factor and severin.

Introduction

Since the first use of genetic methods to generate a cytoskeletal mutant in *Dictyostelium discoideum* [1], the genome of this organism has been deprived of more than 20 genes coding for actin-binding proteins. The mutants can be

[1]To whom correspondence should be addressed.

Table 1. Mutants of *Dictyostelium* lacking actin-binding proteins

Protein(s) missing	Phenotypic effects	References
F-actin cross-linking and bundling		
α-Actinin	Viscoelastic response	[2]
120 kDa gelation factor (GF)	Cell shape and motility (dependent on the parent strain); slug phototaxis	[3–7]
α-Actinin/GF double and triple mutants in combination with severin	Cell–substratum contact; cell shape and motility; phagocytosis; multicellular development	[8–10]
34 kDa bundling protein	Cell shape and motility	[11]
Cortexillin I/cortexillin II double mutant	Cell size and shape; cell motility; cytokinesis	[12]
Fimbrin	No obvious phenotypic change	[13]
F-actin severing and capping		
Severin	No obvious phenotypic change	[14]
GRP 120 (gelsolin-related protein of 120 kDa)	Slug phototaxis	[15]
Cap 32/34	Cell motility	[16]
G-actin binding and nucleating		
Profilin I/profilin II double mutant	Cell motility; cytokinesis; multicellular development	[17]
Binding to the plasma membrane		
Ponticulin	Positional stabilization of pseudopods	[18]
Talin	Phagocytosis and adhesion to substrata	[19]
Hisactophilin I/hisactophilin II double mutant	Response to changes in intracellular pH	[20]

Table 1. (contd.)

Protein(s) missing	Phenotypic effects	References
Signal transduction		
Coronin	Phagocytosis; motility; cytokinesis; actin depolymerization	[21–23]
Myosins		
Myosin II	Cell shape and motility; retraction at the rear of a cell; cytokinesis in suspension; multicellular development	[24–28]
Myosin IA	Cell motility	[29]
Myosin IB	Cell motility; movement of intracellular particles	[30]
Myosin IC	No obvious phenotypic change	[31]
Double and triple mutants lacking combinations of myosins IA, IB, IC and ID	Cell motility; phagocytosis; pinocytosis	[32,33]
Myosin myoJ	No obvious phenotypic change	[34]

roughly divided into six groups, according to the known or putative function of the missing proteins: (1) actin-cross-linking and actin-bundling proteins; (2) proteins that cap or/and sever actin filaments; (3) proteins that nucleate actin polymerization; (4) proteins involved in binding of actin filaments to the plasma membrane; (5) proteins that play a role in signalling to the actin system; and (6) myosins. A list of mutants and their phenotypes is compiled in Table 1. The phenotypes of some mutants are strong and well defined, whereas others are weak and can be seen only under special conditions. Some mutants acquire specific defects that can be deduced from the function of the missing protein, some are pleiotropic, and others show no apparent phenotypic change. A number of double mutants and even mutants lacking three or four actin-binding proteins have been generated in *Dictyostelium*, which offer a unique opportunity to study the combined effects of elimination of multiple cytoskeletal proteins in motile cells.

The life cycle of *Dictyostelium* consists of a unicellular and a multicellular phase. When the cells are deprived of food, they aggregate and build a multicellular structure, a slug, which migrates and finally culminates in a fruiting body carrying the spores. Several cytoskeletal mutants show significant aberrations in specific stages of multicellular development, for instance in the aggregation stage [35], slug phototaxis [6,7] and fruiting body morphology [17]. It is likely that anomalies in the motility of cells inside this primeval tissue lie at the basis of such aberrations. However, it is difficult to deduce the effects of the specific mutations at the single-cell level from the effects that they have on the complex morphogenetic processes that take place in a conglomerate of around 10^5 cells. In this overview, we will discuss primarily the consequences that a lack of actin-binding proteins have for the motility of individual, pre-aggregative cells.

We focus our attention on two events where motility plays a central role in *Dictyostelium* cells: chemotaxis and cytokinesis. In the course of each of these two processes, the organization and dynamics of the actin cytoskeleton are optimized for a cell either to move towards the source of a chemoattractant or to divide. This makes the motile response of cells during chemotaxis and cytokinesis specific and fairly reproducible. This is in contrast with the highly variable behaviour of a cell in the course of random movement, which also varies from cell to cell. Our discussion will concentrate primarily on changes that occur in cytoskeletal mutants in cell shape, in protrusive activities at the leading edge and in cell–substratum interactions during the chemotactic response and during cell division.

For illustration, three mutants lacking proteins from three different groups of actin-binding proteins have been chosen. All three of these cytoskeletal mutants were recently isolated in our laboratory and have marked and specific defects in cell shape and motility. The missing proteins are: (1) talin, which is one of the links between the actin cytoskeleton and the plasma membrane; (2) coronin, an actin-associated protein which is also a member of the WD-repeat family of proteins; and (3) cortexillins, two isoforms belonging to a novel class of actin-bundling proteins. Also, extensive and somewhat controversial studies of mutants that lack two actin-cross-linking proteins, α-actinin and 120 kDa gelation factor (GF), will be reviewed.

Chemotaxis and cytokinesis in *Dictyostelium* wild-type cells

The motile mechanisms involved in the chemotactic movement and in the division of *Dictyostelium* cells attached to a substratum have a lot in common. This is due, in the first instance, to the important role that protrusions at the leading edge of a cell play in both processes. Protrusion, coupled with attachment to a substratum, makes normal cytokinesis possible even in cells lacking myosin II [28]. Establishment of cell polarity, and the spatio–temporal co-ordination of activities at the front and the back of a cell, are central to both chemotaxis and cytokinesis. Although the molecular controls of initiation, maintenance and spontaneous changes of cell polarity in *Dictyostelium* are mostly unknown, cytoskeletal proteins are probably their ultimate targets. A difference between the two processes is that, in chemotaxis, one leading front is made, whereas in cytokinesis two regions of intense protrusive activity are formed at the leading fronts of incipient daughter cells. This is not of crucial importance, however, since cells sometimes extend two pseudopods simultaneously during a chemotactic response, and cytokinesis in large, multinucleated cells is characterized by multiple fronts [36]. Another difference between cells performing chemotaxis and those undergoing cytokinesis is in their shape. Chemotactic, aggregation-competent cells are elongated and have a reduced contact with the substratum, whereas cells that are about to divide have a rounded appearance and are well spread on a substratum (compare Figs. 1 and 2).

By using a micropipette filled with a chemoattractant (cAMP), movement of an individual cell can be directed and manipulated at will [37,38]. Cells shown in Fig. 1 were at first moving towards the tip of a micropipette positioned about 10 μm to the right of the frame border. At the beginning of the sequence shown, the pipette was repositioned to the top of the frame (indicated by an arrow). To create these images, a double-view microscope was utilized, which enables simultaneous imaging using bright-field microscopy and reflection interference contrast microscopy (RICM) [9]. Contours of the cell-body projections, as seen in bright-field images, were extracted by an image-processing routine [39], and superimposed on to RICM images. Extension of a pseudopod in the direction of a chemotactic gradient takes place either by spreading over the substratum surface (Fig. 1A) or by protrusion into the liquid medium without making contact with the substratum (Fig. 1B).

A cell dividing on an albumin-coated glass surface is shown in Fig. 2. Images obtained by RICM are displayed side by side with phase-contrast micrographs in order to demonstrate the extensive contact a cell has with a substratum during cytokinesis, besides showing the cell shape. Two symmetrically positioned cell fronts exhibit protrusive activity in the form of ruffling lamellipodia and occasional filopodia. The cleavage furrow stays in contact with a substratum until a relatively late stage of cytokinesis (Fig. 2; 120 s). It should be noted that, after an initial elongation (Fig. 2; 70 s), a dividing cell does not elongate any further (cf. [36]).

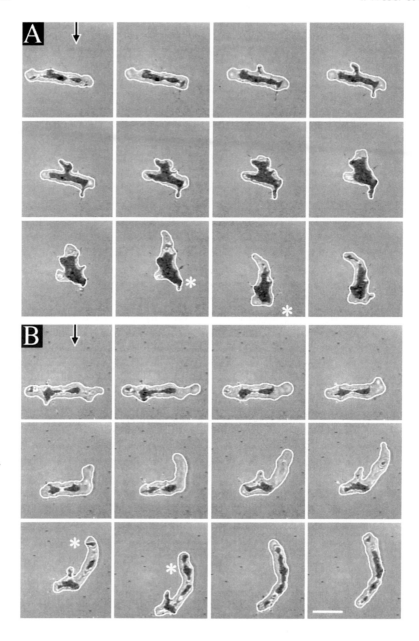

Fig. 1. Turning of two wild-type AX2 cells in response to a chemotactic signal. (A) A cell turning with a pseudopod that spreads over the substratum; (B) a cell that turns by bending the leading pseudopod, without attaching to the substratum. The cells were moving for 2 min in the direction of a micropipette filled with cAMP, which was placed to the right of the frame. At the beginning of each of the sequences shown, the pipette was placed at the top of the frame (arrows). Asterisks mark stationary points on the substratum. The substratum was albumin-coated glass (in this and all Figs.). The interval between frames is 10 s. Bar = 10 μm.

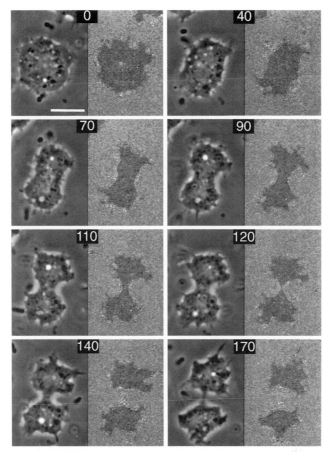

Fig. 2. Pairs of phase-contrast and RICM images of a dividing wild-type cell. Cell–substratum contacts appear dark grey in RICM images. The time that has passed from the beginning of the sequence is designated in seconds on the top of each double-frame. Bar = 10 μm.

Talin links actin to the plasma membrane and supports adhesion of cells to external surfaces

Evidence has accumulated that talin plays an important role in interactions between the cell plasma membrane and the submembranous cytoskeleton. In several mammalian cell types, talin has been localized to focal adhesion complexes [40], phagocytic cups [41] and cell–cell adhesion areas [42], which are all sites of direct contact of a cell with the environment. Talin has also been shown to bind to actin, vinculin, β integrins and membrane lipids *in vitro*. *D. discoideum* is the first non-metazoan known to contain a full-length talin homologue. However, *Dictyostelium* cells do not form elaborate adhesion complexes resembling focal adhesion plaques in fibroblasts. Moreover, immunofluorescence data do not show prominent localization of talin in the cell–substratum contact areas (I. Weber and J. Niewöhner, unpublished work). Instead, tips of

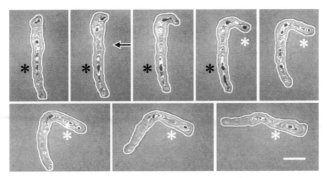

Fig. 3. Turning of a talin-null cell in response to a chemotactic signal.
Interference fringes specify the topology of the ventral cell surface. The vertical
distance between a minimum and a maximum of the intensity pattern is about
135 nm. A micropipette filled with cAMP was placed to the right side of the cell
in the second frame (arrow). Before that, the cell was moving in the direction
of a pipette placed at the top of the frame. Asterisks mark stationary points on
the substratum. The interval between frames is 10 s. Bar = 10 μm.

filopodia in vegetative cells and leading edges in aggregation-competent cells are
enriched in talin [43].

Gene replacement of *Dictyostelium* talin results in cells that are strongly
impaired in adhesion to external surfaces [19]. This is demonstrated by the
defects that these cells have in phagocytosis, in cell-to-substratum contact during locomotion, in the cohesion of cells and in cytokinesis. Greatly decreased
contact of talin-null cells to an albumin-coated glass surface during chemotactic movement is shown in Fig. 3. The cell moves over stationary point-like
contacts which, in contrast with the contacts of the wild-type cells (Fig. 1), do
not increase in size significantly. Spreading of the leading pseudopod over a
surface in the forward direction, characteristic of the chemotactic response in
the wild-type, is also absent in talin-null cells. Cell–substratum adhesion is
affected in a similar manner when cells lacking talin divide on a solid surface.
Most mitotic talin-null cells are more spherical than their wild-type counterparts, and they remain so throughout cytokinesis (Fig. 4).

Given the evidence presented, it is clear that talin influences decisively the
interaction of *Dictyostelium* cells with external surfaces. However, there are as
yet no biochemical or ultrastructural data that would help to elucidate the molecular mechanism of its action. One possibility is that the coupling of talin to a
membrane protein regulates cell adhesion by inducing clustering of the membrane protein, as is known for integrins in other cells. However, a candidate
binding partner for talin is not known in *Dictyostelium*, and immunofluorescence work has provided no evidence for clustering. An alternative explanation
is that a less specific mechanism, based on the suppression of thermal membrane fluctuations by coupling of the plasma membrane to the cortical
cytoskeleton, is responsible for talin's mode of action in *Dictyostelium* cells.

Soft two-dimensional membranes exhibit thermal out-of-plane fluctuations, also called undulations. When a membrane approaches a flat surface, the

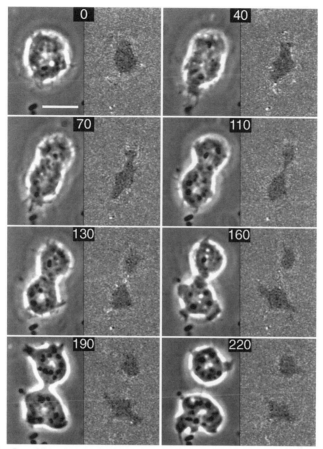

Fig. 4. Cytokinesis of a talin-null cell. See the legend to Fig. 2 for details. Note the reduced cell–substratum contact in comparison with the wild-type cell shown in Fig. 2. The bright halo in the phase-contrast images indicates that the cell is round. Bar = 10 μm.

undulations are suppressed due to a reduction in the configuration space available for fluctuations perpendicular to the plane of a membrane. The situation is analogous to the pressure exerted by an ideal gas when the available volume is reduced [44]. This effect leads to a repulsive entropic force, often referred to as the undulation force. It is known that the amplitudes of thermal undulations are inversely proportional to the bending stiffness of the membrane, and also to its lateral tension. Coupling of the membrane to the cortical cytoskeleton increases its resistance to bending deformations that arise during undulations, thus decreasing the undulation amplitudes, and in turn decreasing the entropic repulsive force. The distance-dependence of this force is the same as for the Van der Waals attractive force, and thus these forces can compete at large membrane–substratum distances. Since talin is considered to participate in membrane–cytoskeleton coupling, it could regulate strength of the cell–substratum adhesion through the undulation-mediated mechanism. Indeed, it has been shown recently that the bending stiffness and lateral tension of the plasma membrane are significantly reduced in talin-null cells [45]. The point-like contacts which sometimes form in talin-deficient cells could be due to a local increase in membrane tension, but the mechanism responsible for their formation is not known. Since these impinging centres occur increasingly during chemotactic stimulation, they could be coupled to local actin polymerization.

Coronin controls actin dynamics in cell protrusions

In contrast with talin, coronin was first discovered in *Dictyostelium* and only later was its homologue identified in two mammalian cell lines [46]. Coronin co-precipitates with an actin–myosin cytoplasmic complex *in vitro*, and co-localizes with F-actin at the leading edge and in the surface projections of *Dictyostelium* cells [47]. Coronin-null cells are impaired in cytokinesis, movement and phagocytosis [21,23]. A possible rationale for such a phenotype is that coronin, besides being associated with the actin–myosin cytoskeleton, also belongs to the WD-repeat family of proteins. These proteins are thought to participate in multiple protein–protein interactions, and are thus implicated in the assembly of protein complexes [48]. Coronin could play a role in transmitting information that controls the dynamics of actin turnover from signalling pathways to actin itself.

Since coronin is enriched in cell extensions, the regions in which the most intense actin reshuffling occurs, it was not a great surprise when mutants that lacked coronin showed anomalies in the dynamics of cell protrusions. A sequence showing a coronin-null cell during chemotactic turning illustrates this point (Fig. 5). The cell is remarkable because of the exaggerated hyaline

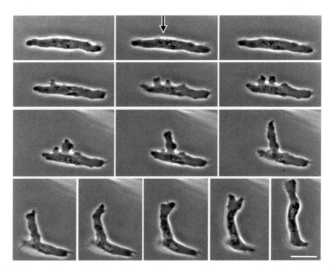

Fig. 5. Coronin-null cell responding to a chemotactic signal. The leading pseudopods are free of intracellular particles. These hyaline zones are filled with polymerized actin. The interval between frames is 10 s, except for the last three frames, which were recorded 20 s apart. Bar = 10 μm.

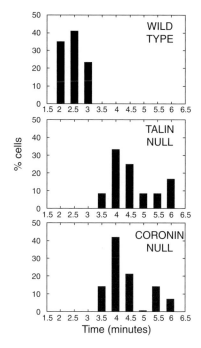

Fig. 6. Histograms showing duration of cytokinesis in wild-type, talin-null and coronin-null cells. Cytokinesis in both mutants takes longer than in the wild-type, and the distributions for the two mutants are statistically indistinguishable. The duration of cytokinesis was measured from the inception of a cleavage furrow to the separation of the two daughter cells (cf. [28]).

zone extending way beyond the leading edge, which is typical for coronin mutants. In wild-type cells, this organelle-free region is restricted to the leading cell margin [22]. It has also been shown that the hyaline zone is full of actin filaments. Taken together, these findings indicate that coronin is probably involved in the control of the actin polymerization/depolymerization cycle. Since it binds to F-actin and its absence leads to over-polymerization of actin *in vivo*, it seems reasonable to assume that coronin regulates the depolymerization of filamentous actin.

One characteristic that coronin and talin have in common is their localization in the newly formed cell protrusions. The two mutants, coronin-null and talin-null, share a tendency to form multinucleate cells when growing in axenic medium and also exhibit a similar delay in the completion of cytokinesis (Fig. 6). In talin-null cells, this delay seems to be caused by reduced cell–substratum adhesion at the opposite fronts of a dividing cell. In coronin-null cells, the character of the protrusions at the fronts changes. Lamellipodia, which are predominant in wild-type cells (Fig. 2), are replaced by numerous filopodia and hyaline pseudopodia (Fig. 7). It is plausible to account for this change by a defect in actin depolymerization, since the characteristic ruffling of lamellipo-

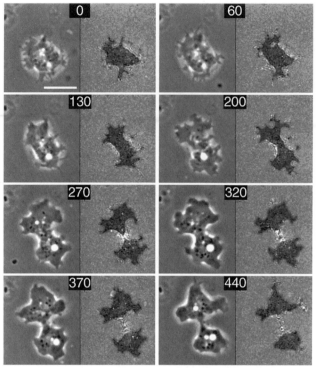

Fig. 7. Cytokinesis of a coronin-null cell. See the legend to Fig. 2 for details. Numerous filopodia are present in the early phase of cytokinesis, and hyaline pseudopodia predominate later. The irregular outline of the dividing cell is typical for this mutant. Bar = 10 μm.

dia is known to be a hallmark of dynamic rearrangements in the actin cytoskeleton.

It is a long-standing opinion, based predominantly on data stemming from experiments on non-motile cells such as echinoderm and amphibian eggs, that cell cleavage is utterly dependent on the myosin-based contraction localized to the contractile ring. Recent experiments performed with motile cells have shown, however, that traction-based mechanisms of cytokinesis are able to contribute to, or even fully substitute for, contraction-mediated cell cleavage [36,49]. The examples of mutants that lack talin and coronin demonstrate that events on the leading edges of incipient daughter cells have a significant influence on the dynamics of cell division.

Cortexillins bundle actin filaments in an anti-parallel fashion and stabilize the cleavage furrow

Two cortexillin isoforms found in *Dictyostelium* are members of a new group of proteins that bundle actin filaments and belong to the α-actinin/spectrin superfamily of actin-binding proteins. Cortexillin double-mutant cells are

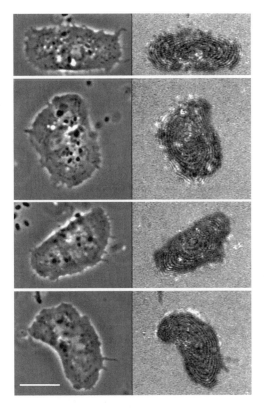

Fig. 8. Interphase of cortexillin-double-null cells. RICM images show the topology of the dorsal cell surface. The fringes result from interference of the light reflected from the glass/buffer interface with the light being reflected from the lower and upper cell membrane/buffer interface. The height difference between the two successive dark fringes is about 270 nm. The thickness of the cells illustrated is approx. 3 μm in the middle of a cell. Such flattened cells are typical for this mutant. Bar = 10 μm.

severely impaired in cytokinesis when grown in shaking suspension, thus resembling the phenotype of myosin-null cells [12]. However, cells lacking both cortexillins also display unique features that distinguish them from myosin-null cells, as well as from other mutant and wild-type *Dictyostelium* cells. First, these cells are unusually flat. This phenomenon is particularly clearly seen in large multinucleated cells, but is also evident in smaller cells that have only one nucleus (Fig. 8). Secondly, cytokinesis is disturbed in cortexillin double-mutant cells dividing on an albumin-coated glass surface. The first phase of cytokinesis proceeds normally, but then the cleavage furrow starts to stretch out as it gets thinner (Fig. 9). This extremely elongated cell morphology in the final phase of cell division is often accompanied by an apparent imbalance of forces that causes the cells to divide unevenly. Furthermore, about one-third of cortexillin-null cells fail to complete cytokinesis when attached to

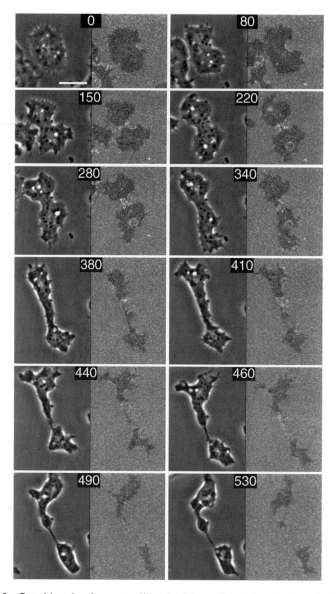

Fig. 9. Cytokinesis of a cortexillin-double-null cell. See the legend to Fig. 2 for details. After the normal initiation of cytokinesis, the cell elongates and the cleavage furrow stretches out. A small cytoplasmic body forms in the middle of the furrow and stays connected to the daughter cells by thin membranous threads (final frame). Eventually, one of the threads broke and the mid-body fused with one daughter cell. Bar = 10 μm.

a substratum, which is significantly more than the one-tenth that fail in the case of myosin-null cells [28].

Other mutants of *Dictyostelium* that lack one or two proteins that bundle or cross-link actin filaments show much milder defects in cell shape and motility [5,10]. The question arises as to why the absence of the two cortexillin iso-forms has such a marked effect at the level of a whole cell. The singular molecular architecture of cortexillin I may provide some clues. All other known actin-bundling proteins in *Dictyostelium* build predominantly anti-par-allel homodimers or remain monomeric *in vitro*. Recombinant cortexillin I forms parallel homodimers *in vitro*, in which N-terminal heads bearing actin-binding sites are brought into juxtaposition [12]. Such an arrangement of actin-binding sites, together with their apparently flexible linkages to the cen-tral coiled-coil domains, could be responsible for the preference of cortexillin I to bundle actin filaments in an anti-parallel fashion. This tendency to connect actin filaments of opposite polarity is another feature that distinguishes cortex-illins from other actin-bundling proteins. Therefore cortexillins could potentially play an important role in determining the relative polarity of single filaments within actin bundles, which are mainly located in the cortex of *Dictyostelium* cells.

A surprisingly elaborate distribution of actin filaments, with respect to their relative polarity, was discovered recently in motile fibroblasts ([50]; see also Chapter 11). Bundles comprising actin filaments with opposite polarity were found predominantly in the middle region of a cell, whereas, at front and rear ends of the cell, bundles of parallel filaments were dominant. Interestingly, in regions with different relative polarity of connected actin filaments, the fila-ments also move in opposite directions relative to the substratum during cell locomotion. The filaments move backward in the lamellipodium and remain stationary or move forward elsewhere in the cell. It has been inferred by the authors [50], but not yet proven, that the distinct patterns of F-actin movement are diagnostic of different mechanisms of motility operating in the lamel-lipodium (protrusion) compared with the lamella and cell body (traction) (see also Chapter 18). The relative orientation of actin filaments within a bundle should also influence its interaction with myosin motors. Bundles composed of anti-parallel actin filaments would be prone to disruption by myosin, which could pull two subpopulations of filaments within a bundle in opposite direc-tions [51]. In the course of cytokinesis, the cortical material has to be transported away from the cleavage furrow as the furrow disassembles [52]. At the same time, the furrow retains its resistance to longitudinal deformation in wild-type *Dictyostelium* cells, but not in cortexillin double-mutant cells (Fig. 9). The geometry of actin filaments in the mid-zone of a dividing cell is likely to determine the proper symmetry and direction of forces that are responsible for cell cleavage. By controlling this geometry, cortexillins seem to be important components of the molecular machinery that executes cytokinesis in *Dictyostelium* cells.

The actin-cross-linking proteins GF and α-actinin have subtle effects on cell shape and motility

The first two mutants that lacked the actin-cross-linking proteins α-actinin and GF revealed basically no phenotype [3,53]. Soon after this, GF-minus strains generated in another laboratory displayed substantial alterations in cell shape and motility [4,54]. Both outcomes have been confirmed in more extended studies, leading to the conclusion that differences in the parental strains from which the two groups of mutants were derived (AX2 versus AX3) are responsible for the observed phenotypic differences [5,10]. In the latter paper, a double mutant lacking both α-actinin and GF was also analysed [10]. In contrast with single mutants, the double-mutant cells showed alterations in phagocytosis, responsiveness to osmotic stress, cell shape and motility, and were also not able to complete multicellular development. On the basis of the phenotypes of single and double mutants, the concept of redundancy of actin-binding proteins in *Dictyostelium* has been proposed [55]. In this view, the removal of one protein does not lead to any discernible changes in cell behaviour, since its role can be taken over by a functionally related protein.

Later studies, however, have cast some doubt on the concept of functional redundancy of α-actinin and GF. Rheological measurements performed *in vitro* demonstrated that the two proteins affect the viscoelastic properties of F-actin solutions in different ways [56]. Whereas α-actinin strongly dampens high-frequency oscillations, GF has a much milder effect. An analogous result has been obtained by using preparations consisting of whole *Dictyostelium* cells, thus strengthening the conclusion that α-actinin enables a cell to resist deformations imposed by external impacts [2]. Further evidence favouring distinct roles for the two actin-binding proteins has been provided recently: only mutants lacking GF displayed anomalies in slug phototaxis [6,7].

When *Dictyostelium* cells are deprived of their food source, they gradually change shape in the first hours of starvation. At the end of this process, the cells attain an elongated morphology and become responsive to chemotactic stimulation. Such aggregation-competent cells also have drastically reduced contact with a substratum when compared with vegetative cells [9]. The dynamics of these changes differ in various cytoskeletal mutants, and differences in early development have also been found between strains that are used as standard, or 'wild-type', cells in different laboratories, i.e. AX2 and AX3 [57]. However, measurements of motility are mostly performed using aggregation-competent cells at a fixed time point after the beginning of starvation. Therefore the difference in the dynamics of early development could have been a significant factor that led to different results in mutant strains derived from the two reference strains. The differences in the adhesion of cells to various substrata that were used in the two sets of experiments might also have played an important role. This conclusion is reinforced by the results, described below, that demonstrate the importance of cell–substratum interactions for the motility of *Dictyostelium* cells.

By using RICM to investigate cell–substratum adhesion in *Dictyostelium*, David Gingell and co-workers were able to show numerous situations in which the physical and chemical properties of substratum surfaces had a marked effect on cell motility [58,59]. In the later experiments on the motility of cytoskeletal mutants, two types of substrata were mostly used: simple glass surfaces and glass surfaces coated with albumin. Plain glass is a rather poorly defined substratum, because its adhesiveness for cells depends on the way it was cleaned. The glass surface is generally more adhesive when it is uncoated than when it is coated with albumin [8]. A recent comparison of cell locomotion on the two substrata in cells lacking the 34 kDa actin-bundling protein showed that the persistence of locomotion is greater on albumin-coated surfaces [11].

The possibility that weakly adhesive surfaces may reveal alterations in cell motility not normally seen on more adhesive substrata led to the utilization of the strongly hydrophilic surface of freshly cleaved mica as a substratum for *Dictyostelium* cells [8,9]. Indeed, triple-mutant cells that lacked α-actinin, GF and the actin-severing protein severin had smaller areas of contact with mica than did their wild-type counterparts. Also, locomotion of mutant cells was strongly impaired on mica, whereas it was nearly normal on albumin-coated surfaces. Locomotion of wild-type AX2 cells was much less sensitive to differences in substratum adhesiveness.

Analysis of the correlation between protrusion and retraction of the cell body, and between attachment to and detachment from a substratum, was also performed in the course of these investigations [9]. To achieve this, a double-view microscopy technique was combined with computer-assisted image

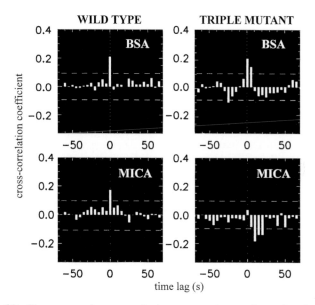

Fig. 10. Cross-correlograms between vectors of protrusion and retraction during random motility of wild-type cells, and of a triple-mutant, on BSA-coated glass and on mica. The triple-mutant cells lack α-actinin, GF and severin. In the wild type, protrusion and retraction are positively correlated on both substrata. This shows that protrusion of pseudopodia at the leading edge occurs simultaneously, and in the same direction, as retraction of the rear end of a cell. In the triple mutant the behaviour changes on the weakly adhesive surface (mica), where the correlation is negative. Moreover, retraction lags behind protrusion by approx. 10–20 s, indicating that cells retract pseudopodia 10–20 s after extending them. A similar effect arises in the correlation between the attachment to and the detachment from the mica surface [9]. For technical details of the cross-correlation analysis, see [9,67–69].

analysis (see the section on chemotaxis and cytokines in *Dictyostelium* cells). The principal result was that wild-type cells co-ordinate protrusion and attachment at the leading edge of a cell with retraction and detachment at the rear in a regular manner. Cross-correlograms can be constructed as footprints of this characteristic spatio–temporal pattern (Fig. 10). In contrast with wild-type cells, cross-correlograms change drastically when triple-mutant cells are placed on a mica surface. This occurs because mutant cells are unable to attach and stabilize newly protruded pseudopods, which are then frequently retracted. From these results it was concluded that α-actinin and GF play a role in the stabilization of the newly formed protrusions at the front of a cell. On the other hand, myosin II is involved in retraction and in detachment of the rear part of a cell. Consequently, glass surfaces coated with poly-L-lysine were used to show that an impairment of movement on such a strongly adhesive substratum is more pronounced in myosin-null cells than in the wild type [27].

Conclusions and outlook

When the first cytoskeletal mutants were isolated in *Dictyostelium*, it was surprising that the absence of these cytoskeletal proteins, which have a strong actin-cross-linking or actin-severing activity *in vitro*, caused little or no change in the motility of the mutant cells. As more and more mutants were generated, it became clear that the motile machinery of a cell is remarkably resistant to the elimination of its constituents. It appeared that no combination of these mutations could immobilize these cells. Logically, however, even small anomalies in phagocytosis, cytokinesis or locomotion would have serious consequences for a defective strain in the natural habitat of *Dictyostelium* amoebae, the forest soil. Wild-type cells not only outperform mutants under standard laboratory conditions, but they are also able to adapt optimally to the wide range of conditions that exist in their natural environment. It is a task for experimentalists to recreate this variety of conditions in the laboratory, a task that has been pursued in several recent studies. It was shown that myosin-null cells are vulnerable to osmotic shock [60], and cannot detach from a sticky surface [27]. Cells that lack two actin cross-linkers cannot spread and stabilize pseudopods on a weakly adhesive surface [8,9]. Talin-null cells cannot attach even to a moderately adhesive substratum, and are not able to phagocytose certain bacterial strains, depending on the composition of their surface carbohydrates [19]. With the use of imaginative experimental approaches, the weak points of each cytoskeletal mutant can and will be eventually found.

The construction and testing of multiple mutants has proved to be useful in clarifying whether the missing proteins can partially substitute for each other. However, this approach was previously limited to the combined deletion of homologous or closely related proteins (Table 1). It would be interesting to knock out two proteins that are genetically unrelated, but have similar or even opposed functions in the control of cell motility. For instance, the absence of coronin, or of the two profilin isoforms, leads to increased polymerization of actin. However, the consequences for cell motility of the corresponding mutants were relatively mild [17,21]. The question is whether these effects would also be apparent in a triple mutant, or would the defects be more predominantly expressed? Talin is involved in the attachment of cells to weakly adhesive substrata during locomotion [19], whereas myosin II is important for detachment from strongly adhesive substrata [27]. In a double mutant, the two defects might compensate for each other. Another possibility is that these cells would behave like talin-null cells on a weakly adhesive surface, but on a strongly adhesive substratum they may behave like myosin-null cells. Also, since the two single mutants already show relatively marked defects, the co-ordination of activities at the front and the rear of a double-mutant cell could prove to be seriously impaired. Such studies would provide fruitful insights into the interactions and interdependencies of the investigated proteins in a living cell.

To learn about functions of actin-binding proteins by examining anomalies in the motility of cytoskeletal mutants is not a goal in itself. Locomotion and cytokinesis are also affected in several mutants lacking components of elaborate signal transduction pathways that operate in *Dictyostelium*. Examples are small GTPases belonging to the Ras superfamily [61,62], proteins that control their action [63], protein phosphatases [64] and protein kinases [65], and other proteins [66]. Thorough characterization of both groups of mutants may help to establish missing links between the signalling networks and some of their targets that reside in the actin cytoskeleton. Comparison of phenotypes should then complement biochemical data, and help to elucidate the molecular regulatory mechanisms that control the microfilament system in *Dictyostelium*.

We thank Dr. Günther Gerisch for his ongoing support and guidance, and Dr. Ewa Wallraff for help with the cell culture.

References

1. Wallraff, E., Schleicher, M., Modersitzki, M., Rieger, D., Isenberg, G. and Gerisch, G. (1986) EMBO J. **5**, 61–67

2. Eichinger, L., Köppel, B., Noegel, A.A., Schleicher, M., Schliwa, M., Weijer, K., Witke, W. and Janmey, P.A. (1996) Biophys. J. **70**, 1054–1060

3. Brink, M., Gerisch, G., Isenberg, G., Noegel, A.A., Segall, J.E., Wallraff, E. and Schleicher, M. (1990) J. Cell Biol. **111**, 1477–1489

4. Cox, D., Condeelis, J., Wessels, D., Soll, D.R., Kern, H. and Knecht, D.A. (1992) J. Cell Biol. **116**, 943–955

5. Cox, D., Wessels, D., Soll, D.R., Hartwig, J. and Condeelis, J. (1996) Mol. Biol. Cell **7**, 803–823

6. Wallraff, E. and Wallraff, H.G. (1997) J. Exp. Biol. **200**, 3213–3220

7. Fisher, P.R., Noegel, A.A., Fechheimer, M., Rivero, F., Prassler, J. and Gerisch, G. (1997) Curr. Biol. **7**, 889–892

8. Schindl, M., Wallraff, E., Deubzer, B., Witke, W., Gerisch, G. and Sackmann, E. (1995) Biophys. J. **68**, 1177–1190

9. Weber, I., Wallraff, E., Albrecht, R. and Gerisch, G. (1995) J. Cell Sci. **108**, 1519–1530

10. Rivero, F., Köppel, B., Peracino, B., Bozzaro, S., Siegert, F., Weijer, C.J., Schleicher, M., Albrecht, R. and Noegel, A.A. (1996) J. Cell Sci. **109**, 2679–2691

11. Rivero, F., Furukawa, R., Noegel, A.A. and Fechheimer, M. (1996) J. Cell Biol. **135**, 965–980

12. Faix, J., Steinmetz, M., Boves, H., Kammerer, R.A., Lottspeich, F., Mintert, U., Murphy, J., Stock, A., Aebi, U. and Gerisch, G. (1996) Cell **86**, 631–642

13. Prassler, J. (1995) Ph.D. Thesis, Technische Universität Munich

14. André, E., Brink, M., Gerisch, G., Isenberg, G., Noegel, A., Schleicher, M., Segall, J.E. and Wallraff, E. (1989) J. Cell Biol. **108**, 985–995

15. Stocker, S. (1997) Ph.D. Thesis, Technische Universität Munich

16. Hug, C., Jay, P.Y., Reddy, I., McNally, J.G., Bridgman, P.C., Elson, E.L. and Cooper, J.A. (1995) Cell **81**, 591–600

17. Haugwitz, M., Noegel, A.A., Karakesisoglou, J. and Schleicher, M. (1994) Cell **79**, 303–314

18. Shutt, D.C., Wessels, D., Wagenknecht, K., Chandrasekhar, A., Hitt, A.L., Luna, E.J. and Soll, D.R. (1995) J. Cell Biol. **131**, 1495–1506

19. Niewöhner, J., Weber, I., Maniak, M., Müller-Taubenberger, A. and Gerisch, G. (1997) J. Cell Biol. **138**, 349–361

20. Stöckelhuber, M., Noegel, A.A., Eckerskorn, C., Köhler, J., Rieger, D. and Schleicher, M. (1996) J. Cell Sci. **109**, 1825–1835

21. de Hostos, E.L., Rehfuess, C., Bradtke, B., Waddell, D.R., Albrecht, R., Murphy, J. and Gerisch, G. (1993) J. Cell Biol. **120**, 163–173

22. Gerisch, G., Albrecht, R., Heizer, C., Hodgkinson, S. and Maniak, M. (1995) Curr. Biol. **5**, 1280–1285

23. Maniak, M., Rauchenberger, R., Albrecht, R., Murphy, J. and Gerisch, G. (1995) Cell **83**, 915–924

24. Wessels, D., Soll, D.R., Knecht, D., Loomis, W.F., De Lozanne, A. and Spudich, J. (1988) Dev. Biol. **128**, 164–177

25. Fukui, Y., De Lozane, A. and Spudich, J.A. (1990) J. Cell Biol. **110**, 367–378

26. Jay, P.Y. and Elson, E.L. (1992) Nature (London) **356**, 316–320

27. Jay, P.Y., Pham, P.A., Wong, S.A. and Elson, E.L. (1995) J. Cell Sci. **108**, 387–393

28. Neujahr, R., Heizer, C. and Gerisch, G. (1997) J. Cell Sci. **110**, 123–137

29. Titus, M., Wessels, D., Spudich, J.A. and Soll, D. (1993) Mol. Biol. Cell **4**, 233–246

30. Wessels, D., Murray, J., Jung, G., Hammer, III, J.A. and Sol, D.R. (1991) Cell Motil. Cytoskeleton **20**, 301–315

31. Peterson, M.D., Novak, K.D., Reedy, M.C., Ruman, J.I. and Titus, M.A. (1995) J. Cell Sci. **108**, 1093–1103

32. Novak, K.D., Peterson, M.D., Reedy, M.C. and Titus, M.A. (1995) J. Cell Biol. **131**, 1205–1221

33. Jung, G., Wu, X. and Hammer, III, J.A. (1996) J. Cell Biol. **133**, 305–323

34. Peterson, M.D., Urioste, A.S. and Titus, M.A. (1996) J. Muscle Res. Motil. **17**, 411–424

35. Shelden, E. and Knecht, D.A. (1995) J. Cell Sci. **108**, 1105–1115

36. Neujahr, R., Heizer, C., Albrecht, R., Ecke, M., Schwartz, J.-M., Weber, I. and Gerisch, G. (1997) J. Cell Biol. **139**, 1793–1804

37. Gerisch, G. and Keller, H.U. (1981) J. Cell Sci. **52**, 1–10

38. Swanson, J.A. and Taylor, D.L. (1982) Cell **28**, 225–232

39. Weber, I. and Albrecht, R. (1997) Comput. Programs Biomed. **53**, 113–118

40. DePasquale, J.A. and Izzard, C.S. (1991) J. Cell Biol. **113**, 1351–1359

41. Greenberg, S., Burridge, K. and Silverstein, S.C. (1990) J. Exp. Med. **172**, 1853–1856

42. Kupfer, A., Swain, S.L. and Singer, S.J. (1987) J. Exp. Med. **165**, 1565–1580

43. Kreitmeier, M., Gerisch, G., Heizer, C. and Müller-Taubenberger, A. (1995) J. Cell Biol. **129**, 179–188

44. Sackmann, E. (1994) FEBS Lett. **346**, 3–16

45. Simson, R., Faix, J., Niewöhner, J., Wallraff, E., Gerisch, G. and Sackmann, E. (1998) Biophys. J. **74**, 514–522

46. Suzuki, K., Nishihata, J., Arai, Y., Honma, N., Yamamoto, K., Irimura, T. and Toyoshima, S. (1995) FEBS Lett. **364**, 283–288

47. de Hostos, E.L., Bradtke, B., Lottspeich, F., Guggenheim, R. and Gerisch, G. (1991) EMBO J. **10**, 4097–4104

48. Neer, E.J., Schmidt, C.J., Nambudripad, R. and Smith, T.F. (1994) Nature (London) **371**, 297–300

49. Burton, K. and Taylor, D.L. (1997) Nature (London) **385**, 450–454

50. Cramer, L.P., Siebert, M. and Mitchison, T.J. (1997) J. Cell Biol. **136**, 1287–1305

51. Verkhovsky, A.B., Svitkina, T.M. and Borisy, G.G. (1997) J. Cell Sci. **110**, 1693–1704

52. Schroeder, T.E. (1987) in Biomechanics of Cell Division (Akkas, N., ed.), pp. 209–230, Plenum Press, New York and London

53. Schleicher, M., Noegel, A.A., Schwarz, T., Wallraff, E., Brink, M., Faix, J., Gerisch, G. and Isenberg, G. (1988) J. Cell Sci. **90**, 59–71

54. Cox, D., Ridsdale, J.A., Condeelis, J. and Hartwig, J. (1995) J. Cell Biol. **128**, 819–835

55. Witke, W., Schleicher, M. and Noegel, A. (1992) Cell **68**, 1–10

56. Janssen, K.-P., Eichinger, L., Janmey, P., Noegel, A.A., Schliwa, M., Witke, W. and Schleicher, M. (1996) Arch. Biochem. Biophys. **325**, 183–189

57. Beug, H., Katz, F.E. and Gerisch, G. (1973) J. Cell Biol. **56**, 647–658

58. Gingell, D. and Vince, S.M. (1982) J. Cell Sci. **54**, 255–285

59. Owens, N.F., Gingell, D. and Trommler, A. (1988) J. Cell Sci. **91**, 269–279

60. Kuwayama, H., Ecke, M., Gerisch, G. and Vanhaastert, P.J.M. (1996) Science **271**, 207–209

61. Daniel, J., Bush, J., Cardelli, J., Spiegelman, G.B. and Weeks, G. (1994) Oncogene **9**, 501–508

62. Larochelle, D.A., Vithalani, K.K. and De Lozanne, A. (1996) J. Cell Biol. **133**, 1321–1329

63. Faix, J. and Dittrich, W. (1996) FEBS Lett. **394**, 251–257

64. Gamper, M., Howard, P.K., Hunter, T. and Firtel, R.A. (1996) Mol. Cell. Biol. **16**, 2431–2444

65. Mann, S.K., Brown, J.M., Briscoe, C., Parent, C., Pitt, G., Devreotes, P.N. and Firtel, R.A. (1997) Dev. Biol. **183**, 208–221

66. Alexander, S., Sydow, L.M., Wessels, D. and Soll, D.R. (1992) Differentiation **51**, 149–161

67. Dunn, G.A. and Brown, A.F. (1987) J. Cell Sci. Suppl. **8**, 81–102

68. Dunn, G.A. and Zicha, D. (1995) J. Cell Sci. **108**, 1239–1249

69. Mandeville, J.T.H., Ghosh, R.N. and Maxfield, F.R. (1995) Biophys. J. **68**, 1207–1217

Biochem. Soc. Symp. **65**, 267–280
Printed in Great Britain

16

Cell crawling two decades after Abercrombie

Thomas P. Stossel, John H. Hartwig, Paul A. Janmey and David J. Kwiatkowski

Brigham & Womens Hospital, Harvard Medical School, Boston, MA 02115, U.S.A.

Abstract

In response to extracellular signals, cells remodel actin networks. Monomeric actin subunits at the cell's leading edge assemble into linear polymers that are cross-linked by accessory proteins into three-dimensional structures that are contracted by myosins to generate hydraulic force; elsewhere in the cell, actin networks dismantle. Actin subunit sequestering proteins prevent spontaneous actin nucleation, but not the growth of actin subunits on to fast-growing filament ('barbed') ends, and at least half of the actin in most cells is filamentous. Therefore regulation of cellular actin assembly also requires proteins that block ('cap') actin filament barbed ends. Members of the capping protein gelsolin family also sever actin filaments mechanically. Calcium and protons activate gelsolin for severing and capping. Phosphoinositides reverse such capping, and a pathway has been defined in which receptor perturbation operates through GTP-Rac1 to stimulate the synthesis of endogenous phosphoinositides that uncap actin filaments. Other GTPases (and other signalling pathways) target phosphoinositide synthesis where other protrusions (e.g. filopodia) emerge. Cells maintain adequate, albeit compromised, locomotion in the absence of some, but not all, important machine parts. For example, gelsolin-null fibroblasts crawl using predominantly filopodia rather than lamellae. However, ABP-280 (actin-binding protein of 280 kDa), which promotes orthogonal branching of short actin filaments, seems to be necessary for membrane stability and translational locomotion. ABP-null cells hardly crawl at all, although they are viable and engage in surface movements.

Introduction

We joined the cell motility research programme so importantly furthered by Michael Abercrombie just about the time he died. This milestone marked the time at which modern molecular tools were being applied to the problem of how external instructions elicit cell protrusions, shape changes and locomotion. At about the same time, investigators were appreciating that the machinery working these mechanical cell functions also accounts for the cell shape changes of blood platelets during haemostasis and thrombosis, of vascular cells accommodating leucocyte diapedesis, and of all cells undergoing fission and cytokinesis. Recently, the way in which certain micro-organisms enter, cavort within and then exit parasitized cells by subverting at least some of the host cell's crawling machinery has become a popular way to address this molecular problem. Despite its general relevance, many aspects of cell crawling movements are likely to be quite cell specific. Therefore results on one cell type are unlikely to explain everything about how cells crawl. Nevertheless, a reasonable short-term goal may be to pool information from diverse cell studies: some cell types are amenable to biochemical investigations of purified components, some to detailed morphological scrutiny, and others to molecular perturbations that test specific models. Signal transduction mechanisms are also better known in some cell types than in others. One of our principal justifications for studying the molecular basis of cell motility is to identify aspects of cellular molecular machinery that predictably or unexpectedly impact on the prevention or treatment of human diseases.

Based on the entitlement of having persevered in this field for a quarter of a century, we take the liberty of sticking our necks out with a few explicit assertions concerning the mechanism by which at least some cells crawl. The importance of actin remodelling in cell crawling was just recognized at the time Abercrombie completed his work, and today we believe even more strongly that the remodelling and contractility of actin filaments builds and deconstructs various types of cell surface protrusions. The only known exception concerns the crawling movements of nematode sperm, and even the sperm follows principles that underlie the crawling movements of other eukaryotic cells, namely polymer assembly and branching at the leading edge regulated by signals generated at the cell surface [1]. In the last few years, rapid advances in signal transduction research have begun to integrate with previous biochemical information concerning the regulation of actin remodelling by actin-binding proteins, and this summary reviews some of that integration, emphasizing studies on the blood platelet.

Our work on cell motility originated with investigations on mammalian phagocytic leucocytes, macrophages and neutrophils, out of which we mined two previously unknown proteins, actin-binding protein of 280 kDa (ABP-280) and gelsolin, which we have examined for a long time and to which we ascribe important roles in cell motility (and other cell functions). We based our approaches to the functions of these proteins on ideas about cell consistency changes, quaintly known as 'sol–gel transformations', that were invoked for two centuries and which we still believe are useful. Simply put, we believe that ABP-280 is important for stabilizing actin filaments in the cell periphery in a cross-linked gel state that is optimal for cell locomotion. Gelsolin is the prototype of a family of proteins that regulates actin polymerization, and provided some of the first clues as to how cellular actin remodelling relates to signal transduction.

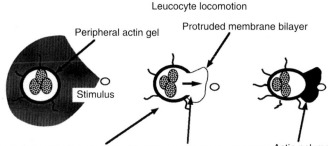

Fig. 1. Proposal for one mechanism of cell protrusion and locomotion.
See the text for details.

Mechanism of crawling

With no claims to originality, here is our simple picture as to how, say, a leucocyte might move, based on rudimentary ideas about an interplay between sol–gel transformations and contractility in the cell periphery (Fig. 1). Among numerous mechanisms proposed for how actin induces protrusion of pseudo-podia during spreading, locomotion and other functions [2], we favour the idea that the cell contains a peripheral elastic rim (the cell cortex). A contraction of the cell cortex induces a hydrostatic force within the cell, which is resisted by the rigidity of the cortical substance. Membrane protrusion occurs in surface locations where the cortex weakens as a localized response to surface stimulation. As the lipid membrane protrudes, tightly coupled strengthening of the submembrane cortex regulates its extended shape and how far it moves. The filling in of cortical substance stabilizes the extension, and it also provides anchoring of transmembrane adhesion molecules that establish the traction force for movement. Contraction of the cell body, from which the protrusion extends, draws it forward in the direction of locomotion. Obviously, adhesion must be reversible to accommodate translational locomotion.

Detail 1: contractility

Many, if not most, actin filaments in crawling cells orient with their fast-growing (barbed with respect to the angle at which myosin head fragments bind to the sides of the filaments) ends facing the plasma membrane. This arrangement is consistent with a mechanism resembling muscle contraction, in which myosin filaments can pull on actin filaments and thereby are positioned to be responsible for the proposed cortical contraction. Both *in vitro* and in cells, such actin–myosin complexes seem to organize themselves even in the relatively messy isotropic actin arrays that do not remotely resemble striated muscle sarcomeres [3]. Clearly co-ordination between actin filament membrane attachment and contraction is necessary, and, while not reviewed here, excellent progress is taking place in understanding this co-ordination [4,5].

Compelling links to cell signal transduction are in hand to account for the regulation of myosin-based contractility. Increases in intracellular Ca^{2+} activate myosin light-chain kinase and the phosphorylation of myosin subunits, leading to a myosin assembly competent to hydrolyse ATP and to move along actin filaments. A recently defined alternative pathway has GTP-charged Rho stimulating a different protein kinase (p160 Rho kinase) that both activates myosin light-chain kinase [4] and inhibits the corresponding phosphatase [6]. These redundant signalling pathways possibly explain why some (but not all) cells can crawl normally sometimes without increasing their intracellular Ca^{2+} [7]. This hypothesis is partially testable by determining whether Ca^{2+}-clamped cells undergoing locomotion are able to phosphorylate myosin in the presence or absence of toxins that inhibit Rho function. To our knowledge, the experiment has not yet been done.

Since myosin isoforms are so numerous, genetic knockout approaches have not been able to establish an absolute requirement for myosins in cell crawling behaviour, although genetic knockout experiments [8], transfection of cells with dominant-negative constructs [9] and use of pharmacological inhibitors suggests that at least some cell surface movements need myosin activity [10]. Worth recalling is the fact that motor proteins ratcheting on actin or tubulin is not the only way to contract polymers. Cross-linking of polymers *per se* can cause them to aggregate (synerese), and possibly the spontaneous cross-linking of actin or of the non-actin polymer of nematode sperm can generate the contractile forces responsible for the mechanism of membrane protrusion that we have proposed.

Detail 2: the cortical gel

The membrane protruded by crawling cells encases an organelle-excluding substance, the 'cortical gel'. This gel is the structure that is proposed to fill in and stabilize hydrostatically extended lipid membranes. When viewed with different kinds of microscopes, the cortex contains predominantly actin filaments organized in diverse ways. Our experiences with neutrophils and pulmonary macrophages led us to be interested in an actin filament architecture within extended lamellae consisting of actin filaments of approx. 1 μm length (or shorter) branching in three dimensions at large angles. This ultimately 'isotropic' actin network appears to be characteristic of the dynamic leading edge of neutrophils and at least some other cell types. Near to perpendicular branching is the optimal way to create an extensive three-dimensional gel out of rod-like actin filaments. In order for local fluid flows to be responsible for protrusive activity, the cell must be able to prevent rapid equilibration of water pressures throughout the cell. Immobilized actin filaments can resist such pressures [11]. We concluded that this type of actin filament organization requires a specific bivalent actin filament cross-linking protein, which we call ABP-280 [12]. Filamin, a variant of ABP-280 encoded by a gene on chromosome 7 (the ABP-280 gene is on the X-chromosome) [13], lacks an important internal motif

[14] of ABP-280 [15] that allows it to promote orthogonal cross-linking of actin filaments. ABP-280, in addition to serving as an actin filament cross-linking protein responsible for efficient actin gelation and high-angle filament branching, links actin filaments to particular membrane proteins. Actin networks cross-linked by ABP-280 resist osmotic forces efficiently [16].

Cells contain numerous actin filament cross-linking proteins. An isotropic architecture of actin filaments, while prevalent, is frequently embellished in tissue culture cells by axial actin filament bundles [17]. However, evidence that ABP-280-based isotropic actin filament architecture is vital for translational locomotion amidst the redundant repertoire of actin filament cross-linking mechanisms resides in the robustly abnormal phenotype of human malignant melanoma cell lines lacking ABP-280 mRNA and protein, and the normal protrusive and locomotory behaviour of these cells when repleted with the protein by stable genetic transfection [18]. Even when ABP-280-deficient cells stabilize their surfaces by increasing their content of polymerized actin, which presumably causes the interpenetration of long actin filaments which confers gel-like properties on actin [19], they are nearly incapable of translational locomotion. Understanding the rudimentary movement of these cells will be of interest.

ABP-280-deficient cells extend blebs in response to stimulation by growth factors. The rate of bleb extension is too high to be caused by actin polymerization, and cytochalasins (which block actin assembly) do not inhibit bleb formation; however, they completely prevent the recovery of these protrusions back to the cell body, suggesting that this latter step requires actin assembly. Monomeric actin enters expanding blebs, and only as the blebs stabilize and contract does filamentous actin appear. All of the results suggest that ABP-280, by cross-linking the assembling network, counteracts hydrostatically induced swelling, leading to the formation of pleats and pseudopodia rather than blebs. The extent to which ABP-280-deficient cells can retract blebs presumably reflects the action of other cross-linking proteins or else simply the effects of long actin filaments becoming interpenetrated, thereby accomplishing in a rudimentary fashion what ABP-280 does with greater efficiency. In either case, the initial stimulus for protrusion must be dissipative, namely a local breakdown of the cortical actin network as well as of its attachments to the plasma membrane [20].

Detail 3: remodelling the actin gel and signal transduction

We propose that the way in which actin fills into an extending structure is by the elongation of linear filaments through the addition of subunits; ABP-280 and other actin filament cross-linking proteins stabilize the filaments into particular three-dimensional organizations. 'Nucleation' is the conventional term for the site where cellular actin filaments elongate. The classic work of Fumio Oosawa, who demonstrated that actin monomers induced to polymerize by salt must first form aggregates of two or three subunits, has powerfully influenced thinking about nucleation. This spontaneous nucleation process discovered by Oosawa is very inefficient, although once a nucleus forms,

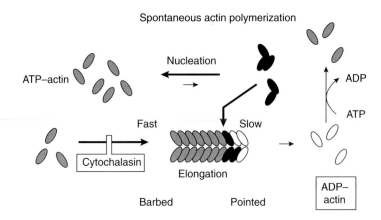

Fig. 2. Kinetics of pure actin assembly. The shading of actin monomers indicates the form of adenine nucleotide bound: ATP, shaded; ADP, unshaded.

elongation is rapid. Subsequent work identified the importance of actin filament polarity and of ATP turnover in actin assembly kinetics, in particular
recognizing that barbed ends are the fast-growing ends (Fig. 2). As a result of
this information, many investigators have fixated on the idea that the stimulated cell presents interfaces that promote the aggregation of actin monomers
into nuclei that then elongate in the barbed direction.

    This intuitively beguiling idea has to contend with three important but
often overlooked facts. First, half of the nearly millimolar concentration of
actin in motile cells is unpolymerized, meaning that the other half is in the form

(1) Actin subunit binding by β-thymosin (⬤) inhibits the spontaneous nucleation of actin subunits

(2) Thymosin has a lower affinity for actin than for free barbed (but not pointed) filament ends

(3) Thus, together with thymosin, reversible barbed-end capping (⬤) regulates actin assembly

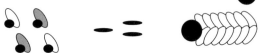

Fig. 3. Regulation of cellular actin assembly. As in Fig. 2, the shading of
actin monomers indicates the form of adenine nucleotide bound: ATP, shaded;
ADP, unshaded. Profilin (black lozenge) accelerates subunit addition on to
barbed filament ends.

of filaments which have barbed ends. Secondly, Oosawa originally used actin filament fragments as nuclei in establishing the concepts of nucleation and elongation in the kinetics of actin assembly. Thirdly, only micromolar concentrations of pure monomeric actin under physiological conditions at steady state remain unpolymerized. If free actin barbed ends are present [21], even in the presence of nearly equivalent molar amounts of β-thymosin, which binds monomeric actin and further impairs its spontaneous nucleation, still only micromolar concentrations or less of actin are unpolymerized. This situation exists especially in the presence of the abundant protein profilin, which accelerates barbed-end elongation by catalysing adenine nucleotide exchange. Hence we have come to believe that actin filament barbed ends are where nucleation takes place. This formulation requires that cells control the exchange of monomers at these barbed ends (Fig. 3).

Our route to this idea arose from our discovery of gelsolin, which provided the first connection between a well-established signalling intermediate, Ca^{2+}, and actin assembly (Fig. 4). Ca^{2+}, at micromolar concentrations, activates gelsolin, which severs the non-covalent bonds between actin monomers in filaments and then binds tightly to their barbed ends. Protons also activate gelsolin, independently of calcium [22]. The rapid actin polymer shortening action caused by severing suggested an efficient mechanism for producing a transition between actin gels and mobile diffusible filament fragments ('sols'), leading to the naming of the protein as 'gelsolin' [23]. Kurth and Bryan showed that the regulation of gelsolin by Ca^{2+} was not strictly reversible, and that gelsolin forms a tight complex with actin that is not dissociated by lowering the Ca^{2+} concentration [24]. This complication emerged as an experimental advantage, as it permitted us to document dynamic gelsolin–actin associations in cells responding to stimulation and to show that cells could dissociate these gelsolin–actin complexes [25]. The experimental approach was to immunoprecipitate gelsolin from extracts of resting and stimulated cells chelated to remove free Ca^{2+} and to document whether gelsolin had sound actin associated with it. Around the same time, we discovered that phosphoinositides are the only

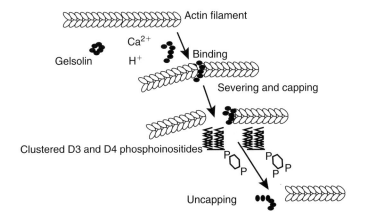

Fig. 4. Action and regulation of gelsolin. See the text for details.

physiological agents capable of separating actin from gelsolin activated by Ca^{2+} or protons [26]. Now it is clear that phosphoinositides reduce the affinity of nearly every actin filament barbed-end capping protein for actin *in vitro* [27]. In addition, phosphoinositides bind to many proteins that cross-link actin filaments [28] and can regulate the function of proteins that link them to the plasma membrane [29]. By concentrating actin filaments at sites of phosphoinositide synthesis or aggregation, these proteins could thereby bring barbed actin filament ends into proximity of uncapping reactions for filament elongation (see Fig. 6) [30].

Studies with the fungal metabolite cytochalasin B established the power of actin filament barbed end capping in controlling actin filament elongation. But the idea of uncapping as a source of cellular actin filament nucleation, especially if it required induced capping associated with severing followed by uncapping, was just too complicated at face value as a way to work the machine. However, studies by John Condeelis and his colleagues with extracts of chemoattractant-stimulated *Dictyostelium* amoebae did imply that the timing of actin polymerization was correlated with the appearance of actin-filament-associated nucleating activity and the release of a capping protein [31], supporting the mechanism inferred from the *in vitro* biochemical studies with gelsolin. Nevertheless, as mentioned above, cell crawling movements and actin assembly occur in cells that do not show increased cytoplasmic Ca^{2+} levels, and bulk changes in phosphoinositide concentrations did not correlate temporally with actin assembly in tissue culture cells or leucocytes [32,33]. Such inconsistencies become targets for vigilant sceptics dedicated to the extermination of silly theories.

Our best case in evidence: actin remodelling in blood platelets

The anucleate platelet has evolved over millenia to protect higher organisms from blood loss following injury without causing inappropriate thrombosis. Derived from a giant polyploid bone marrow precursor, the megakaryocyte, the platelet circulates in trillions of numbers in human blood as flat discs. This discoid shape presumably facilitates access to the vessel wall, where it anticipates thrombogenic stimuli. In our evolutionary heritage these stimuli arose from trauma; at present they all too frequently come from atherosclerosis, and the consequence of platelet shape changes and adhesion is potentially lethal coronary or carotid thromboses, leading to heart attacks and strokes. The advantages of platelets as subjects of studying actin-based cell functions include: (1) their relative simplicity: they remodel actin predominantly sequentially in time rather than in space, so that steps can be more easily separated from one another; nevertheless, they produce the same types of protrusions, lamellae and filopodia characteristic of spreading and locomoting nucleated cells; (2) their abundance of actin and actin-related proteins amenable to biochemical analysis; (3) extensively studied platelet signal transduction mechanisms; (4) they are drawn directly from blood, uninfluenced by cultivation *in vitro*; and (5) they have direct relevance to human physiology and disease.

The unstimulated circulating platelet is rigid, densely packed with chemicals needed for its multiple thrombotic functions, and resistive of high flows and pressures occurring in arterial circulations. This discoid structure derives from a core of about 2000 cross-linked actin filaments investing into a membrane skeleton morphologically indistinguishable from the erythrocyte spectrin-based submembrane lamina. The actin filaments in the platelet core, representing 40% of the total platelet actin (millimolar in concentration and 20% of the total cell protein), are cross-linked and have their fast-growing ends capped [34]. This capping is important. The rest of the platelet actin is monomeric and is bound to thymosin β_4, which prevents its spontaneous nucleation, a step required for elongation. Thymosin β_4, however, unloads actin monomers on to the barbed ends of actin filaments, and would do so in the resting platelet were the filament ends not capped. Attesting to the importance of barbed-end capping, it has been estimated that 40 barbed-end capping proteins exist for every barbed end in a platelet [35].

Stimulation of platelets by thrombin, acting on a heterotrimeric G-protein-linked seven-membrane-spanning receptor, leads to a series of signal transduction events. These include influx of Ca^{2+} from the extracellular fluid and hydrolysis of phosphatidylinositol 4,5-bisphosphate (PtdInsP_2) by phospholipase Cβ, leading to the production of inositol 1,4,5-trisphosphate, which releases Ca^{2+} from intracellular stores. It also generates diacylglycerol, which activates various protein kinase C isoforms. Ca^{2+} initiates the phosphorylation of myosin light chains by myosin light-chain kinase, which is activated by calmodulin-bound Ca^{2+}. Contractility exerted by activated myosin appears to be important for drawing cell surface receptors to the centre of the platelet [36], possibly increasing the internal hydrostatic pressure to provide force for protrusion, while actin filament severing facilitates the transition of the platelet from a discoid to a spherical shape [37].

Following the Ca^{2+}-dependent events described above, actin elongation takes place in two discrete compartments. In one, actin filaments elongate from the barbed ends of the severed filaments, and these filaments constitute an orthogonal network that comprises the substance of an extending circumferential lamella when the platelet is adherent to a surface. At the vertices of this network resides ABP-280. In the platelet, ABP-280 also binds to Gp1b/9, the von Willebrand factor receptor complex. In the second main actin filament compartment appearing in activated platelets, long axial filament bundles form filopodia. This actin filament assembly step is Ca^{2+}-independent. If the Ca^{2+} transient induced by thrombin is prevented by internal chelation, long actin filament coils form and distort the platelet, presumably representing a distribution of monomers on to a relatively small population of uncapped actin filaments rather than on to the larger repository of ends normally formed by Ca^{2+}- and gelsolin-dependent severing [37].

The uncapping of actin filament barbed ends in platelets is driven by formation of membrane phosphoinositides of both the D3 and D4 types (where D3 and D4 are used to denote the presence of phosphate groups at the 3- and 4-

positions respectively of the inositol ring of phosphoinositides). All known actin filament barbed-end capping proteins decrease affinity for actin filament barbed ends in the presence of these phospholipids, although the chemistry of lipid presentation to the protein is important for this effect. In platelets, the rise and fall of intracellular Ca^{2+} and of phosphoinositide levels correlates well with the times at which gelsolin binds and comes off actin filaments, and the expression of free actin filament barbed ends and of actin assembly parallels the decline in Ca^{2+} and the rise in phosphoinositides. Phosphoinositides uncap actin filament barbed ends in permeabilized platelets, and phosphoinositide-binding peptides inhibit this effect. In a permeabilization scheme that preserves signal pathways from the thrombin receptor to actin filament barbed-end uncapping, the active form of the GTPase Rac, previously documented to induce actin-based ruffling activity in fibroblasts [38], promotes PtdInsP_2 synthesis and actin filament barbed-end uncapping [39]. During platelet activation, both gelsolin and capping protein dissociate from the cytoskeletal fraction of platelets, indicating that these two (and presumably additional) capping proteins work co-operatively to regulate actin filament remodelling [40].

Phorbol myristate acetate, which activates protein kinase C, directly promotes filopodial but not lamellar extension, and the steps in the signalling pathway leading to this slower and more limited actin assembly require β_3 integrin activation and D3 phosphoinositide biosynthesis [40]. The thrombin pathway connects with this chain of events, since thrombin-induced hydrolysis generates diacylglycerol, which stimulates protein kinase C. The requirement of D3 phosphoinositides for protein kinase C- and integrin-induced filopodial actin assembly contrasts with thrombin-based lamellar actin assembly, which depends principally on D4 phosphoinositides [41]. Filopodial extension has been linked to activation of Cdc42 [42], and it is not yet clear whether Cdc42 is involved in platelet filopod formation. Since the protein that is mutated in Wiskott–Aldrich syndrome, WASP, interacts with Cdc42, and affected platelets are functionally abnormal, Cdc42 will presumably integrate into our understanding of the pathways of platelet actin assembly.

In summary, the platelet has helped us to link biochemical information about the regulation of actin assembly by known actin-binding proteins with emerging data concerning certain GTPases, and recent results with other cell types further this integration (Fig. 5). Manipulation of phosphoinositides in cells by extracellular instillation or by genetic overexpression of phosphoinositide-synthesizing or -degrading enzymes leads to actin-based ruffling and other actin changes [43–45], although these results do not address whether the lipids are working directly to promote actin assembly. D3 phosphoinositides, for example, are now implicated in the activation of GTPases, and inhibition of D3 phosphoinositide synthesis inhibits ruffling and chemotaxis in fibroblastic cells or in growth-factor- (but not fMet-Leu-Phe)-stimulated neutrophils [46], consistent with results observed in thrombin-stimulated platelets [47].

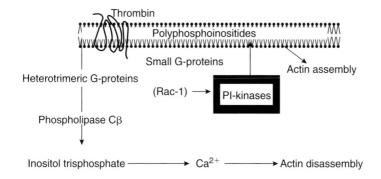

Fig. 5. Signal transduction pathways controlling actin remodelling in platelets. Abbreviation: PI, phosphoinositide. See the text for details.

Relating the platelet results to other crawling cells

There are many members of the gelsolin protein family, and several members reside in the same cell. Moreover, recent evidence indicates that a class of abundant cellular proteins (cofilin, destrin, actin-depolymerizing factor) with weak actin filament severing activity [48] regulated by serine phosporylation [49] is important and possibly essential for cellular actin depolymerization [50]. Therefore it is not surprising that gelsolin knockout mice develop normally (indeed, the mouse embryo does not express gelsolin until day 13), although slower platelet shape changes and leucocyte and fibroblast locomotion *in vitro* correlate with prolonged bleeding times, delayed inflammatory responses and impaired wound healing in adult mice *in vivo*.

The behaviour of gelsolin-null cells in culture is even more interesting [51]. Compared with wild-type fibroblasts, gelsolin-null dermal fibroblasts have markedly reduced ruffling responses to platelet-derived growth factor or epidermal growth factor, which signal through Rac. Bradykinin-induced filopodial formation, attributable to activation of Cdc42, is similar in both cell types. Wild-type fibroblasts exhibit typical lamellipodial extension during translational locomotion, whereas gelsolin-null cells move at 50% lower rates using structures resembling filopodia. Multiple gelsolin-null tissues, as well as gelsolin-null fibroblasts, overexpress Rac by 5-fold, but have normal levels of Cdc42 or Rho. Re-expression of gelsolin, by stable transfection or adenovirus, in gelsolin-null fibroblasts results in reversion of the ruffling response and Rac expression to normal. Rac migrates to the cell membrane in a step requiring D3 phosphoinositide synthesis following epidermal growth factor stimulation in both wild-type and gelsolin-null cells. Therefore gelsolin is an essential effector of Rac-mediated actin dynamics, acting downstream of the recruitment of Rac to the membrane. The gelsolin-null cell is able to use a Cdc42-related filopodial mechanism as a compensatory but suboptimal method of locomotion. Just as cross-communication between structural (gelsolin) and signalling (Rac) molecules is being elucidated in this case, it is very likely that similar interactions will emerge in the system that leads to filopodial extension.

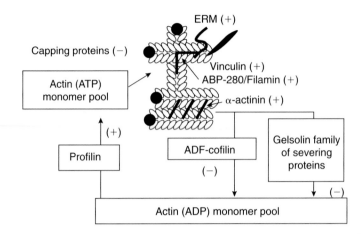

Fig. 6. Phosphoinositide regulation of actin remodelling. + indicates stimulation; − indicates inhibition. Abbreviations: ERM, ezrin-radixin-moesin; ADF, actin-depolymerizing factor.

In summary, in our view the field has progressed markedly in understanding the reactions that control cellular actin remodelling and cellular locomotion. Our own interests emphasize the importance of actin filament barbed-end uncapping in controlling actin assembly in time and space, the organization of elongated actin filaments into distinct three-dimensional structures by actin filament cross-linking proteins, and the co-ordinated control of both processes by phosphoinositide metabolism (Fig. 6). The diversity of cellular functions powered by actin remodelling certainly suggests that other important control mechanisms exist. The ongoing vigorous research programme into host cellular actin remodelling by micro-organisms is identifying these controls [52–54].

Practical spin-offs

Cytoskeletal components are usually intracellular, being viewed as 'structural' or 'housekeeping' constituents of cells and therefore considered as unlikely targets for pharmaceutical development. However, our cell motility research may have clinical ramifications.

Gelsolin has an extracellular isoform that is presumed to participate, along with the vitamin D-binding protein (also called Gc globulin), in scavenging the abundant actin that is released with cell debris during injury or inflammation [55,56]. One site of intense inflammation is the airway of patients with cystic fibrosis, and leucocyte-derived DNA makes airway secretions very viscous, exacerbating the inflammatory damage. Leucocytes also contain abundant actin, and actin and DNA interact to produce large fibre bundles [57]. Gelsolin decreases the viscosity of sputum expectorated by patients with cystic fibrosis or chronic bronchitis, suggesting that inhalational therapy with this protein might have a therapeutic benefit in inflammatory airway disease [58]. This hypothesis is currently being tested in clinical trials.

The discovery that phosphoinositides regulate actin assembly by uncapping actin filament barbed ends suggested a mechanism for and a method of prevention of the mysterious phenomenon of platelet activation in the cold. A membrane phase transition at 15°C, the temperature at which platelets change shape, might cluster phosphoinositides. Chilling also impairs Ca^{2+} extrusion. Thus Ca^{2+} accumulation in the cell, together with lipid aggregation, would mimic the signalling changes occurring during normal platelet activation. We have shown that cooling platelets results in actin assembly and actin filament barbed-end uncapping. These reactions and the platelet shape change can be inhibited by a permeant Ca^{2+} chelator and by cytochalasin B [59]. These observations are being applied to the development of a method for the liquid cryopreservation of platelets, a technology not hitherto possible.

The research described was supported by USPHS (NIH) grants HL19429, HL07680, HL56252, AR38910 and HL54188, and by grants from the Edwin S. Webster Foundation.

References

1. Italiano, J., Roberts, T., Stewart, M. and Fontana, C. (1996) Cell **84**, 105–114
2. Mogliner, A. and Oster, G. (1996) Biophys. J. **71**, 3030–3045
3. Verkhovsky, A., Svitkina, T. and Borisy, G. (1997) J. Cell Sci. **110**, 1693–1704
4. Burridge, K. and Chrzanowska-Wodnicka, M. (1996) Annu. Rev. Cell Dev. Biol. **12**, 463–519
5. Lauffenburger, D. and Horwitz, A. (1996) Cell **84**, 359–369
6. Kimura, K., Ito, M., Amano, M., Chihara, K., Fukata, Y., Nakafuku, M., Yamamori, B., Feng, J., Nakano, T., Okawa, K., et al. (1996) Science **273**, 245–248
7. Alteraifi, A. and Zhelev, D. (1997) J. Cell Sci. **110**, 1967–1977
8. Coluccio, L. Myosin I. (1997) Am. J. Physiol. **273**, C347–C359
9. Burns, C., Reedy, M., Heuser, J. and De Lozanne, A. (1995) J. Cell Biol. **130**, 605–612
10. Cramer, L. and Mitchison, T. (1995) J. Cell Biol. **131**, 179–189
11. Ito, T., Zaner, K.S. and Stossel, T.P. (1987) Biophys. J. **51**, 745–753
12. Hartwig, J. (1994) in Protein Profile (Sheterline, P., eds.), pp. 711–778, Academic Press, London
13. Maestrini, E., Patrosso, C., Mancini, M., Rivella, S., Rocchi, M., Repetto, M., Villa, A., Frattini, A., Zoppè, M., Vezzoni, P. and Toniolo, D. (1993) Hum. Mol. Genet. **2**, 761–766
14. Barry, C., Xie, J., Lemmon, V. and Young, A. (1993) J. Biol. Chem. **268**, 25577–25586
15. Gorlin, J., Yamin, R., Egan, S., Stewart, M., Stossel, T.P., Kwiatkowski, D.J. and Hartwig, J.H. (1990) J. Cell Biol. **111**, 1089–1105
16. Ito, T., Suzuki, A. and Stossel, T. (1992) Biophys. J. **61**, 1301–1305
17. Small, J., Herzog, M. and Anderson, K. (1995) J. Cell Biol. **129**, 1275–1286
18. Cunningham, C., Gorlin, J., Kwiatkowski, D., Hartwig, J., Janmey, P., Byers, H. and Stossel, T. (1992) Science **255**, 325–326
19. Janmey, P., Hvidt, S., Käs, J., Lerche, D., Maggs, A., Sackmann, E., Schliwa, M. and Stossel, T. (1994) J. Biol. Chem. **269**, 32503–32513
20. Cunningham, C. (1995) J. Cell Biol. **129**, 1589–1599
21. Carlier, M.-F. and Pantaloni, D. (1997) J. Mol. Biol. **269**, 459–467
22. Lamb, J., Allen, P., Tuan, B. and Janmey, P. (1993) J. Biol. Chem. **268**, 8999–9004

23.   Yin, H.L. and Stossel, T.P. (1979) Nature (London) **281**, 583–586
24.   Kurth, M. and Bryan, J. (1984) J. Biol. Chem. **259**, 7473–7479
25.   Lind, S.E., Janmey, P.A., Chaponnier, C., Herbert, T. and Stossel, T.P. (1987) J. Cell Biol. **105**, 833–842
26.   Janmey, P. (1994) Annu. Rev. Physiol. **56**, 169–191
27.   Schafer, D., Jennings, P. and Cooper, J. (1996) J. Cell Biol. **135**, 169–179
28.   Fukami, K., Endo, T., Imamura, M. and Takenawa, T. (1994) J. Biol. Chem. **269**, 1518–1522
29.   Hirao, M., Sato, N., Kondo, T., Yonemura, S., Monden, M., Sasaki, T., Takai, Y., Tsukita, S. and Tsukita, S. (1996) J. Cell Biol. **135**, 37–51
30.   Mackay, D., Esch, F., Furthmayr, H. and Hall, A. (1997) J. Cell Biol. **138**, 927–938
31.   Hall, A., Warren, V., Dharmawardhane, S. and Condeelis, J. (1989) J. Cell Biol. **109**, 2207–2213
32.   Eberle, M., Traynor-Kaplan, A., Sklar, L. and Norgauer, J. (1990) J. Biol. Chem. **265**, 16725–16728
33.   Zigmond, S., Joyce, M., Borleis, J., Bokoch, G. and Devreotis, P. (1997) J. Cell Biol. **138**, 363–374
34.   Hartwig, J. and DeSisto, M. (1991) J. Cell Biol. **112**, 407–425
35.   Barkalow, K. and Hartwig, J. (1995) Biochem. Soc. Trans. **23**, 451–456
36.   Kovacsovics, T. and Hartwig, J. (1996) Blood **87**, 618–629
37.   Hartwig, J. (1992) J. Cell Biol. **118**, 1421–1442
38.   Ridley, A. (1994) BioEssays **16**, 321–327
39.   Hartwig, J., Bokoch, G., Carpenter, C., Janmey, P., Taylor, L., Toker, A. and Stossel, T. (1995) Cell **82**, 643–653
40.   Barkalow, K., Witke, W., Kwiatkowski, D. and Hartwig, J. (1996) J. Cell Biol. **134**, 389–399
41.   Hartwig, J., Kung, S., Kovacsovics, T., Janmey, P., Cantley, L., Stossel, T. and Toker, A. (1996) J. Biol. Chem. **271**, 32986–32993
42.   Nobes, C. and Hall, A. (1995) Cell **81**, 53–62
43.   Derman, M., Toker, A., Hartwig, J., Spokes, K., Falck, J., Chen, C.-S., Cantley, L. and Cantley, L. (1997) J. Biol. Chem. **272**, 6465–6470
44.   Sakisaka, T., Itoh, T., Miura, K. and Takenawa, T. (1997) Mol. Cell. Biol. **17**, 3841–3849
45.   Shibasaki, Y., Ishihara, H., Kizuki, N., Asano, T., Oka, Y. and Yazaki, Y. (1997) J. Biol. Chem. **272**, 7578–7581
46.   Thelen, M., Uguccioni, M. and Bösiger, J. (1995) Biochem. Biophys. Res. Commun. **217**, 1255–1262
47.   Kovacsovics, T., Bachelot, C., Toker, A., Vlahos, C., Duckworth, B., Cantley, L. and Hartwig, J. (1995) J. Biol. Chem. **270**, 11358–11366
48.   McGough, A., Pope, B., Chiu, W. and Weeds, A. (1997) J. Cell Biol. **138**, 771–781
49.   Agnew, B., Minamide, L. and Bamburg, J. (1995) J. Biol. Chem. **270**, 17582–17587
50.   Theriot, J. (1997) J. Cell Biol. **136**, 1165–1168
51.   Witke, W., Sharpe, A., Hartwig, J., Azuma, T., Stossel, T. and Kwiatkowski, D. (1995) Cell **81**, 41–51
52.   Gertler, F., Niebuhr, K., Reinhard, M., Wehland, J. and Soriano, P. (1996) Cell **87**, 227–239
53.   Southwick, F. and Purich, D. (1995) Infect. Immun. **63**, 182–190
54.   Welch, M., Iwamatsu, A. and Mitchison, T. (1997) Nature (London) **385**, 265–269
55.   Herrmannsdoerfer, A., Heeb, C., Reustel, P., Estes, J., Keenan, C., Minnear, F., Selden, L., Giunta, C., Flor, M. and Blumenstock, F. (1993) Am. J. Physiol. **265**, G1071–G1081
56.   Lee, W. and Galbraith, R. (1992) N. Engl. J. Med. **326**, 1335–1341
57.   Sheils, C., Käs, J., Travassos, W., Allen, P., Janmey, P., Wohl, M. and Stossel, T. (1996) Am. J. Pathol. **148**, 919–927
58.   Vasconcellos, C., Allen, P., Wohl, M., Drazen, J., Janmey, P. and Stossel, T. (1994) Science **263**, 969–971
59.   Winokur, R. and Hartwig, J. (1995) Blood. **85**, 1796–1804

Biochem. Soc. Symp. **65**, 281–299
Printed in Great Britain

17

Using molecular genetics as a tool in understanding crawling cell locomotion in myoblasts

M. Peckham[*][1]**, C. Wells**[*]**, P. Taylor-Harris**[*]**, D. Coles**[*]**,
D. Zicha**[†] **and G. A. Dunn**[†]

[*]Muscle and Cell Motility Research Centre and [†]MRC Muscle and Cell Motility
Unit, The Randall Institute, King's College London, 26–29 Drury Lane, London
WC2B 5RL, U.K.

Abstract

We have used digitally recorded interference microscopy with automatic phase shifting (DRIMAPS) to investigate the crawling locomotion of normal and mutant mouse myoblasts. Contraction forces that give rise to cell body movement, tail retraction and cell adhesion to the substrate in myoblasts and other locomoting tissue cells arise from the interactions of actin and non-muscle myosin II. The activity of non-muscle myosin II is regulated differently from that of skeletal myosin. Using DRIMAPS, we found that crawling locomotion was altered in myoblasts that heterologously expressed human β-cardiac myosin heavy chain (MHC); the cells moved more slowly and had reduced rates of protrusion and retraction. Immunolocalization demonstrated that MHC and non-muscle myosin II were not co-localized, suggesting that MHC does not compete directly with myosin II, but interferes with cell locomotion by binding inappropriately to actin filaments and possibly cross-linking them. Myosin I may be involved in protrusion of the lamellipodia. However, using DRIMAPS, we found that crawling locomotion was unaltered in myoblasts that heterologously expressed a truncated myosin I which lacked the membrane-binding tail domain. This suggests that, if endogenous myosin I is important for cell locomotion, this mutant was unable to interfere with its action. We conclude that the effects on locomotion of expressing foreign or mutant proteins of the cytoskeleton in vertebrate cells can be subtle and can be swamped by the intrinsic variability of the cells. Their characterization requires automated methods of acquiring data, such as

[1]To whom correspondence should be sent, at present address: School of Biomedical
Sciences, University of Leeds, Leeds LS2 9JT, U.K.

DRIMAPS, and careful statistical analysis in order to take account of other sources of variation.

Introduction: crawling cells and involvement of actin and myosins

A crawling cell moves forward over a surface by (1) protruding processes; (2) forming adhesions to the surface; (3) drawing the cell body towards the new adhesions by traction; and (4) detaching and retracting the tail region. Although highly interdependent, these events can happen separately and discontinuously. In the case of fibroblasts or fibroblast-like cells such as myoblasts, the protrusion is usually a thin lamellar sheet or lamellipodium, 100–150 nm thick, which does not generally advance continuously but undergoes cycles of extension and withdrawal [1]. Withdrawal is often accompanied by a folding backwards to make membranous folds or ruffles in the dorsal surface of the cell that travel rearwards, away from the leading edge [2].

The traction force required for cell locomotion probably results from the interaction between myosin II and actin (see, for example, Chapter 12 in this volume). In fibroblasts, non-muscle myosin II is co-localized with oblique microfilament bundles that contain actin (often called stress fibres), each of which terminates at one end in a focal adhesion [3]. This co-localization results in an approximately sarcomeric organization [4]. Contraction produced by the interaction of actin and myosin in these structures could draw the cell body towards the site of adhesion to the substratum [5,6].

However, while the stress fibres doubtless exert force on the substratum, they can hardly be considered to be the sole engine of tractive force, since the most motile cells tend to have fewer or even no such microfilament bundles [7]. Moreover, the most motile cells tend to exert the least force on the substratum [8], and an alternative function of stress fibres is to generate intracellular tension in fibroblasts and hence to exert the forces that fibroblasts use in their primary role of remodelling the extracellular matrix (see Chapter 3 in the present volume). Thus the tension generated by stress fibres is probably far greater than that required to pull the cell forward against external resistance; however, the possibility cannot be ruled out that the stress fibres do take over this function when they are present in fibroblasts.

There are now known to be several different types of organization of microfilament bundles other than stress fibres in moving fibroblasts, with different patterns of polarity and of association with adhesions, and at least some of these may contribute to the force of cell traction. For example, there are ventral actin bundles with a non-sarcomeric, graded polarity that span the whole length of the cell [9], and transverse arc-shaped bundles [10] that are commonly seen in fibroblasts. Nevertheless, the focus of attention has now largely turned away from bundles towards the more open meshwork of microfilaments that pervades the cell cytoplasm and is particularly rich in the cortex underlying the cell surface. The actin/myosin-mediated contraction of this meshwork may exert traction via close contacts between the cell and the substratum which are more amorphous and changeable than the focal contacts associated with micro-filament bundles [11,12].

Recent work has revealed that the open meshwork of microfilaments in fish keratinocytes has quite a high degree of organization, with myosin II clusters embedded in loose transverse ribbons of actin filaments ([13]; Chapter 12 in the present volume). In fact, it seems that the contraction of this meshwork in keratinocytes drives the forward translocation of the cell body and generates a transverse arc-shaped bundle of microfilaments at the rear of the leading lamella (Chapter 12). Thus it is probable that the actomyosin meshwork and the bundle co-exist in a state of dynamic equilibrium, contracting more or less continuously and constantly being assembled and disassembled, but maintaining a more or less stable pattern. Fibroblasts can appear to behave somewhat differently, and certainly do not maintain a stable pattern of organization, but the transverse arc-shaped bundles commonly seen in fibroblasts [10] strongly suggest that a similar mechanism of traction based on actin and myosin II may be operating in these cells.

Retraction of the tail region is known to involve active contraction which is equally likely to depend on actin and myosin II [14]. In fact, it may be misleading to treat tail retraction as a separate phase of locomotion from forward traction of the cell body. Fibroblasts are highly variable in morphology and often develop long tails, which are not prevalent in keratinocytes. When a fibroblast tail detaches from the substratum, either spontaneously or as a result of experimental manipulation, there is a rapid shortening of the tail as tension is released, but this is followed by an active retraction of the tail and a simultaneous traction of the nuclear region of the cell towards its foremost adhesions.

The mechanism of lamellipodial protrusion is not so well understood, and it is much less clear whether the interaction of non-muscle myosin II and actin contributes to it. The evidence against is that non-muscle myosin II is mostly absent from lamellipodia, although towards the edge of keratinocytes there are punctate areas of myosin II in which the filaments have a zig-zag appearance and are possibly the precursors of the ribbon structure in the leading lamella of the cell [13]. In *Dictyostelium*, furthermore, mutant amoebae which do not express myosin II can still form pseudopodia, but do so more slowly and less efficiently than wild-type cells [15].

Lamellipodial protrusion may not require myosin at all, but may depend simply on the polymerization of actin [16]. Certain specialized instances of protrusion, such as the formation of the acrosomal process in *Thyone* and other invertebrates, are known to depend simply on actin polymerization. In the case of vertebrate tissue cells, it has been shown that actin filaments at the leading edge are oriented with their barbed ends (also known as plus or fast-growing ends) pointing forward into the membrane, which raises the problem of how actin monomers push themselves in between the end of the growing filament and the membrane [17]. A thermal ratchet mechanism has been proposed to account for this [18], but the highest theoretical rates of operation of a thermal ratchet seem to fall short of some measured rates of protrusion in tissue cells [19].

An alternative view is that, although actin polymerization does occur at the leading edge of the cell, the driving force for protrusion comes from a positive internal hydrostatic pressure within the cell ([20]; see also Chapter 16 in the present volume) which could involve non-muscle myosin II and actin. The idea that internal hydrostatic pressure may be required for protrusion has been around for some time, and it is known that the protrusion of blebs and cell locomotion can be prevented by reverse osmotic pressure caused by hypertonic media [20,21]. There is evidence that hydrostatic pressure in epidermal cells is dependent on Ca^{2+}-mediated contractions of the actomyosin system and that this internal pressure is linked to cell locomotion [22].

Evidence for the possible involvement of non-muscle myosin II in the mechanism of protrusion of tissue cells is largely circumstantial, and the issue remains an open question. However, non-muscle myosin II is not the only myosin found in crawling cells. During the last 10 years many new myosins, in addition to non-muscle myosin II, have been discovered. The functions of most of these myosins in the cell, and whether they have roles in cell locomotion, are still largely unknown. Whereas these myosins share a reasonably conserved head domain which binds to actin and hydrolyses ATP, the tail domains are highly variable, and only class II myosins (which include non-muscle myosin II) are able to form thick filaments. Based on their sequence diversity, there are now thought to be at least 13 different classes of myosin [23–26]. A single cell type can express at least five or six different classes of myosin [27,28].

This raises the possibility that another myosin molecule might be involved in extending the leading edge of the cell [17]. Candidate myosins that could be involved in protrusion are the myosin I and myosin V isoforms, as they are both found at the leading edge of crawling cells [28–31] (Fig. 1). Myosin I is the most likely myosin isoform to play a role in protrusion. It is a single-headed molecule which has a membrane-binding domain in its tail. It is a barbed-end-directed motor, so if it pulls on actin while it is attached to a fixed point on the membrane, it will tend to pull the end of the actin filament away from the membrane. This gap would enable a new actin monomer to add on to the actin filament. Alternatively, if myosin I pulls on the actin filament while it is attached to the membrane, this could result in the membrane associated with myosin or actin being displaced, causing the membrane to bend, as may happen when the lamellipodia retracts.

It is not clear how myosin I is involved in cell protrusion. Knocking out the expression of myosins IA and IB in *Dictyostelium* suggested that neither of these myosins is involved in protrusion [32,33]. Knocking out the function of myosin IB in nerve-growth cones by microchromophore-assisted laser inactivation [34] expanded the lamellipodia, and overexpression of myosin IB in *Dictyostelium* reduced both the rate of cell migration (to 35% of wild-type levels) and actin-based protrusive structures [35]. Both of these experiments suggest that myosin I may even inhibit or reduce the rate of protrusion. Following up this idea, myosin inhibitors such as butanedione monoxime stimulate protrusion, but also block the process of retrograde flow, i.e. the transport of actin away from the leading edge of the cell [36]. This process is not blocked by cytochalasin, which inhibits actin polymerization.

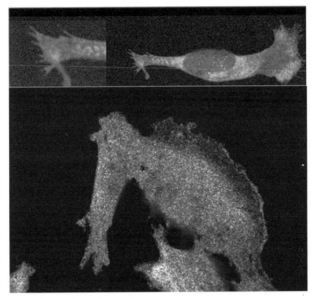

Fig. 1. Immunolocalization of myosin Iα and myosin V in myoblasts.
The upper panel shows a myoblast immunostained for myosin Iα; the inset to the left-hand side is an enlarged image of the tail region of the myoblast to show that myosin Iα can be found in the rear microspikes of the cell. Myosin Iα can clearly be found in the leading edge of the cell in the main picture (upper panel). The lower panel shows a myoblast immunostained for myosin V. The majority of the staining is in the cell body; however, some staining can be found at the leading edge of the cell. The antibody to myosin Iα was kindly provided by Martin Bähler (University of Munich, Munich, Germany); the antibody to myosin V was kindly provided by Richard Cheney (University of North Carolina at Chapel Hill, NC, U.S.A.). Confocal micrographs were taken with the expert guidance of Alan Entwistle. Magnification, 20 μm = 40 mm.

Myosin V is less likely to play a direct role in protrusion, as it is thought to be an organelle motor. It is a two-headed molecule, which is known to transport vesicles containing melanin (melanosomes) [37] and must therefore be able to bind to membranes as well as to actin. Its localization at the leading edge could simply arise because it has transported a cargo there, although it is not known what this cargo is in myoblasts. The cargo could be vesicles, protein or RNA. The kinetics of the actomyosin ATPase cycle are different for myosin V compared with myosin I or myosin II, which means that it tends to remain attached to actin for a much greater proportion of the ATPase cycle. This could enable it to act more like a processive motor such as kinesin [38], enabling it to act as an organelle motor.

Molecular genetic approaches to investigating cell locomotion

Molecular genetics has been used to mutate myosin isoforms to try to learn more about their contribution to cell locomotion, either by knocking out expression of the protein of interest by gene targeting or by expressing mutant isoforms that interfere with the function of the endogenous protein (dominant-negative approach). Most of this work has been undertaken in *Dictyostelium*. Gene targeting is relatively straightforward in *Dictyostelium* because this organism has a haploid genome, which means that it has only a single copy of each gene, and hence a single homologous recombination event will result in a targeted gene. Further, the frequency at which homologous recombination occurs is high. For example, the expression of myosin II and several myosin I isoforms has been knocked out by gene targeting and antisense approaches in *Dictyostelium* [32,33,39–41]. Overexpression and expression of dominant-negative mutants have also been used to address questions about the function of myosin isoforms in *Dictyostelium*.

While these experiments are relatively easy to carry out, the results may be difficult to interpret. Often when the expression of a protein is knocked out, this is lethal to the cells and no mutant cells will be recovered. Alternatively, the cells may be able to recruit another protein or protein pathway to compensate for the loss of the targeted gene, and the mutant cells will not have a distinctive phenotype. For example, *Dictyostelium* expresses many different isoforms of myosin I, and knocking out these isoforms appears, disappointingly, to have very little effect on cell locomotion. Only when three isoforms were knocked out together did cell behaviour change subtly ([32,33]; reviewed in [42]). Similarly, when a dominant-negative protein is expressed, it too can potentially be lethal to the cells, or it may not produce a distinctive phenotype. Expression of dominant-negative mutants of myosin IB in *Dictyostelium* had little effect on the cells, but overexpression of the full-length myosin IB decreased the instantaneous velocity to 35% of that in wild-type cells [35]. Conversely, myosin IB-null cells had increased numbers of pseudopodia [31].

Although genetic manipulations of the motor proteins in *Dictyostelium* can be informative about cell locomotion, the results are not necessarily directly applicable to the crawling cell behaviour of mammalian cells such as fibroblasts, due to differences in their locomotion [43–45]. These differences arise mainly from differences in the organization of the actin cytoskeleton, which is much simpler in *Dictyostelium* amoebae compared with fibroblasts. As described above, fibroblasts have a higher order of cytoskeletal organization, an actin-rich lamellipodium, stress fibres and high orders of F-actin assemblies; the latter two are often absent from fast-moving *Dictyostelium* amoebae. Part of the function of the cytoskeleton in fibroblasts is to attach the cells to the substratum.

Molecular genetics should therefore also be used to answer questions about cell locomotion in mammalian cells such as fibroblasts. However, it is much more difficult to carry out equivalent gene-targeting experiments, because these somatic cells are diploid. Both alleles of the gene must be targeted in two consecutive experiments in order to knock out expression of a gene. Even more problematical is the fact that most cultured cell lines are no longer diploid, but have lost karyotype. This means that if gene targeting were to be attempted in these cells, upwards of four alleles would have to be targeted, a pretty hopeless task! However, some somatic cells have been used successfully in gene targeting experiments when they have a normal karyotype, and gene targeting using embryonic stem cells (undifferentiated cells derived from mouse embryos), which can be used to make transgenic mice, is now relatively straightforward.

So far, this type of approach has not been used to investigate the functions of the non-muscle myosins in crawling cell locomotion in mammalian cells. Dominant-negative mutants of myosin I and myosin IX have been expressed in mammalian cells [46,47], but potential alterations in cell locomotion were not investigated. However, gene targeting and overexpression have been used, with some success, to examine the functions of proteins that bind to actin and are involved in focal adhesions, and their effects on locomotion in mammalian cells. The locomotory activity of fibroblasts is increased when gelsolin, an actin-binding protein which severs actin filaments, is overexpressed [48], and inhibited when gelsolin expression is knocked out using fibroblasts derived from gelsolin-null transgenic mice [49]. Locomotory activity is decreased when vinculin, an actin-binding protein involved in focal adhesion formation, is overexpressed [50], but is increased when its expression is knocked out either by antisense (in 3T3 cells [51]) or by gene targeting in embryonic stem cells [52].

Measuring parameters of cell locomotion

Using molecular genetics to mutate proteins that may play a role in crawling cell locomotion will only be informative if there is a good method of analysing any resultant changes in cell behaviour. In general the effects of many of the mutations described above have been limited to the investigation of cell speed and speed of spreading, or to the morphology of fixed cells. Potential changes to the rate of lamellipodial extension, cell body movement or tail retraction have not usually been investigated. Furthermore, cell speed has not always been measured directly but has been estimated by measuring, for example, the rate at which cells repopulate an artificial 'wound' introduced into a culture, or the rate at which they invade a filter membrane in the Boyden chamber. Such techniques are unlikely to reveal small or subtle effects of mutations, although they can be effective for screening large numbers of experimental populations.

A major problem in characterizing the locomotion of freely moving, isolated cells in culture is the large variability between individual cells, even from the same clone, and the large variation in motile characteristics with time. Thus the detection of subtle experimental effects requires the analysis of large numbers of cells. This necessitates that data be gathered automatically by computer, but conventional forms of light microscopy for living cells are not ideally suited to automatic image analysis. In phase-contrast or differential interference contrast images, for example, cellular material can appear both brighter and darker than the cell-free background, which makes automatic recognition of cell outlines very difficult. However, high-speed, optical-sectioning differential interference contrast microscopy has enabled changes in the three-dimensional space occupied by a moving cell to be reconstructed [53], and interference reflection microscopy has been used successfully to investigate locomotion in *Dictyostelium* [54]. The case of vertebrate fibroblasts is more difficult, because their lamellipodial protrusions can be very thin, but a technique which can give accurate and detailed information about their locomotion is digitally recorded interference microscopy with automatic phase shifting (DRIMAPS) [55,56].

The DRIMAPS system produces digital images in which the intensity at each point is directly proportional to the amount of non-aqueous cellular material in the light path. It depends on the fact that all materials within a cell increase its refractive index in direct proportion to the concentration of material present. Proteins, lipids and nucleic acids all have very similar effects. This means that, when a beam of light passes through a cell, it is slowed down or retarded in direct proportion to the dry mass of cellular material encountered. In the interference microscope, the illuminating beam of light is split into two; one beam passes through a chamber containing cells, while the other (reference beam) passes through a dummy chamber without cells. When the two beams of light are recombined, they interfere with each other so that differences in intensity in the interference image depend, among other things, on how much the different parts of the cells have retarded the light. Phase-shifting is a method of accurately quantifying this interference effect. Three or more interference images are acquired while the reference beam is progressively retarded to one wavelength. These interference images contain sufficient information to construct a phase-shifted image in which the intensity at each point is directly proportional to the amount by which the light has been retarded by the cells. This final phase-shifted image is thus a contour map of the distribution of non-aqueous cellular material.

In the DRIMAPS system, eight interference images are captured within a fraction of a second while phase-shifting is carried out under computer control, and a single phase-shifted image is calculated from these during the interval between frames. Typically, 1200 such images are acquired at 1 min intervals and stored during a recording session. After further processing to remove drifts and distortions in the wavefront settings of the interference microscope, the sequence of images is tracked by operating a mouse to identify individual cells throughout the sequence. The mass and area moments which describe size, location, orientation, shape and mass distribution are then calculated for each identified cell throughout the sequence, and differences between consecutive frames are used to calculate the positions and areas of the sites of protrusion and retraction of the cell margin. Clusters of identified cells in contact are automatically recognized by the computer; in the analysis presented below, only data from single isolated cells were used. A continuous sequence or run of data from a single cell is interrupted when the cell either contacts another cell or leaves the recording field. Only runs longer than 60 min were used in the following analysis, and these were partitioned into 60-min runs, discarding any remainder. For each parameter of cell behaviour, the values obtained throughout each 60-min run were averaged to give a single value to be entered into the analysis.

The DRIMAPS system can provide sufficiently large data sets to alloow the detection of quite subtle changes in cell behaviour, but the analysis of such data is strewn with pitfalls. In analysing the effects of an experimental treatment on a parameter's values, it is critically important to take account of random variations between cultures and between cells within a culture. Simply applying a t-test to all the data from control and treated cells, as is often done in published analyses, can yield very misleading significance levels. In fact, with large volumes of data, most parameters yield highly significant differences when tested in this way, irrespective of whether the experimental treatment has a true effect! The reason is that the t-test assumes that all values within each group are drawn independently from a population, whereas values drawn at different times from the same cell, or from different cells in the same culture, are not generally independent. The correct way to handle such data is to nest them, first by cells and then by cultures and by treatments etc., and to apply a hierarchical analysis of variance (ANOVA), which tests the significance of the variation at each level of nesting. The critical null hypothesis to be tested is usually that the treatments do not introduce a significant difference in the parameter's values in addition to the random variations between cultures, between cells and within cells. It is especially important when transfecting cells to take additional account of random variation between cell clones, since any two clones may be significantly different in some aspects of cell behaviour regardless of any different treatments that they may have received. Standard ANOVA tests require balanced data (equal numbers of observations within groups at any level) but, fortunately, methods are now available for applying ANOVAs to unbalanced data [57,58].

Using DRIMAPS to investigate changes in crawling cell locomotion in myogenic cells transfected with human β-cardiac myosin II

Locomoting myoblasts express non-muscle myosin, but they do not start to express skeletal myosin until they have become post-mitotic and have begun terminal differentiation into muscle myotubes. One reason for this may be because the activity of skeletal myosin is regulated differently from that of non-muscle myosin II [59]. Skeletal myosin is inhibited from interacting with actin by the regulatory proteins tropomyosin and troponin present on the actin-containing thin filament in muscle, whereas non-muscle myosin II is regulated by phosphorylation of the heavy chain, which controls its ability to form filaments, and by phosphorylation of the light chains, which controls its ability to interact with actin. Skeletal myosin cannot be regulated by troponin/tropomyosin in the locomoting myoblasts, since, although they express tropomyosin isoforms, they do not express troponin.

Expression of foreign skeletal myosin in locomoting myoblasts might be expected to interfere with the function of non-muscle myosin II, due to this difference in regulation. To test this idea, we heterologously expressed full-length human β-cardiac myosin II heavy chain (MHC) and a truncated fragment of this protein (t-MHC; containing the first 954 amino acids encoding the subfragment-1 region and 146 amino acid residues of the rod) in proliferating myogenic cells in culture, using a pCMV5 expression vector [60]. Permanently transfected clones were analysed for expression of β-cardiac myosin by Western analysis and immunofluorescence using antibodies specific for this isoform. Immunofluorescence of myogenic cells expressing full-length MHC showed that it had a filamentous appearance, but was not co-localized with non-muscle myosin II, whereas t-MHC had a punctate distribution in the cytoplasm similar to that observed previously in COS cells [61].

In total, 27 cultures of clones expressing t-MHC, MHC, the pCMV5 vector alone (NULL) or the original wild-type clone (W/T) were recorded for 20 h for the DRIMAPS analysis (Table 1). The results were analysed by ANOVA tests, and this analysis demonstrated that ANOVAs are essential for distinguishing random sources of variation from any variation due to the transfections. In the ANOVA tests the upper level of data nesting compared the two groups expressing β-cardiac myosin sequences (MHC and t-MHC) with those not expressing β-cardiac myosin (W/T and NULL). Significant differences at this level would indicate a significant effect of expression of human β-cardiac myosin II (whether MHC or t-MHC). Significance at the next level of nesting indicates either that W/T and NULL differ significantly from each other, or that MHC and t-MHC differ significantly from each other, or both. No parameter tested revealed significance at this level. At lower levels, significant differences between cultures often arose, and differences between cells were usually highly significant; however, these are likely to be random sources of variation.

More than 20 parameters of cell growth, spreading, shape and motility were tested by ANOVA, but we will discuss only three in detail in the following brief summary of the results.

An important feature of the DRIMAPS system is that it allows the growth of individual cells to be monitored in order to ascertain whether they remain healthy throughout the recording period. It is clear that all four transfection groups were growing at a mean rate of around 4%·h^{-1}, which is equivalent to a mass doubling time of about 17 h, indicating that culture conditions were optimal (Fig. 2A). The ANOVA test of cell growth revealed no significant difference due to the expression of β-cardiac myosin sequences, and no significant differences between W/T and NULL or between MHC and t-MHC. There were highly significant differences between cultures and marginally significant differences between cells within a culture.

The speed of cell movement in the two transfection groups expressing β-cardiac myosin was suppressed compared with cells in the W/T and NULL groups (Fig. 2B) [mean values (μm·min^{-1}): W/T, 1.05; NULL, 0.94; MHC, 0.61; t-MHC, 0.67]. The ANOVA confirmed that this difference between the two pairs of groups was highly significant ($P < 0.0001$), but again there were no

Table 1. Breakdown of the number of cultures, cells and 1-h runs within each transfection group for wild-type (W/T), null-transfected (NULL), MHC-transfected and t-MHC-transfected clones.

	W/T	NULL	MHC	t-MHC
Cultures	6	5	7	9
Cells	55	45	95	136
1-h runs	122	85	296	311

significant differences between W/T and NULL or between MHC and t-MHC. We analysed this suppression of cell speed by β-cardiac expression further, and found it to be due both to a decrease in the areas of protrusion and retraction of the cell margin and to a decrease in their separation, although further details will not be given here.

The area/mass ratio was not significantly affected by β-cardiac myosin expression, and this was confirmed in an ANOVA test (Fig. 2C). We have found previously that fibroblast spread area is closely related to cell mass, and have therefore proposed that the area/mass ratio is a better index of cell spreading than is cell area [56]. Again there were no significant differences between W/T and NULL or between MHC and t-MHC. Further analysis of cell shape also revealed no significant effects of β-cardiac myosin expression.

In conclusion, the only major effect to emerge was the suppression of cell speed and of closely related parameters such as protrusion and retraction. It is unclear how the heterologous expression of full-length or truncated MHC is able to affect cell locomotion in this way. The transfected MHC probably does not interfere directly with endogenous non-muscle myosin II function, as it

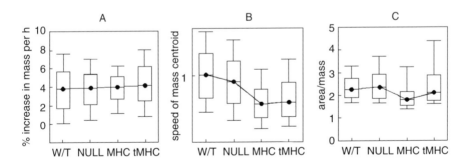

Fig. 2. DRIMAPS analysis of myogenic cells expressing MHC and t-MHC. The distribution of values within each transfection group is represented by a 'box-and-whiskers', in which the box covers 50% of the data values and the whiskers span 80% of values. (A) Growth of the cells, expressed as percentage increase in dry mass per hour. (B) Speed of cell locomotion, measured as the magnitude of the displacement of the cell's mass centroid during 5 min and divided by 5 to give units of $\mu m \cdot min^{-1}$. A 5 min rather than a 1 min interval was chosen to smooth out any track noise [58]. (C) Spreading of the cells (area/mass ratio).

does not co-localize with it. It is also unlikely to be interfering with the expression levels of endogenous non-muscle myosin II since, using Western analysis, we could not detect any change in the levels of this protein as a result of MHC or t-MHC transfection (results not shown). It is possible that the transfected MHC is sequestering light chains that normally bind to non-muscle myosin II, but again it is unlikely that this would have the observed effect, even if the expression levels of light chain are not up-regulated, since we believe that the levels of expression of MHC are too low for this to be a major factor. For the same reason, it is equally unlikely that the transfected MHC competes significantly for actin sites with other actin-binding proteins and thus disrupts the dynamics of actin assembly/disassembly cycling.

The most probable explanation is that, as MHC is regulated differently from non-muscle myosin II, it can bind to actin inappropriately, possibly cross-linking actin filaments in an unregulated manner. The t-MHC is similar in length to heavy meromyosin and, like this protein, will probably form two-headed dimers. If so, its actin-cross-linking potential could be similar to that of full-length MHC, which would explain the similar effects of these two transfections. Even small numbers of unregulated cross-links might significantly 'seize up' the dynamics of the actin cytoskeleton. We need to do further work to test this idea, possibly by transfecting with subfragment-1 of MHC, which should remain single-headed and have no cross-linking potential.

Effect of expressing a dominant-negative mutant of myosin I on cell locomotion

Myogenic cells express five different classes of myosin: class I (homologue to rat myr3, and mammalian Iα), class II (IIA and IIB), class V, class IX and class X [28] (Table 2). Using antibodies to all of these isoforms except class X, for which no antibody was available, we investigated their localization in crawling myoblasts and found a staining pattern similar to that observed for fibroblasts [28]. Of interest here is the fact that two isoforms, myosin I and myosin V, could be found at the leading edge of lamellipodia (Fig. 1), whereas myosins IIA and IIB and myosin IX were excluded. Myosins IIA and IIB co-localized with actin filament bundles in the cell body, but myosin IX showed a punctate distribution in the cytoplasm.

The localization of mammalian myosin Iα at the leading edge of crawling myoblasts suggests that it could play a role in protrusion of the lamellipodia, as described above. The head domain of myosin Iα binds to actin, whereas the tail domain is thought to contain the membrane-binding domain. Expression of the tail domain in NRK cells [46] showed that this domain alone could not localize myosin I to the plasma membrane as well as the full-length molecule. This suggests that the head domain is also important in localization, and perhaps by a more complex method than by simply attaching to actin.

Expression of a truncated myosin Iα head fragment might therefore be expected to affect protrusion of the lamellipodia. Such a fragment should be able to at least partially localize to the plasma membrane, and bind to actin. However, it will not be able to move the plasma membrane while attached to

Table 2. Myosin isoforms discovered by reverse transcription–PCR using mouse myoblast cDNA. The majority of the PCRs were carried out using degenerate primers to the conserved sequences LEAFGNA and GES-GAGKT [28]. These sequences are highly conserved across all classes of myosin so far described, and are separated by about 30–50 amino acid residues which are highly variable. Therefore sequence from these amplified reverse transcription–PCR fragments can be used to classify the myosin fragment into a class of the myosin superfamily based on identity with known myosin sequences. Five different classes of myosin were found: I, II, V, IX and X. The sequence between LEAFGNA and GESGAGKT is given in the table where these two primers were used in the reverse transcription–PCR. Where the amino acid sequence was different from that of the myosin with the best identity, the altered residues are shown in bold face.

Name of myosin in myoblast	Myosin with best matching sequence	Sequence between LEAFGNA and GESGAGKT
Myosin I	Rat myr3	VAAKYIMSYVSRVSGGGP KVQHVKDIILQSNPL
Myosin Iα	Mammalian Iα	
Myosin IIA	Human IIA	
Myosin IIB	Human IIB	
Myosin V	'Dilute'	VSAKYAMRYFATVSGSASEAN VEEKVLASNPIME
Myosin IX	Rat myr5	QSTNFLIHCLTALSQKGYAS GVERTILGAGPV
Myosin X	Bovine X	LILKFLS**V**ISQQTLDL**GL**QEK TSSVEQAIL**Q**SSPIME

actin, as it has lost its membrane-binding domain. To test this idea, we expressed in myogenic cells a myosin Iα head fragment (t-MIα) that was truncated just subsequent to the calmodulin-binding domains or IG motifs using the pCMV5 vector [60]. In this case a total of 29 cultures were each recorded for 20 h for the DRIMAPS analysis. Our experience with the β-cardiac myosin analysis had suggested that it might be important to keep a track of the separate transfection clones during the ANOVA tests, and so a new level was introduced, the Clones level, to replace the Vectors level which was now redundant (Table 3). In ANOVA tests, the Expression level compared the three clones expressing t-MIα (MC8, MC9 and MC15) with the three clones not expressing it (the parent W/T clone and two clones, PC10 and PC12, expressing the null vector). The remaining two levels, Cultures and Cells, were as before.

We found that several parameters did indeed reveal highly significant clonal differences, notably the mean mass of the cells and cell spreading (expressed as the area/mass ratio). In the case of cell mass, when we examined the mean masses of individual cultures within each clone, we found that the five transfected subclones each had a smaller range of masses than the parent wild-type clone, whose range of masses spanned all those of the subclones. This demonstrates that it is important to include several subclones in a properly designed transfection experiment, and a more balanced type of ANOVA test

Table 3. New nesting structure and sample sizes. Two clones were transfected with the null vector, whereas three were transfected with the t-MIα vector.

	No t-MIα			t-MIα		
	W/T	PC10	PC12	MC8	MC9	MC15
Cultures	10	4	2	3	5	5
Cells	62	19	12	19	32	48
1-h runs	210	67	19	74	111	128

would compare the transfected clones only with null-transfected clones. If there are significant mass differences between clones, as in this case, it is also important to check that any other clonal differences found are not simply due to this difference in masses.

In contrast with the transfections with β-cardiac myosin, the transfection with t-MIα produced no significant effects in ANOVAs at the Expression level in any of the parameters studied. It can be seen that the responses are essentially flat in all three cases (Fig. 3). As with the β-cardiac myosin experiment, each transfection group was growing at a mean rate of around $4\% \cdot h^{-1}$ (Fig. 3A). The ANOVA tests revealed no significant differences at the Expression level or at the Clones level. There was no significant difference in the speed of cell locomotion at the Expression level (Fig. 3B). In this case it was particularly important to be sure that clonal differences were not masking any differences due to t-MIα expression and, fortunately, the Clones level also revealed no significant difference. Furthermore, we ordered the cultures within each clone in order of increasing mass, and it was clear that the mean cell speed of the cultures was not related to their mean mass, which explains why the clonal differences in mass did not give rise to clonal differences in speed. One point to note, however, is that the parent wild-type clone used for the t-MIα transfection had a much lower speed of cell locomotion that the one used for the β-cardiac myosin transfection. This emphasizes the importance of clonal variation and the need to use the same parent clone for all transfections within any one experiment. For the spreading of the cells expressed as the area/mass ratio, there was a significant difference at the Clones level. Therefore we cannot be sure that the t-MIα transfection did not affect cell spreading; all we can say is that the clonal variation in spreading was so large that no significant effects of t-MIα expression emerged.

In summary, we observed no effect of transfection with truncated myosin Iα on cell behaviour. We did detect significant differences between different transfected subclones, however, and so these could mask any differences due to myosin Iα expression. Nevertheless, in the case of cell speed, the clonal variation was not significant, and we can be sure that the transfection had little or no effect on this parameter. However, as with β-cardiac myosin, we were unable to confirm expression of t-MIα by Western analysis, although expression of the same construct in COS cells did give a band of the predicted size. Thus the

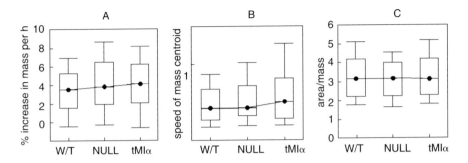

Fig. 3. DRIMAPS analysis of wild-type, null-transfected and t-MIα-transfected myoblasts. 'Box-and-whisker' plots (see legend to Fig. 2) comparing the data from the wild-type with that from the pooled null transfections and the pooled t-MIα transfections for the three selected parameters are shown. (A) Growth of the cells expressed as percentage increase in dry mass per hour; (B) speed of cell locomotion; (C) spreading of the cells (area/mass ratio).

expression levels were low, and it is possible that they were too low for us to observe a significant effect. However, we confirmed the presence of mRNA by PCR analysis, and it is likely that the expression levels were as high as those for the β-cardiac myosin constructs which did have a significant effect. We conclude that, if the expression of t-MIα has any effect on myoblast behaviour, it is much less conspicuous than the effect of the β-cardiac myosin constructs, which were also expressed at low levels.

An explanation for the relative ineffectiveness of t-MIα transfection might be that, even if t-MIα does interfere with the function of native myosin Iα, full functionality of the native protein might not be essential for locomotion in these cells. Even if it is, noticeable suppression of this function might require very high expression levels of the non-functional form of myosin Iα if its only action is merely to compete with the native form for functional sites on actin. It may be that the only hope of detecting effects on cell locomotion of a foreign protein, especially at low expression levels, is if it actively disrupts the locomotory machinery of the cell. We were hoping that this might happen with t-MIα in view of its localization at active protrusion sites. In comparison with β-cardiac myosin, however, it seems that t-MIα may be ineffective for two reasons. Firstly, its binding may be regulated in the same way as that of native myosin Iα, and thus it would not accumulate abnormally at the protrusion sites and permanently paralyse them. Secondly, being monomeric, it has no potential for cross-linking actin sites, and so it may be unlikely to be seriously disruptive even if it did bind in an unregulated manner. It is noteworthy in this respect that a truncated mutant of myosin I expressed in *Dictyostelium* did not affect protrusive behaviour, whereas overexpression of the full-length myosin I did [35].

Conclusions

We have discussed the much greater difficulties of using molecular genetics to increase our understanding of the crawling locomotion of vertebrate tissue cells in comparison with those presented by haploid, free-living cells or even by lower metazoan cells. Nevertheless, the mechanism of vertebrate cell locomotion will need to be studied in its own right, since potentially important differences in the locomotion of the different classes of cells are emerging as our understanding increases. Despite the great technical difficulties, progress is being made in obtaining viable null mutants in vertebrate cells that fail to express components of the locomotory machinery. Our approach of expressing foreign or defective components of the cytoskeleton, while being technically less difficult, has the disadvantage that the results are harder to interpret. Even so, the information that can be obtained by over- or under-expressing native proteins is limited, and the additional information that can be obtained by expressing other proteins that may interfere with the functioning of the native components of the cytoskeleton will doubtless contribute to our understanding. Such experiments are potentially at least as informative as those using drugs to disrupt the functioning of the cytoskeleton, an approach that has yielded most of our current knowledge of the system.

When levels of expression of foreign proteins are low, competition with native proteins for binding sites or other resources is unlikely to be effectively disruptive, whereas relatively few abnormal cross-links of native filamentous systems could be highly disruptive. This is the explanation that we have suggested for the high effectiveness of the double-headed myosin II in reducing the speed of cell locomotion, whereas the single-headed truncated myosin I, which would be expected to compete with native myosin I, had no detectable effect. Many more experiments will be needed to test this interpretation.

Finally, the analysis of our results has revealed an unexpectedly high variability in behavioural parameters between different clones taken from the same parent clone, irrespective of any transfection treatment. Similarly, there can be a highly significant variation between different cultures of the same clone, and there is almost always highly significant variation between cells within any one culture. These observations emphasize the necessity of performing an ANOVA or equivalent statistical procedure before attributing any observed differences in cell behaviour to an experimental treatment. A consequence of this is that large quantities of data from many cultures need to be collected if subtle effects are to be detected, and this necessitates an automatic method of data acquisition such as the DRIMAPS system described here.

References

1.   Abercrombie, M., Heasyman, J.E.M. and Pegrum, S.M. (1970) Exp. Cell Res. **59**, 393–398
2.   Abercrombie, M., Heasyman, J.E.M. and Pegrum, S.M. (1970) Exp. Cell Res. **60**, 437–444
3.   Lazarides, E. and Weber, K. (1974) Proc. Natl. Acad. Sci. U.S.A. **71**, 2268–2272
4.   Sanger, J.W., Sanger, J.M. and Jockusch, B.M. (1983) J. Cell Biol. **96**, 961–969
5.   Abercrombie, M., Heaysman, J.E. and Pegrum, S.M. (1971) Exp. Cell Res. **67**, 359–367
6.   Heath, J.P. and Dunn, G.A. (1978) J. Cell Sci. **29**, 197–212
7.   Couchman, J.R. and Rees, D.A. (1979) Cell Biol. Int. Rep. **3**, 431–439

8.   Harris, A.K., Stopak, D. and Wild, P. (1981) Nature (London) **290**, 249–251
9.   Cramer, L.P., Siebert, M. and Mitchison, T.J. (1997) J. Cell Biol. **136**, 1287–1305
10. Heath, J.P. (1983) J. Cell Sci. **60**, 331–354
11. Abercrombie, M., Dunn, G.A. and Heath, J.P. (1977) Soc. Gen. Physiol. Ser. **32**, 57–80
12. Dunn, G.A. (1980) Symp. Br. Soc. Cell Biol. **3**, 409–423
13. Verkhovsky, A.B., Svitkina, T.M. and Borisy, G.G. (1995) J. Cell Biol. **131**, 989–1002
14. Chen, W.-T. (1979) J. Cell Biol. **81**, 684–691
15. Jay, P.Y., Pham, P.A., Wong, S.A. and Elson, E.L. (1995) J. Cell Sci. **108**, 387–393
16. Tilney, L.G. (1975) Soc. Gen. Physiol. Ser. **30**, 339–388
17. Small, J.V., Anderson, K. and Rottner, K. (1996) Biosci. Rep. **16**, 351–368
18. Peskin, C.S., Odell, G.M. and Oster, G.F. (1993) Biophys. J. **65**, 316–324
19. Keller, H.U. and Bebie, H. (1996) Cell Motil. Cytoskeleton **33**, 241–251
20. Fedier, A. and Keller, H.U. (1997) Cell Motil. Cytoskeleton **37**, 326–337
21. Harris, A.K. (1973) Ciba Found. Symp. **14**, 3–26
22. Strohmeier, R. and Bereiter-Hahn, J. (1987) J. Cell Sci. **88**, 631–640
23. Mooseker, M.S. and Cheney, M.E. (1995) Annu. Rev. Cell Dev. Biol. **11**, 633–675
24. Cope, M., Jamie T.V., Whisstock, J., Rayment, I. and Kendrick-Jones, J. (1996) Structure **4**, 969–987
25. Hasson, T. and Mooseker, M.S. (1996) J. Biol. Chem. **271**, 16431–16434
26. Sellers, J.M., Goodson, H.V. and Wang, F. (1996) J. Muscle Res. Cell Motil. **17**, 7–22
27. Bement, W.M., Hasson, T., Wirth, J.A., Cheney, R.E. and Mooseker, M.S. (1994) Proc. Natl. Acad. Sci. U.S.A. **91**, 6549–6553
28. Wells, C., Coles, D., Entwistle, A. and Peckham, M. (1997) J. Muscle Res. Cell Motil. **18**, 501–515
29. Conrad, P.A., Giuliano, K.A., Fisher, G., Collins, K., Matsudaira, P.T. and Taylor, D.L. (1993) J. Cell Biol. **120**, 1381–1391
30. Fukui, Y., Lynch, T.J., Brzeska, H. and Korn, E.D. (1989) Nature (London) **341**, 328–332
31. Wessels, D., Murray, J., Jung, G., Hammer, J.A. and Soll, D.R. (1992) Cell Motil. Cytoskeleton **20**, 301–315
32. Novak, K.D., Petersen, M.D., Reedy, M.C. and Titus, M.A. (1995) J. Cell Biol. **131**, 1205–1221
33. Jung, G., Fukui, Y., Martin, B. and Hammer, J.A. (1995) J. Cell Biol. **168**, 14981–14990
34. Wang, F.S., Wolenski, J.S., Cheney, R.E., Mooseker, M.S. and Jay, D.G. (1996) Science **273**, 660–663
35. Novak, K.D. and Titus, M.A. (1997) J. Cell Biol. **136**, 633–647
36. Lin, C.H., Espreafico, E.M., Mooseker, M.S. and Forscher, P. (1996) Neuron **16**, 769–782
37. Wu, X., Bowerx, B., Wei, Q., Kocher, B. and Hammer, J.A. (1997) J. Cell Sci. **110**, 847–859
38. Howard, J. (1997) Nature (London) **389**, 561–567
39. DeLozanne, A. and Spudich, J.A. (1987) Science **236**, 1086–1091
40. Knecht, D. and Loomis, W.F. (1987) Science **236**, 1081–1086
41. Manstein, D.J., Titus, M.A., Delozanne, A. and Spudich, J.A. (1989) EMBO J. **8**, 923–932
42. Ostap, E.M. and Pollard, T.D. (1995) J. Cell Biol. **133**, 221–224
43. Lackie, J.M. (1986) Cell Movement and Cell Behaviour, Allen and Unwin, London
44. Heath, J.P. and Holifield, B.F. (1993) Symp. Soc. Exp. Biol. **47**, 35–56
45. Harris, A.K. (1994) Int. Rev. Cytol. **150**, 35–68
46. Ruppert, C., Godel, J., Müller, R.T., Kroschewski, R., Rheinhard, J. and Bähler, M. (1995) J. Cell Sci. **108**, 3775–3786
47. Kalhammer, G., Bähler, M., Schmitz, F., Jockel, J. and Block, C. (1997) FEBS Lett. **414**, 599–602
48. Cunningham, C.C., Stossell, T.P. and Kwiatkowski, D.J. (1991) Science **251**, 1233–1236
49. Witke, W., Sharpe, A., Hartwig, I., Azuma, T., Stossel, T. and Kwiatkowsi, D. (1995) Cell **81**, 41–51

50.   Fernandez, J.L.R., Geiger, B., Salomon, D. and Benzeev, A. (1992) Cell Motil. Cytoskeleton **22**, 127–134

51.   Fernandez, J.L.R., Geiger, B., Salomon, D. and Benzeev, A. (1993) J. Cell Biol. **122**, 1285–1294

52.   Coll, J.L., Benzeev, A., Ezzell, R.M., Fernandez, J.L.R., Baribault, H., Oshima, R.G. and Adamson, E.D. (1995) Proc. Natl. Acad. Sci. U.S.A. **20**, 9161–9165

53.   Murray, J., Vawter-Hugart, H., Voss, E. and Soll, D.R. (1992) Cell Motil. Cytoskeleton **22**, 211–223

54.   Weber, I., Wallraff, E., Albrecht, R. and Gerisch, G. (1995) J. Cell Sci. **108**, 1519–1530

55.   Dunn, G.A. and Zicha, D. (1993) Symp. Soc. Exp. Biol. **47**, 91–106

56.   Dunn, G.A. and Zicha, G.A. (1995) J. Cell Sci. **108**, 1239–1249

57.   Milliken, G.A. and Johnson, D.E. (1992) Analysis of Messy Data. Volume 1: Designed Experiments, Chapman and Hall, New York and London

58.   Dunn, G.A., Zicha, D. and Fraylich, P.E. (1997) J. Cell Sci. **110**, 3091–3098

59.   Wilson, A.K., Gorgas, G., Claypool, W.D. and de Lanerolle, P. (1991) J. Cell Biol. **114**, 277–283

60.   Andersson, S., Davis, D.N., Dahlback, H., Jornvall, H. and Russell, D.W. (1989) J. Biol. Chem. **264**, 8222–8229

61.   Moncman, C.L., Rindt, H., Robbins, J. and Winkelmann, D.A. (1993) Mol. Biol. Cell **4**, 1051–1067

Biochem. Soc. Symp. **65**, 299–314
Printed in Great Britain

18

Forces in cell locomotion

E.L. Elson[*][1]**, S.F. Felder**[†]**, P.Y. Jay**[‡]**, M.S. Kolodney**[§] **and C. Pasternak**[*]

[*]Department of Biochemistry and Molecular Biophysics, Box 8231, Division of Biology and Biomedical Sciences, Washington University School of Medicine, 660 South Euclid Avenue, St. Louis, MO 63110, U.S.A., [†]Neotech, P.O. Box 64326, Tucson, AZ 85728, U.S.A., [‡]Department of Cardiology, Children's Hospital, 300 Longwood Avenue, Boston, MA 02115, U.S.A., and [§]Division of Dermatology, Department of Medicine, UCLA School of Medicine, 200 UCLA Medical Plaza, Suite 550, Los Angeles, CA 90095, U.S.A.

Abstract

The molecular mechanisms that drive animal cell locomotion are partially characterized, but not definitively understood. It seems likely that actin polymerization contributes to the forward protrusion of the leading edge of a migrating cell. Both myosin-dependent contractile forces and selective detachment of adhesive interactions with the substratum seem to contribute to release of the posterior of an extended cell. It is probable, but not certain, that a separate 'traction' force advances the cell body towards the forward anchorage sites formed by the advancing lamellipodium. The molecular mechanism of this force is unknown. Determining the role of traction forces in migrating fibroblasts and keratocytes is complicated by the fact that the primary functions of the relatively strong forces exerted on the substratum by these cells may be to establish tissue 'tone' and to remodel tissue matrices, rather than to drive locomotion. In accordance with this notion, rapidly moving cells such as neutrophils and *Dictyostelium* amoebae exert weaker forces on the substratum as they migrate. The traction force in cell migration may be distinct from traction forces with tissue functions. Ultimately, the mechanism may be revealed by using molecular genetics to disrupt the motors that provide this force. Reconstituted tissues provide systems in which to investigate the regulation of cell forces and their contribution to tissue mechanical properties and development.

[1]To whom correspondence should be addressed.

Introduction

Cells migrate over planar substrata using a repetitive sequence of steps. These include: (1) advancing a leading process (a pseudopodium or lamellipodium) and anchoring it to the substratum, (2) drawing the main body of the cell forward, and, finally, (3) detaching the tail from the substratum and drawing it forward to prepare the way for further advance. Each step requires the exertion of force on the substratum to drive forward motion or on the cell itself to change cell shape, and it appears that the mechanisms for force development might be different for each step. It is thought that actin polymerization drives protrusion of the lamellipodium [1]. Detachment can be assisted by a myosin II-driven contraction. The origin of the force that advances the cell body, called the 'traction force' is unknown, and poses a basic problem in the field. Observations of the migration of diverse cell types provide different perspectives on the nature and co-ordination of these steps. Although the style of locomotion varies from one type of cell to another, e.g. from smooth and gliding by fish epidermal keratocytes to halting and pulsatile by fibroblasts, it has been supposed that there are underlying features that are common to locomotion among diverse cell types. Then, it would be legitimate to carry over some mechanistic interpretations from studies of one type of cell to others that have a different style of locomotion. The range over which this assumption is valid represents another important question.

A cell exerts forces on itself to change shape and on the substratum to propel itself forward. For example, the lamellipodium can be projected forward while out of contact with the substratum [2–4]. Then the forces that drive protrusion act only to change cell shape and not against the substratum. To make a net advance, however, an adherent cell, which we are supposing cannot 'swim', must push or pull against the substratum. Two kinds of forces, described as *contraction* and *traction*, could drive this motion [5]. As its name suggests, a contraction force causes the cell or a portion of it to reduce its degree of extension, as when a contracting muscle cell shortens. As the cell contracts, the region of the substratum to which it is attached comes under tension. Of course, this can occur with no change in cell volume. Traction describes the contrary process, in which the cell increases its degree of extension, i.e. it spreads over the substratum. To extend itself, the cell pulls outward at its margins against the substratum and is also thought to stretch structural components (presumably mainly cytoskeletal) to generate an elastic tension, which the cell exerts on the substratum [6]. Similar inwardly directed force vectors (i.e. forward from the back and backward from the front) towards the nucleus of the cell are exerted against the substratum for both processes. Both traction and contraction forces could supply force both to advance portions of the cell and to detach regions of the cell from its adherence to the substratum. Whether contraction and traction are both simply consequences of the same molecular mechanism, or whether they differ in the motors that drive them, their mechanisms of regulation or their physiological functions, are important questions that remain to be answered [5].

Three types of cells with different styles of locomotion and different properties advantageous for investigation have been studied intensively during recent years. In their initial work, which provided a systematic foundation for this research area, Abercrombie and co-workers characterized the properties of primary chick embryo fibroblasts. These cells move in a jerky, sporadic fashion and are strongly retarded by firm adhesion to the substratum. Despite their low rate of migration and the complexity of their shape changes during locomotion, these and similar cells remain the focus of much attention [5]. The forces that a fibroblast exerts on its substratum as it migrates are far greater than are required to propel the cell [7], may be irrelevant for locomotion and might have a role in tissue development [7,8] and wound healing [9].

Epidermal keratocytes from fish and amphibia move much more rapidly and, moreover, do so in a smooth gliding fashion, projecting forward a thin, broad, optically clear lamellipodium [10,11]. Their retention of constant shape and the favourable optical properties of their lamellipodia make these cells advantageous for studies of locomotion using optical microscopy. The preponderant forces exerted by these cells on the substratum are transverse to the direction of migration [12,13], are again greater than are required for cell propulsion, and so may also have a tissue function other than cellular migration. The physiological function of these forces and the way in which they are co-ordinated to produce lamellipodial extension and the smooth gliding of the cell body remain important questions.

Neutrophil leucocytes are rapidly moving cells that have been studied extensively because of the importance of locomotion to their physiological function. The small amoeba, *Dictyostelium discoideum*, moves with a style and speed similar to neutrophils. Despite its small size, non-uniform optical properties and the complexity of its shape changes during migration, *Dictyostelium* cells are the subject of extensive research because of their suitability for genetic manipulation [14]. Among a host of *Dictyostelium* mutants, deletion of conventional myosin [15,16], diverse myosin I isoforms [17–19], a number of other cytoskeletal proteins [20–22] and the small GTP-binding protein Ras [23] are of particular interest. The forces exerted on the substratum by neutrophils and *Dictyostelium* amoebae, cells specialized for locomotion, are much smaller than those exerted by fibroblasts and epidermal keratocytes, lending further support to the idea that the forces exerted by the latter have tissue functions other than locomotion [5].

Advance of the forward process

The force that drives protrusion of the leading cell edge could be exerted either by motor molecules such as myosins or by actin polymerization, and either locally at the leading edge or more remotely in the cell body [1]. Protrusive force generated in the cell body would have to be transmitted to the lamellipodium via a mechanical linkage, presumably the actin cytoskeleton, which is the principal structural component of the lamellipodium [24,25]. According to this view, actin filaments pushed into the lamellipodium would drive its extension, and so forward motion of the edge would be co-ordinated with comparable motion of actin filaments in the lamellipodium. Fluorescence photoactivation and photobleaching experiments indicate, however, that actin filaments in the lamellipodium are either stationary or move rearward in the laboratory frame of reference, and always move rearward relative to the leading edge [1,26,27]. Therefore it seems likely that the force that drives the advance of the leading edge is generated near the edge [1,28].

Myosin II is not detectable at the leading cell edge and, furthermore, *Dictyostelium* myosin II-null mutants retain the ability to migrate over a planar substratum [15,29]. Although a number of myosin I isoforms are present at the leading edge of *Dictyostelium* cells [30,31], none has been shown to contribute to their advance.

The tight coupling of actin polymerization with the advance of the leading edge suggests, but does not prove, that such polymerization contributes to the driving force. The feasibility of this mechanism was demonstrated by showing that polymerization of actin within a lipid vesicle deformed the vesicle [32]. The simple lipid membrane in these vesicles could, however, have been substantially more compliant than a cell membrane. Additional evidence that actin polymerization can provide a driving force arises from studies of the locomotion of certain bacteria and viruses within the cytoplasm of animal cells [33–35]. Inhibition of actin polymerization prevents protrusion, but this does not prove that polymerization supplies the driving force. To prove that polymerization is necessary and sufficient, one must also demonstrate that it can drive protrusion in the absence of other forces. Until all myosin types and isoforms have been identified and either inhibited or knocked out, it will not be possible to conclude definitively that none of these motors contributes to the protrusion force. The observation that the myosin inhibitor butanedione monoxime prevents post-mitotic cell spreading, but not the motion of *Listeria* driven by actin polymerization, suggests an avenue by which to approach this subject [36].

Another approach is to characterize the balance between the protrusion force that can be supplied by actin polymerization and the resistance of the membrane to outward deformation. This could be accomplished by analysing quantitatively the deformation of a membrane system by actin polymerization. To provide a relatively realistic model, the actin in the form of capped filaments and monomer bound to a sequestering protein such as thymosin β_4 could be contained within large lipid vesicles and triggered to polymerize by displacing the capping and sequestering proteins. The effect of polymerization on the shape of the vesicle can be followed by light microscopy [32]. The polymerization driving force can be represented simply as the free energy change for polymerization divided by distance (d) that the actin filament is extended by the addition of a monomer [37]. The polymerization reaction could be represented as follows:

$$A_1S + A_nB \rightleftharpoons A_{n+1} + B + S$$

where S is the monomer-sequestering protein; B is a barbed-end capping protein, A_1 is actin monomer and A_n is actin polymer. The free energy change per mol for this process is:

$$\Delta G = RT \ln (Q/K)$$

where R is the gas constant and T is the absolute temperature. Q is given by:

$$Q = [A_{n+1}] \cdot [B] \cdot [S]/[A_1 S] \cdot [A_n B]$$

where concentrations are denoted by square brackets. K is the equilibrium constant. To obtain this, concentrations in the equation defining Q are replaced by equilibrium concentrations, $[X]_{eq}$. Thus:

$$K = K_p \cdot K_b \cdot K_s$$

where:

$$K_p = [A_{n+1}]_{eq}/[A_1]_{eq} \cdot [A_n]_{eq}$$
$$K_b = [A_n]_{eq} \cdot [B]_{eq}/[A_n B]_{eq}$$
$$K_s = [A_1]_{eq} \cdot [S]_{eq}/[AS]_{eq}$$

Then, the protrusion force due to actin polymerization is $-\Delta G/d$, which is balanced by the resistance to extension of the leading edge, Fr, i.e. $Fr = -\Delta G/d$. Under static conditions the driving force for polymerization is insufficient to deform the leading edge. Polymerization could be initiated by uncapping of the rapidly polymerizing end of the filaments, e.g. by increasing the concentration of polyphosphoinositides [38] and/or by releasing actin monomers. A change in the deformability of the membrane at the leading edge of the cell could also promote protrusion. In principle, one could also apply this kind of model to evaluate the balance of driving force and resistance to deformation in a cell, but most of the relevant parameters remain to be determined.

In a simpler model system, one could dispense with the capping and monomer-sequestering proteins and initiate polymerization by changing the salt concentration in the vesicle [32]. For example, in a system containing actin monomers bound to Ca^{2+} ions, increasing the concentration of Mg^{2+} would cause this ion to replace the Ca^{2+} bound to the actin and initiate polymerization. Overall this reaction can be represented as:

$$2Mg^{2+} + A_1 \cdot Ca^{2+} + (A \cdot Mg^{2+}_2)_{n-1} \longleftrightarrow Ca^{2+} + (A \cdot Mg^{2+}_2)_n$$

A detailed analysis of this system in which polymerization is coupled to Mg^{2+} has been presented [39]. Increasing the Mg^{2+} concentration by an infinitesimal change, $\delta[Mg^{2+}]$, would yield an infinitesimal change in the free energy of the system, i.e. $\delta\Delta G = -2RT(\delta[Mg^{2+}]/[Mg^{2+}])$, which, for a finite change in Mg^{2+} concentration, becomes $\Delta\Delta G = -2RT\ln([Mg^{2+}]_f/[Mg^{2+}]_i)$, where f and i denote the final and initial concentrations of Mg^{2+}, respectively. Hence a 10-fold increase in Mg^{2+} concentration would yield a free energy change of $\Delta\Delta G = -4.6k_B T$ per actin filament, where K_B is Boltzmann's constant. Taking the distance by which one monomer lengthens the filament to be $d = 3$ nm [37], the force exerted by the lengthening filament is $\Delta G/d = 6.3$ pN. Hence, by varying the concentration of Mg^{2+} ions in the vesicle, one could correspondingly vary the driving force deforming the vesicle. A bundle of 150 filaments would produce a force close to 1 nN in response to a 10-fold change in $[Mg^{2+}]$. Indentation measurements yield a stiffness for the leading lamella in the range 1–2 nN/μm [40]. This is 5–10-fold higher than stiffnesses measured by indentation and micropipette aspiration on neutrophils [41,42]. Hence the number of polymerizing filaments needed to extend the cell front may vary from one type of cell to another, but is in the range 10–200 filaments for neutrophils and *Dictyostelium* amoebae. Of course, a smaller effective concentration change would necessitate a larger number of filaments to produce the same force. Moreover, in a cell, effectors other than the concentration of bivalent cations (e.g. polyphosphoinositides) are likely to be important.

Contraction forces

On glass substrata, *Dictyostelium* myosin-null mutant cells migrate at less than half the rate of wild-type cells [29]. Although myosin II is not essential for locomotion, it does contribute. Some clues to its function include the observations that depletion of ATP by azide caused wild-type *Dictyostelium* cells to stiffen and to become more likely to detach from the substratum [43]. Moreover, microscopy and mechanical measurements suggest that myosin II functions preferentially at the rear of *Dictyostelium* cells [40,44]. These results suggest the hypothesis that one function of myosin II in locomotion is to provide a contractile force that detaches the tail of the cell from its adhesion to the substratum, as had been proposed previously for both fibroblasts and *Dictyostelium* cells [45,46]. This hypothesis was tested by placing wild-type and myosin-null mutant *Dictyostelium* cells on to substrata coated with different amounts of polylysine to vary the adhesiveness [6]. Although the velocities of wild-type cells were only slightly reduced on the adhesive substrata, the null mutant cells barely moved from sites on the substratum, to which they appeared to be tethered. These results demonstrate that one function of myosin II in the locomotion of *Dictyostelium* cells is to detach the cell from the substratum. Using interference reflection microscopy, it was also shown that a myosin II-dependent contractile force is exerted specifically at the rear of the wild-type cells. The ability of the myosin II-null mutant cells to migrate over a glass surface despite their weakened ability to detach from the substratum suggests the operation of other myosin II-independent force-generating systems. The myosin II-independent transport of beads discussed below may be an indicator of this force.

The role of contractile force in detachment of the tails of cells migrating over solid substrata is probably not unique to *Dictyostelium*. It appears that both passive elasticity and active contractile force contribute to the detachment and retraction of the stretched tails of fibroblasts [45]. In fish epidermal keratocytes the large, mostly transverse, contractile forces at the rear of the cell are also likely to contribute to their detachment as they migrate [12]. In

Dictyostelium cells, myosin II-dependent contractile forces also contribute to cell shape changes other than those associated with detachment of the tail, e.g. to the 'cringe' response, during which the amoebae stiffen and round up in each locomotory cycle and in response to cAMP (C. Pasternak and E.L. Elson, unpublished work).

Traction

Minimal model: no active traction force

The traction force is the most elusive of the forces contributing to locomotion. It is possible to construct a minimal model for locomotion in which there is no active traction force. Suppose that protrusion of the leading edge by actin polymerization generates an elastic tension within the cell. Although this tension must initially be supported by the actin filaments that drive the protrusion, the tension can be shifted to the cell body after attachment of the protruded edge to the substratum by weakening the force supplied by the driving filaments. The tension pulls equally on the front and the back of the cell, and so, for the cell to have net forward progress, it must detach selectively at the rear, allowing the body of the cell to move forward towards the leading anchorage sites [47]. It has been suggested that this mechanism could be sufficient to drive the migration of cells that are weakly adherent to the substratum, but might be insufficient to detach strongly adherent cells [1]. If detachment from the substratum were accomplished by means other than a traction force, e.g. either by contraction forces and/or by selective disruption of adhesive linkages [48], then the elastic traction force provided by the minimal model might also be sufficient to drive migration even of strongly adherent cells. We cannot at present disprove this minimal model. Nevertheless, several lines of evidence suggest that there is an active traction force.

Indications of a traction force in fibroblasts

During the migration of fibroblasts, a lamellipodium is projected forward above the substratum from the leading lamella. Some of these protrusions attach to the substratum and provide anchorage for forces that draw the cell body forward. Others fail to adhere and are swung back over the cell surface to form 'ruffles'. It is plausible that the forces responsible for pulling the cell body forward towards advanced anchorage sites are also responsible for elevating those lamellipodia that fail to anchor themselves. Studies of the mechanical characteristics and evolution in time of these ruffling lamellipodia in fibroblasts provide clues about their mechanism of formation [2]. Reconstruction of optical microscope images taken in different focal planes to provide three-dimensional views of migrating fibroblasts has shown that the lamellipodia swing stiffly upward, retaining a straight cross-sectional profile, about a hinge 2–4 µm behind the leading edge, consistent with earlier studies [49]. Force applied to the elevated lamellipodia by means of a fine flexible glass beam causes them to swing stiffly about the hinge. When the lamellipodia are released from the probe, they quickly spring back to their original position, indicating that resistance to rotation about the hinge is mainly elastic. Because the probe causes the lamellipodium to rotate as a unit, it is appropriate to characterize the stiffness of the hinge about which the rotation occurs in terms of the torque required to generate a defined angular displacement of the ruffle. Measuring the bending of the glass beam as it applies force to the lamellipodium indicates that the stiffness of the hinge is roughly in the range 1–2 mdyn·μm/rad (1 dyn = 10^{-5} N). Hence, to a crude approximation, the gravitational force exerted by 1 μg applied 1–2 μm from the hinge would be required to rotate a ruffle through 1 rad. Both the actin filament matrix at the hinge and the traction force which elevates the ruffle provide the structural basis for this substantial stiffness. Further experimental work is needed to determine the relative contributions to this stiffness of the passive elasticity of the actin matrix and the active retraction force.

The ruffle continues to lengthen as it is swung upward. This observation indicates that protrusion of the leading lamellipodium can operate in parallel with the force that elevates the ruffle, presumably the traction force. Furthermore, it is difficult to reconcile this continued extension after the lamellipodium is at a sharp angle relative to the substratum with hypotheses which propose that the force for lamellar extension arises from motor molecules acting in a more central region of the cell. On the contrary, the results are most simply interpreted by supposing that the force for lamellipodial extension is developed within the lamellipodium itself, presumably by actin polymerization [1,2]. The stiffness of the lamellipodium is likely to be due to the dense matrix of cross-linked actin filaments with which it is filled and the membrane skeleton.

In order to exert a protrusive force against the leading edge, there must be structures within the cell to provide a foundation against which either the polymerizing actin filaments or the frontward-operating motors can push. The nature of this foundation is unknown, but several observations suggest that it might be related to rearward-moving structures that contain actin. During fibroblast locomotion it was observed that, as the lamellipodium swung upward, the hinge was retracted towards the nucleus. This suggests that the foundation that supports the force generated by the polymerizing actin filaments recedes rearward, potentially diminishing the force that could be exerted at the leading edge. Evidence for this supposition was obtained in studies of amphibian growth cones, which showed that advance of the growth cone was inversely proportional to the rate of rearward flow of actin within it [50]. In these studies the rearward transport of beads bound to the surface of the growth cone was shown to occur at the same rate as that of the underlying actin filaments. Then, under conditions whereby regions of the growth cone experienced different rates of forward protrusion, it was shown that the greater the rate of protrusion, the slower the rearward motion of the beads and actin filaments.

In summary, fibroblasts exert a force on unanchored lamellipodia which elevates the lamellipodia to form ruffles which swing back over the body of the cell. If this same force acted on a newly extended lamellipodium anchored to the substratum, it would draw the cell forward. The motors that supply this force and their regulation are unknown and are difficult to determine in mammalian cells, which have several isoforms of myosin II as well as a number of other forms of unconventional myosins. Studies of the rearward motion of surface-bound beads on *Dictyostelium* amoebae provide a different perspective on the traction force.

Bead transport on *Dictyostelium* amoebae as an indicator of traction force

The rearward transport of particles attached to the surfaces of migrating cells has long been seen as an important clue to the mechanism of traction in cell migration [5,51]. Studies of this process in mutants of *Dictyostelium* amoebae lacking conventional myosin provide additional information. Somewhat surprisingly, the myosin II-null mutants retain their ability to migrate on glass substrata, albeit at one-half the rate of wild-type cells [29]. The null cells are also impaired in their ability to execute conventional cytokinesis although, when adherent to a substratum, they can form cleavage furrows [22] and can also use a traction-mediated process to separate into mononuclear cells [52].

One form of rearward transport is the process known as capping, which has been studied extensively in lymphocytes as well as in *Dictyostelium* amoebae [53–55]. In both lymphocytes and *Dictyostelium* cells, capping is initiated by the cross-linking of surface proteins by multivalent ligands and their aggregation into 'patches'. These patches are then actively transported to the rear of the cell in a process that involves the actin cytoskeleton. This process has been interpreted in terms of the myosin-dependent cortical flow of actin filaments to which the aggregated patches are presumed to be attached [56]. Apparently, the aggregation process triggers not only the attachment of the aggregate to the actin cortical cytoskeleton, but also the activation of myosin. The myosin, which is presumed to be randomly distributed throughout the actin cortex, generates an isotropic cortical tension, which is then presumed to relax at the front of the cell, causing the tensed cortex to be drawn up the tension gradient to the rear of the cell, carrying with it the attached aggregates to form the cap. From this model one can predict that the tensing of the cortex in the early stages of the capping process should cause the cell first to stiffen and then, as the cortex is drawn to the rear of the cell, to soften. This stiffening and subsequent softening was seen in lymphocytes [57], and the dependence of the process on myosin was demonstrated using the *Dictyostelium* myosin II-null mutants [43]. These studies demonstrate that, like lymphocytes, normal *Dictyostelium* cells stiffen and then soften during capping of the multivalent lectin concanavalin A (ConA). On the myosin-null mutant cells, ConA molecules remain uniformly distributed over the cell surface and are not capped. Neither do the cells stiffen. Hence both capping and stiffening require myosin II. These data can be interpreted most simply by supposing that cross-linked surface proteins and the actin cytoskeleton to which they are linked are transported to the rear of the cell due to a myosin-dependent tension, as suggested by the cortical relaxation model.

In further studies, the surprising observation was made that ConA-coated fluorescent beads are transported to the rear even of *Dictyostelium* myosin II-null mutants [58]. Careful examination revealed, however, that this rearward transport is distinct from capping. The transported beads at the back of the null cells are distributed in a line, rather than as a compact cap as in wild-type cells. This difference results from the fact that the trajectories of the beads on the null mutant cells are parallel, rather than convergent as in wild-type cells. As had been shown previously on fibroblasts [59], the particles move fastest at the front of the cell, and they slow progressively as they move rearwards. A 3–4-fold decrease in the velocity of rearward particle transport was seen on both null mutant and wild-type *Dictyostelium* cells, but the rate of transport on null cells was significantly lower than on wild-type cells [58]. These results demonstrate that there is a myosin II-independent transport process in *Dictyostelium* cells, which might be responsible for the traction force that drives locomotion and traction-mediated cytofission. If this were so in *Dictyostelium* cells, then in other types of cells as well the traction force might depend on motors that could differ from myosin II in both mechanism and regulation. For example, myosin I might be responsible for this force. Several myosin I isoforms are concentrated at the leading edge of *Dictyostelium* cells [30,31]. The contribution of several myosin I isoforms to the rearward transport of ConA-coated beads was tested by comparing transport rates of wild-type and null mutants, but the motors that drive the myosin II-independent transport remain unknown (P.Y. Jay, unpublished work). As another example, the hypothetical actin pointed-end-directed motor recently proposed [60] could also be the generator of the traction force.

Are there actin tracks upon which the traction motors run?

In contrast with capping, during which a large amount of surface-bound ligand can be conveyed to the back of the cell, the myosin-independent transport has a limited capacity. On the myosin II-null mutant cells, rearward transport of ConA is seen only at low ConA concentrations; at high concentrations the ConA remains apparently uniform over the surface (P.Y. Jay, unpublished work). Beads are always limited to only a few per cell and therefore are also at low concentration and so are transported. (This resolves the apparent paradox that no rearward transport of ConA was seen in the earlier studies of capping on myosin II-null mutant cells [43]. In these studies the cells were exposed to relatively high concentrations of ConA, as used in earlier studies of capping on lymphocytes and *Dictyostelium* cells.) The limited capacity for transport, and the observation that the myosin II-independent transport process apparently occurs without an isometric contraction (detectable as a cell stiffening) and subsequent rearward flow of the cortical cytoskeleton, suggests that this transport is mediated by a more direct interaction between the motor and its cargo (the bead) than is seen in capping, in which the cargo is transported because it is attached to the cytoskeleton, which is under a myosin-dependent tension. This apparently more direct interaction and the straight, parallel trajectories of the beads on the myosin-null cells suggest that transport is driven by a motor which travels on actin tracks from the front to the back of the cell. This would be a specific example of a transport model for force production, as previously proposed [1]. Therefore we suggest that there are two cortical actin filament systems. One of these, which includes most of the actin filaments, participates in rearward flow on locomoting cells and in capping. The other, which includes only a small fraction of the filaments, provides tracks for myosin in a low-capacity transport system which can drive the motion of surface-attached beads or provide the traction force for locomotion if the motors are attached to immobile anchorage sites on the substratum. In contrast with the myosin molecules interacting with the bulk of the actin filaments, the myosin molecules on the tracks would be coupled more or less directly to their cargo. The force that they generate would drive the transport of the cargo or supply traction directly, and would not cause a general cortical contraction or cellular stiffening as occurs during capping. These hypothetical tracks have not been observed in *Dictyostelium* or other animal cells, but only a relatively small number of filament tracks would be required for the low-capacity transport process compared with the entire actin filament contents of the cortical cytoskeleton. Hence these tracks would be difficult to find, although they might be detected using a sensitive optical method such as fluorescence-detected linear dichroism, which has been used to reveal a minority fraction of actin filaments oriented perpendicular to the cleavage furrow in NRK cells [61].

Is there any indication of track motion in other cell types? A recent study of bead transport on fish epidermal keratocytes has shown that the nucleus, internal particles and organelles such as mitochondria in the cell body, as well as beads applied to the exterior surface of the cell, rotate about an axis parallel to the substratum and normal to the direction of motion of the cell [62]. (Beads on the dorsal surface of the lamellipodium either diffused randomly or were transported to the boundary between the lamellipodium and the cell body, in agreement with earlier work [63].) These results were interpreted as indicating that the cell body rolls over the substratum, thereby minimizing resistance to migration, as it follows the lamellipodium, which is presumed to drive the migration of the cell. There was some slipping as well as rolling, however, because only one rotation was accomplished as the cell moved a distance three times the circumference of the cell body.

An alternative interpretation of the apparent rolling motion of the cell body is that actin filaments oriented circumferentially on the inner surface of the membrane of the cell body provide tracks for the rotational motion of myosin molecules, similar perhaps to the rotation of myosin seen in algal cells such as *Nitella* [64]. The motion of these motor molecules could entrain and drive the rotational motion both of the nucleus and internal organelles and of the surface particles [65]. Furthermore, if the rotating surface molecules dominated the interaction between the cell body and the substratum, the effect of this rotational transport in diminishing the resistance of the cell body to migration would be similar to that proposed for rolling. This explanation for the experimental observations also avoids the difficulty of understanding how the rolling of the cell body would be compatible with its connection to the lamella at the front of the cell. One way to test this hypothesis is to identify the myosin motors that drive the rotational motion. It would also be interesting to determine, using photoactivation of fluorescence [66], whether the entire cage of actin filaments in the cell body is rotating as predicted by the 'rolling' hypothesis. If, in contrast, there is tracked motion, a fraction of the cytoskeletal actin containing the tracks might appear to turn at a different rate than the bulk of the rotating actin filaments.

Traction force in fibroblasts has been discussed in terms of a recently discovered and surprising pattern of actin filament organization ([26]; see also Chapter 11 in the present volume). The polarity of actin filaments in migrating chick embryo heart fibroblasts was determined by decoration with the S1 fragment of myosin. It was shown that the majority of the actin filaments, present in overlapping bundles parallel to the direction of migration in the ventral region of the cell, were organized according to a 'graded-polarity' pattern. That is, most of the filaments near the leading edge had their barbed (plus) ends forward, whereas most of the filaments at the back of the cell had their pointed (minus) ends forward. Between the front and the rear, the polarity of the filaments in the bundles varied continuously between these two extremes so that, for example, at the centre of the cell 50% of the filaments were oriented with barbed ends forward and 50% with barbed ends to the rear. In addition, experiments using photoactivation of fluorescence showed that actin filaments flowed rearward with respect to the substratum in the lamellipodium, were immobile in the lamella, and were present as two populations in the cell body, one of which moved forward with the cell body and the other of which remained stationary with respect to the substratum. If, as in growth cones, there is an inverse dependence of protrusion on the velocity of rearward actin transport [50], the rearward transport of lamellipodial actin might also provide a sort of 'clutch' mechanism for regulating the rate of protrusion in fibroblasts. The forward flow of a fraction of actin in the cell body might facilitate the recycling of actin to permit the sustained polymerization of filaments at the front of the cell.

Based on the abundance of actin filaments that are present in bundles with graded polarity, their location at the ventral surface of the cell and their specificity to locomoting fibroblasts, it was argued that these filaments are important for generating motile force during fibroblast locomotion [26]. Although this is a plausible and reasonable suggestion, it is worth considering another possibility. Because the force exerted by a fibroblast on its substratum is so much greater than what is needed to propel it, Harris and colleagues have suggested that these large tractional forces are developed for morphogenetic functions rather than to drive locomotion [7], and have demonstrated how such forces might operate in the formation of ligaments, tendons and muscles [67]. From this perspective, the graded polarity of fibroblast actin filaments could be considered as an adaptation to serve this tissue developmental function. The preferential orientation of filaments with their barbed ends pointed to the adjacent cell edge at the front and rear of the cell is well suited to generating static centripetal traction, as measured originally on deformable substrata [68] and more recently and quantitatively using a micromechanical device which can register the local tractional force of migrating cells [69]. The large transverse centripetal forces generated by fish epidermal keratocytes provide a further illustration of this idea [12,13]. The orientation of these forces indicates that they are unlikely to contribute directly to cell propulsion, and so it is plausible to suggest that their normal physiological function is in tissue development or to provide mechanical stabilization of tissue (in this case, skin).

The large adhesive forces that bind fibroblasts to tissue-culture substrata are an impediment to their migration and are overcome with the help of the large tractional forces that these cells exert. This does not necessarily mean, however, that the capability of the cell to develop these forces has evolved for this purpose. On the contrary, it is possible that, as for *Dictyostelium* amoebae and neutrophils, the propulsive forces are much smaller. These forces would be difficult to detect over the background of the observed large tractional forces. Nevertheless, it has been observed that, on fibroblasts, beads are transported to a ring-shaped region around the nucleus (in contrast with the transport of beads to the posterior of amoebae) [5]. Arguing as above for *Dictyostelium* cells, we propose that the specifically propulsive forces in fibroblasts might arise from motor molecules operating individually on centripetal tracks towards the nucleus, in contrast with the much larger collective cytoskeletal contraction which gives rise to the observed tractional forces. Whether the motors that operate on these hypothetical tracks are isoforms of myosin II or, as in *Dictyostelium* cells, are different from conventional myosin is an important question. The tracks could be oriented with their barbed ends towards the nucleus, or a pointed-end-directed motor could supply the propulsive force. Identification of these motor molecules and their visualization in migrating cells could provide the best way to test this proposal.

Traction force in tissue modelling

One way to study the proposed effects of fibroblast traction on tissue development [5,7] is with reconstituted tissue models, which can provide useful systems in which to study cell and tissue mechanical properties and force regulation. These model tissues can be constructed from extracellular-matrix components and cells. For example, model connective tissues called fibroblast-populated collagen lattices (FPCLs) are formed by suspending fibroblasts in liquid collagen and then allowing the collagen to gel, entrapping the cells. Over a period of time the cells compress the matrix, reducing its volume by more than 10-fold and increasing the mechanical stiffness of the tissue. FPCLs provide valuable models of connective tissue and wound healing [70], and have been studied extensively (e.g. [71]; see also Chapter 3).

As the cells compress the tissue, they generate a contractile force, which can be measured using an isometric force transducer [72–74]. This force is presumably related to the traction force with which the cells are thought to remodel the matrix. Measurements of the development of this force can be followed over the time period of matrix compression, and the effects of various perturbations (e.g. inhibition of selected kinases or the interaction of the cells with the collagen) can be observed (T. Wakatsuki, unpublished work). This method should be useful for characterizing the mechanism and regulation of the forces responsible for tissue compression. It will also be important to determine whether this force is, as suggested, related to the traction that fibroblasts exert on substrata as they migrate.

In addition to developing a 'basal' force during matrix compression, the model tissues can further increase force in response to mitogenic agonists that trigger activation of myosin by phosphorylation of the myosin regulatory light chain of 20 kDa (LC_{20}) [73].

The force exerted by FPCLs is also regulated by microtubules [75]. When an FPCL is exposed to the microtubule-disrupting drug nocodazole, the force that it exerts on the isometric force transducer increases. This increase can be blocked by the microtubule stabilizer taxol. This result is predicted by the 'force counterbalance' or 'tensegrity' model of cytoskeletal structure in terms of a transfer of contractile load from internal struts (microtubules) to the sites at which the cells are anchored to the matrix [76,77]. Not expected from this model, however, was the observation that disruption of microtubules by noco-dazole also activates myosin, as indicated by phosphorylation of LC_{20}. Moreover, the effects of nocodazole and taxol on the phosphorylation of LC_{20} closely parallel their effects on force. Hence it appeared equally likely that the nocodazole-induced increase in force could result from force transfer or from activation of myosin. It was possible to distinguish the contributions of force transfer and myosin activation by determining the effect on the nocodazole-induced increase in force of pre-activating myosin prior to nocodazole treatment. This experiment showed that myosin activation, not force transfer, makes the dominant contribution to the nocodazole-induced force response [75].

Conclusions and questions

Although there has been much progress in the analysis of the mechanism of cell locomotion, the structural basis, mechanism and regulation of none of the steps of the process have been completely characterized as yet. Tail retraction with the aid of a myosin-dependent contractile force as well as, for fibroblasts, selective disruption of interactions between cellular integrins and the matrix [48] may be the best understood. Much evidence points to the importance of actin filament polymerization in the frontal protrusion step, but a detailed molecular characterization of this force and the mechanical resistance against which it operates remains to be determined. Least understood is the traction force that draws the body of the cell towards frontal anchorage sites. In *Dictyostelium* amoebae, motor(s) other than myosin II are expected to make important contributions to this force. Whether a distinct motor is also involved in fibroblast motility is unknown, and this is difficult to determine because of the multiplicity of myosin II isoforms in higher animal cells. The large forces exerted on the substratum by fibroblasts and fish epidermal keratocytes may be directed towards a tissue modelling or wound healing function rather than locomotion. The presence of these forces renders detection of the smaller forces responsible for cell propulsion still more difficult. The tissue modelling traction force is, however, interesting and physiologically important in its own right. Reconstituted model tissues provide one approach to characterizing this morphogenetic force.

References

1. Mitchison, T.J. and Cramer, L.P. (1996) Cell **84**, 371–379
2. Felder, S. and Elson, E.L. (1990) J. Cell Biol. **111**, 2513–2526
3. Wessels, D., Vawter-Hugart, H., Murray, J. and Soll, D.R. (1994) Cell Motil. Cytoskeleton **27**, 1–12
4. Weber, I., Wallraff, E., Albrecht, R. and Gerisch, G. (1995) J. Cell Sci. **108**, 1519–1530
5. Harris, A.K. (1994) Int. Rev. Cytol. **150**, 35–68
6. Jay, P.Y., Pham, P.A., Wong, S.A. and Elson, E.L. (1995) J. Cell Sci. **108**, 387–393
7. Harris, A.K., Stopak, D. and Wild, P. (1981) Nature (London) **290**, 249–251
8. Harris, A.K. (1986) Dev. Biol. **3**, 339–357
9. Ehrlich, H.P. and Rajaratnam, J.B. (1990) Tissue Cell **22**, 407–417
10. Bereiter-Hahn, J., Strohmeier, R., Kunzenbacher, I., Beck, K. and Voth, M. (1981) J. Cell Sci. **52**, 289–311
11. Euteneuer, U. and Schliwa, M. (1984) Nature (London) **310**, 58–61
12. Lee, J., Leonard, M., Oliver, T., Ishihara, A. and Jacobson, K. (1994) J. Cell Biol. **127**, 1957–1964
13. Oliver, T., Dembo, M. and Jacobson, K. (1995) Cell Motil. Cytoskeleton **31**, 225–240
14. Egelhoff, T.T. and Spudich, J.A. (1991) Trends Genet. **7**, 161–166
15. De Lozanne, A. and Spudich, J.A. (1987) Science **236**, 1086–1091
16. Knecht, D.A. and Loomis, W.F. (1987) Science **236**, 1081–1086
17. Novak, K.D. and Titus, M.A. (1997) J. Cell Biol. **136**, 633–647
18. Jung, G., Wu, X. and Hammer, III, J.A. (1996) J. Cell Biol. **133**, 305–323
19. Wessels, D., Murray, J., Jung, G., Hammer, III, J.A. and Soll, D.R. (1991) Cell Motil. Cytoskeleton **20**, 301–315
20. Schindl, M., Wallraff, E., Deubzer, B., Witke, W., Gerisch, G. and Sackmann, E. (1995) Biophys. J. **68**, 1177–1190
21. Niewohner, J., Weber, I., Maniak, M., Muller-Taubenberger, A. and Gerisch, G. (1997) J. Cell Biol. **138**, 349–361
22. Neujahr, R., Heizer, C. and Gerisch, G. (1997) J. Cell Sci. **110**, 123–137
23. Tuxworth, R.I., Cheetham, J.L., Machesky, L.M., Spiegelmann, G.B., Weeks, G. and Insall, R.H. (1997) J. Cell Biol. **138**, 605–614
24. Small, J.V., Rohlfs, A. and Herzog, M. (1993) Symp. Soc. Exp. Biol. **47**, 57–71
25. Small, J.V., Herzog, M. and Anderson, K. (1995) J. Cell Biol. **129**, 1275–1286
26. Cramer, L.P., Siebert, M. and Mitchison, T.J. (1997) J. Cell Biol. **136**, 1287–1305
27. Wang, Y.L. (1985) J. Cell Biol. **101**, 597–602
28. Evans, E. (1993) Biophys. J. **64**, 1306–1322
29. Wessels, D., Soll, D.R., Knecht, D., Loomis, W.F., De Lozanne, A. and Spudich, J. (1988) Dev. Biol. **128**, 164–177
30. Jung, G., Fukui, Y., Martin, B. and Hammer, III, J.A. (1993) J. Biol. Chem. **268**, 14981–14990
31. Fukui, Y., Lynch, T.J., Brzeska, H. and Korn, E.D. (1989) Nature (London) **341**, 328–331
32. Cortese, J.D., Schwab, B., Frieden, C. and Elson, E.L. (1989) Proc. Natl. Acad. Sci. U.S.A. **86**, 5773–5777
33. Cudmore, S., Cossart, P., Griffiths, G. and Way, M. (1995) Nature (London) **378**, 636–638
34. Cossart, P. (1995) Curr. Opin. Cell Biol. **7**, 94–101
35. Theriot, J.A. (1995) Annu. Rev. Cell Dev. Biol. **11**, 213–239
36. Cramer, L.P. and Mitchison, T.J. (1995) J. Cell Biol. **131**, 179–189
37. Cooper, J.A. (1991) Annu. Rev. Physiol. **53**, 585–605

38. Stossel, T.P. (1993) Science **260**, 1086–1094
39. Frieden, C. (1985) Annu. Rev. Biophys. Chem. **14**, 189–210
40. Pasternak, C. and Elson, E.L. (1990) J. Cell Biol. **111**, 7a
41. Worthen, G.S., Schwab, B., Elson, E.L. and Downey, G.P. (1989) Science **245**, 183–186
42. Evans, E. and Yeung, A. (1989) Biophys. J. **56**, 151–160
43. Pasternak, C., Spudich, J.A. and Elson, E.L. (1989) Nature (London) **341**, 549–551
44. Yumura, S., Mori, H. and Fukui, Y. (1984) J. Cell Biol. **99**, 894–899
45. Chen, W.T. (1981) J. Cell Biol. **90**, 187–200
46. Small, J.V. (1989) Curr. Opin. Cell Biol. **1**, 75–79
47. DiMilla, P.A., Barbee, K. and Lauffenburger, D.A. (1991) Biophys. J. **60**, 15–37
48. Lauffenburger, D.A. and Horwitz, A.F. (1996) Cell **84**, 359–369
49. Ingram, V.M. (1969) Nature (London) **222**, 641–644
50. Lin, C.H. and Forscher, P. (1995) Neuron **14**, 763–771
51. Abercrombie, M., Heaysman, J.E. and Pegrum, S.M. (1970) Exp. Cell Res. **62**, 389–398
52. Fukui, Y., De Lozanne, A. and Spudich, J.A. (1990) J. Cell Biol. **110**, 367–378
53. Bourguignon, L.W.W. and Bourguignon, G.J. (1984) Int. Rev. Cytol. **87**, 195–224
54. Schreiner, G.F. and Unanue, E.R. (1976) Adv. Immunol. **24**, 37–165
55. Condeelis, J. (1979) J. Cell Biol. **80**, 751–758
56. Bray, D. and White, J.G. (1988) Science **239**, 883–888
57. Pasternak, C. and Elson, E.L. (1985) J. Cell Biol. **100**, 860–872
58. Jay, P.Y. and Elson, E.L. (1992) Nature (London) **356**, 438–440
59. Harris, A. and Dunn, G. (1972) Exp. Cell Res. **73**, 519–523
60. Cramer, L.P. and Mitchison, T.J. (1997) Mol. Biol. Cell **8**, 109–119
61. Fishkind, D.J. and Wang, Y.L. (1993) J. Cell Biol. **123**, 837–848
62. Anderson, K.I., Wang, Y.L. and Small, J.V. (1996) J. Cell Biol. **134**, 1209–1218
63. Kucik, D.F., Elson, E.L. and Sheetz, M.P. (1990) J. Cell Biol. **111**, 1617–1622
64. Sheetz, M.P. and Spudich, J.A. (1983) Nature (London) **303**, 31–35
65. Nothnagel, E.A. and Webb, W.W. (1982) J. Cell Biol. **94**, 444–454
66. Theriot, J.A. and Mitchison, T.J. (1991) Nature (London) **352**, 126–131
67. Stopak, D. and Harris, A.K. (1982) Dev. Biol. **90**, 383–398
68. Harris, A.K., Wild, P. and Stopak, D. (1980) Science **208**, 177–179
69. Galbraith, C.G. and Sheetz, M.P. (1997) Proc. Natl. Acad. Sci. U.S.A. **94**, 9114–9118
70. Bell, E., Ivarsson, B. and Merrill, C. (1979) Proc. Natl. Acad. Sci. U.S.A. **76**, 1274–1278
71. Grinnell, F. (1994) J. Cell Biol. **124**, 401–404
72. Kolodney, M.S. and Wysolmerski, R.B. (1992) J. Cell Biol. **117**, 73–82
73. Kolodney, M.S. and Elson, E.L. (1993) J. Biol. Chem. **268**, 23850–23855
74. Huang, D., Chang, T.R., Aggarwal, A., Lee, R.C. and Ehrlich, H.P. (1993) Ann. Biomed. Eng. **21**, 289–305
75. Kolodney, M.S. and Elson, E.L. (1995) Proc. Natl. Acad. Sci. U.S.A. **92**, 10252–10256
76. Buxbaum, R.E. and Heidemann, S.R. (1988) J. Theor. Biol. **134**, 379–390
77. Ingber, D.E. (1993) Cell **75**, 1249–1252

Biochem. Soc. Symp. **65**, 315–341
Printed in Great Britain

19

A dozen questions about how tissue cells crawl

Albert K. Harris

Department of Biology, Coker Hall, CB# 3280, University of North Carolina at Chapel Hill, Chapel Hill, NC 27599-3280, U.S.A.

Abstract

A series of long-standing questions about tissue cell locomotion now need to be reconsidered in relation to accumulating molecular information. For example, if ruffling represents localized re-assembly of actin at advancing cell margins, does this mean that contact inhibition of cell locomotion reflects inhibition of actin assembly at cell–cell contacts or adhesions? If so, then to what extent does invasiveness result from abnormal continuation of actin assembly at contacted margins? What mechanisms restrict actin assembly to certain parts of cells, margins, and how do microtubules control this, such that fibroblasts become less directional ('the Vasiliev phenomenon'), but more strongly contractile ('the Danowski phenomenon'), when treated with microtubule inhibitors? What concentrates cell–substratum adhesions into foci, and reduces adhesions beneath cell centres? A related question is whether actomyosin stress fibres are nucleated at focal adhesions, as opposed to being gathered together from lamellipodial actin; in addition, how are myosins added into the fibres in such a way as to contribute to force exertion? Does the orientation of fibroblasts along ridges result from inhibition of actin oriented across ridges, or from stimulation of actin assembly where cell margins are bent? Related questions concern the roles of cellular traction forces in embryonic development, whether cell sorting and aggregate rounding are due more to active contraction or to maximization of adhesion, and why individual cells round-up in mitosis or when treated with very high pressures or deprived of bivalent cations. Alternative answers to each question are discussed in relation to the available evidence.

Introduction

Much information is accumulating about tissue cell locomotion, especially molecular and structural information. If we can synthesize this new information, the result should be a much more complete understanding of the subject, extending from how molecules interact to exert traction forces all the way up to the anatomical level of how these forces are used to create and maintain normal tissue structures. These goals are still some distance off, but it may help to formulate specific questions to which we would most like to know the answers. This article proposes a dozen such questions, together with some alternative possible answers.

To the best of my knowledge, none of these 12 questions has yet been answered conclusively. I suspect, however, that data sufficient to answer several of them may already exist in the published literature. Focusing attention on specific questions may stimulate the application of such existing data, as well as encouraging further research. In addition, several of the questions were formulated specifically to challenge assumptions that many people seem to be making, based more on plausibility than evidence. In particular, the questions about focal adhesion and stress fibre formation, as well as those about rounding-up by cells and cell aggregates, have this motivation. Another goal was to shift thinking towards more geometric and morphogenetic points of view. The cytoskeleton is a self-organizing system, after all. So it is not sufficient to understand which molecules pull on which other molecules, as if motility were simply a physiological problem (given a structure, how does it work?). We also need to ask how the combinations of properties and interactions of these molecules cause them to rearrange spontaneously into the patterns that we observe, not just how they function once these patterns are achieved. Similarly, at the multicellular level, what structure-creating functions are served by the phenomenon that expresses itself in isolated cells as crawling? Just because a cell crawls doesn't mean that the only function of its force-generating mechanisms is to move cells from one location to another. There can be other reasons for cells to exert pulling forces on materials around them, such as the constriction of wound margins and the rearrangement of collagen to form capsules and tendons. As it happens, Michael Abercrombie wrote several of the most seminal research papers in the field of wound repair, and in relation to collagen fibres [1–3]. This aspect of his career seems not to be known by most of us in the field of cell motility; I learned of it only when invited to speak at a convention of plastic surgeons, and asked my hosts what papers I should read concerning the basic cell biology of their field. They told me that the best and most basic papers were by Abercrombie. They, in turn, had not known about his work on locomotion and contact inhibition.

Some of the dozen questions concern the self-organization and self-perpetuation of cell polarity. Although cell motility can be viewed from a purely physiological point of view (given the observed spatial arrangements of actin, myosins, etc., how does this produce propulsive forces?), we also need to ask what mechanisms put these molecules into their correct relative positions, and put them back there when they move. Consider a crawling cell, with a front and a rear, or even with two opposite front ends and a middle being stretched between them: this cell's polarity tends to persist over time, despite continuous rearrangements of the component molecules responsible (I refer especially to actin treadmilling). So how can such polarity persist? And how can this polarity create itself in the first place? Probably only by the use of some kind of positive feedback; in other words, causal effects that run in a circle. Nevertheless, people tend to ask which elements are 'upstream' or 'downstream' of which others, as if causality had to be (or even could be) purely a one-way street. Perhaps it is no accident that such 'upstream/downstream' debates are so often inconclusive, leading to paradoxes.

We especially need to elucidate the normal control functions of the Rho family of GTP-binding proteins. One might regard these as answers in search of questions (or mechanisms in search of functions), in that changing their activities definitely produces drastic changes in cell behaviour, although we don't yet really know which processes of actin assembly and disassembly, force exertion, adhesion and de-adhesion are being controlled.

Question 1: how is cell polarity created and maintained?

Idealized fibroblasts are often drawn with a single front end and with a trailing tail at the opposite end (Fig. 1). Such cells can be found in most cultures, but they are only one among many different shapes. The more common morphology has two or three advancing margins along opposite sides of the cell, each pulling in a different direction and stretching the cell between them. The physical forces responsible for pushing these advancing margins forward have been the subject of much discussion, with regard to what extent the force is osmotic, or caused by the local assembly of actin fibres, or Brownian but then quickly supported by actin fibres. I will take a different approach by focusing attention on the self-perpetuation of each of the different alternative states of behaviour of cell margins, what special properties each one has, what causal relationships exist between the different properties of each, and (most of all) why one set of properties continues to exist at one point location while very different sets of properties exist only a few micrometers away. For this pur-

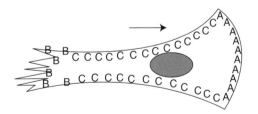

Fig. 1. Idealized fibroblast cell outline. The letters A, B and C are used to indicate the approximate locations of three different types of (apparently self-perpetuating) cytoskeletal activities: A, the advancing margins where actin is assembled; B, retreating end behind the cell; C, the comparatively inactive catenary margins.

pose, I tentatively categorize the margins of cells (spread and crawling in tissue culture) into three distinct kinds, to be called type A margins (A for advancing), type B margins (B for back) and type C margins (C for catenary, following Michael Abercrombie's own use of the term 'catenary margins' for these regions). Future knowledge may suggest more subdivisions than just these three; or possibly fewer subdivisions, for example if type B turns out to be a transition state between A and C.

The special properties of type A margins are forward protrusion, the formation of new adhesions (especially of the dispersed but weak form that appear grey in interference reflection microscopy) and the exertion of centripetally directed traction forces tangentially through their plasma membranes on both upper and lower surfaces. If a fibroblast has three separate parts of its margin undergoing type A behaviour, then its cell body will be stretched between them into an approximately triangular shape. When a crawling cell is made to undergo a small amount of blebbing, for example by making the culture medium hypotonic, then the blebs (temporary membrane herniations) will be almost always concentrated exactly along the type A and type B margins; neither the central areas nor the type C margins will bleb until the treatment is made more drastic. Type C margins sag inward, are not attached to the substratum, and do not protrude outward, undergo ruffling or exert forces. Type B margins are attached to the substratum, but do not protrude outward or form new adhesions.

We need to ask what causes a given part of a cell margin to develop one or other of these three sets of properties and activities, how each property of a given region is related to its other properties (e.g. why do forward-protruding regions also form new adhesions?), and what perpetuates this activity over time at the same approximate location. This question of self-perpetuation is most interesting in relation to the type A margins: for one thing, the basis of contact inhibition seems to be a decrease or termination of type A activity at sites that have touched or adhered to another cell.

Furthermore, if the molecular components of these regions are being swept rearward by the flow of actin, then what keeps the actin assembly process concentrated at the same place over time? When a trypsinized fibroblast is first plated out and begins to spread on a flat surface, the whole cell margin initially shows only type A behaviour: one margin extends 360° all the way around the periphery, producing a 'fried egg' shape. Later, regions with type C properties begin to appear, and the cell develops its typical polygonal shape. The largest and strongest of the type A margins becomes the front end, pulling the rest of the cell around behind itself. Tracks of fibroblast movement often show a series of relatively straight lines (periods of locomotion consistently in one direction), separated by sharp turns. These sharp turns reflect some shift in dominance to a new or different type A margin, and often follow major detachments and partial rounding-up of the cell. As will be discussed further in relation to question 2 below, treatment of cells with vinblastine or other microtubule poisons causes many fibroblasts to lose this polarity, with the entire margin reverting to type A behaviour (the 'Vasiliev phenomenon'). Conversely, the phenomenon named 'contact guidance' by Paul Weiss [4], and recently renamed 'topographic response' [5], seems to reflect some kind of stimulation of type A activity by grooves and ridges in the substratum; this will be the subject of question 9 below.

Another way of posing our first question would be to ask why type A margins are so much more prone to being protruded outward, whether in combination with actin assembly or in bleb formation. Is the membrane less resistant to outward forces? Concentration of blebbing along these margins would seem to imply less resistance. Could this lesser resistance predispose to actin protrusions? A third special property of type A margins is their adhesiveness. What causal interrelationships can be proposed between actin assembly, greater susceptibility to outward protrusion and the formation of new adhesions? For some reason, these three go together. Cell margins that ruffle also bleb, and also form new adhesions. Perhaps one of these peculiarities causes the other two, or all three share a common cause. Some cycle of causation seems likely, to account for the self-perpetuation of each type of behaviour.

Because actin assembly is concentrated along type A margins, the simplest way to pose our question might be to ask why (and whether) actin doesn't assemble elsewhere instead. Here are some possible alternative answers to this question. (A) The barbed ends ('plus' ends, which favour assembly) are concentrated in this region. A variation on this idea is that any barbed ends at all other locations somehow become capped [6]. (B) Actin monomer is concentrated there, in one theory because actin mRNA is localized there by some kind of molecular addressing mechanism. (C) There could be some sort of enzyme that catalyses actin assembly, and this catalyst is concentrated there. (D) The process of addition of ATP nucleotides to actin monomers (allowing them to assemble) is somehow concentrated there. (E) The plasma membrane has some special properties in this area, causing it to protrude outward, or perhaps causing it to form new adhesions. (F) Mechanical tensions concentrated at these areas somehow promote actin assembly, as well as being created by the rearward actin flow.

The second of these possibilities (explanation B) is derived from the work of Singer and co-workers [7], who have shown very elegantly that actin mRNA definitely tends to become concentrated near the leading margins of crawling fibroblasts. Whether this is a cause or a side-effect is less certain. The only way that I can think of for it to be a side-effect would be if newly synthesized actin molecules, still in the process of synthesis on the ribosomes and still attached to their mRNA, were to undergo assembly into fibrous actin at the normal locations. This would tend to trap the messengers at this location, even though the movement of the messenger, and its ribosomes, and its nascent actin proteins, had been random. This possibility might be tested directly, for example in a cell-free system where actin is being synthesized from labelled mRNA and where one could nucleate actin assembly at some arbitrary location. An indirect test would be to see whether cell polarity, or indeed locomotion and centripetal transport, is rapidly destroyed by chemical inhibition of protein synthesis.

The last of the possibilities listed (explanation F), involving mechanical tension, was included because of observations that cells can be 'steered' by pushing on them sideways just behind the leading margin. The advancing cell margin reacts to the pressure by turning in the opposite direction, in other words turning towards the direction in which tension has been increased. Wessells and Nuttall [8] showed this very dramatically in the case of nerve growth cones, which they could steer into any chosen pathway, including letters of the alphabet. Less dramatic equivalents of the same effect can be produced in fibroblasts by using a micromanipulator to push the end of a flexible glass needle against one of a cell's type C margins, so that the needle is bent into a bow. The adjacent type A margins will turn towards the direction from which the tension is being applied. That is exactly what ought to happen if tension were part of the normal feedback cycle that creates and maintains the directionality of locomotion. It would also explain why fibroblasts tend to change direction following major retractions, since that is when tension is temporarily lost.

Question 2: why do microtubule poisons inhibit polarity, but strengthen the contraction and increase the formation of stress fibres?

Treatment of tissue culture cells with microtubule poisons (colchicine etc.) has been found to produce two major effects (Fig. 2). One effect is that type A activity (protrusion, ruffling, adhesion, etc.) gradually tends to spread more broadly around the cells' circumference; I shall call this 'the Vasiliev phenomenon', because it was first described by his group [9]. The other effect is much more rapid, consisting of an increased strength of contractility (doubling or more), with the development or enlargement of actomyosin stress fibres. I shall refer to this increase in contractile strength as 'the Danowski phenomenon', after its discoverer [10], although it has been confirmed by others [11]. An especially dramatic aspect of this phenomenon is the reversal of disruption of stress fibres by phorbol ester tumour promoters and several other drugs. In other words, cells lose their stress fibres when treated with phorbol esters, but quickly re-form them when treated with a microtubule poison in addition, even though the phorbol ester is still present. Analogous changes in cortical contractility have been observed in frog oocyte cytoplasm treated with microtubule poisons [12].

It is tempting to hope that both of these phenomena (reduced polarization and increased contractility) can somehow be explained by the same basic mechanism, and also that this mechanism reflects some normal function of microtubules in controlling actin behaviour during cell spreading. But it is a puzzle to imagine what that single mechanism might be. Why should taking away microtubules favour stress-fibre formation? Why should it also reverse the effects of other drugs that disrupt stress fibres? And how can it also cause type A margins to spread at the expense of type C margins? In addition, why should this latter effect be gradual and only occur in some cell lines, while the former effects are rapid and occur universally?

One type of explanation for the Danowski phenomenon is that microtubules might exert large pushing forces, counterbalancing a substantial

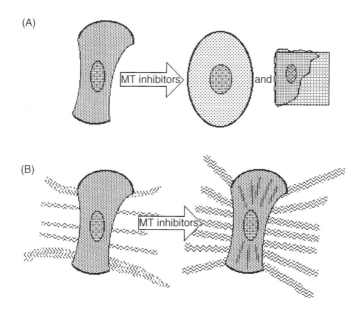

Fig. 2. Two responses of fibroblasts to microtubule (MT) inhibitors.
(A) The Vasiliev phenomenon, including reduced morphological polarity and increased conformity to shapes imposed by adhesive patterns of artificial substrata. (B) The Danowski phenomenon, including stronger traction, increased stress fibres and reversal of effects of phorbol ester tumour promoters.

fraction of the pull exerted by actin; thus poisoning the microtubules would allow the full force of the actin to be transmitted to the substratum. This is one aspect of Ingber's 'tensegrity theory' [13]. Unfortunately, it won't explain why more stress fibres appear in cells treated with microtubule poisons, even where none had been present before; this is especially puzzling in cells whose contraction has been inhibited with phorbol esters and other drugs. Why should this inhibition of actin stress fibres be reversed by reducing the pressure of growing microtubules? Nevertheless, Clare Waterman-Storer's video records of microtubule formation near the type A margins show clearly that microtubules growing towards type A margins are subjected to substantial rearward pressures from actin flow. If the actin is pushing hard on the microtubules, even bending them back and breaking them, then this would reduce the effective contractile force that the actin can exert on the substratum. Whether this can account for an effective doubling or tripling of the strength of traction is more doubtful. If so, it would mean that the actin is actually pulling two or three times harder than the force exerted at the surface, i.e. that microtubules absorb forces at least as large (or sometimes twice as large) as the total forces exerted on the substratum. That would seem to be a huge waste of energy, both for the actin and for the microtubules.

A very different type of explanation is that some material transported along microtubules serves some control function, inhibiting actin in some way.

Thus, when the microtubules are eliminated, the inhibition is reduced, allowing the actin to pull harder and to assemble around a larger proportion of the cell periphery. Rodionov et al. [14] have showed that injection of antibodies against kinesin produces morphological effects similar to those of microtubule poisons. A paradox is that microtubules mostly point towards the type A margins, so if they were transporting some control substance, you might expect the function of this substance to be stimulation (rather than inhibition) of actin assembly and function. Disruption of the microtubules might then allow more of the cell margin to be stimulated; but why should this greatly increase the total force exerted, and why should this increase in force occur very much faster than the loss of polarization, and in a wider variety of cells?

Alternative types of explanation for the Danowski phenomenon are as follows. (A) Large pushing forces result from microtubule assembly (tensegrity). (B) Motor proteins (kinesins) transport something (inhibitory?) towards the type A margins. (C) Microtubule dynamics themselves inhibit actin assembly and/or contraction. (D) Release of tubulin (or some microtubule-associated protein) from disassembly of peripheral ends of microtubules has some controlling effect on actin assembly. Eliot Elson's chapter in this symposium (Chapter 18) indicates that the increased contractility may result largely from phosporylation of myosin light chains.

Question 3: what is the molecular basis of contact inhibition?

One reason why it is so important to determine the molecular mechanism of ruffling and the forward protrusion of cell margins (i.e. is it really rapid localized actin assembly?) is so that we can then go on to determine the mechanism of contact inhibition of cell locomotion. This phenomenon, of course, was discovered and measured in the classic and elegant experiments of Joan Heaysman and Michael Abercrombie 45 years ago [15,16].

As they demonstrated so carefully, when one crawling fibroblast collides with another, it tends to reverse direction, or to turn aside, or at least to slow down. Thus cell–cell overlaps are significantly less frequent than random; thus the average speeds of cells with many contacts are only about half as fast as those of cells not in contact with any other cells; thus outgrowth zones around square or elongate explants nevertheless quickly approximate circles; thus cell densities between adjacent explants are significantly less than twice the densities found at equal distances from explant sides not facing another explant. These are only some of the consequences of contact inhibition that Abercrombie and Heaysman were able to measure. A subsequent paper by Abercrombie and Ambrose [17] showed that this inhibition of locomotion is usually accompanied by an inhibition of ruffling along the parts of the cell margins that touch each other ('contact paralysis'), as if ruffling somehow represented the motor that was pulling the cell margin forward, with contact somehow turning off this motor. The topic of contact inhibition has been extensively reviewed [18–20].

It would be a dream come true to be able to present evidence that actin assembly is inhibited locally where cell margins touch, or to be in a position to test alternative possible mechanisms by which cell–cell adhesions could cause this inhibition. Instead, I propose these questions (does contact inhibit actin assembly? and why?) as being among the most important targets for research today. One might bleach spots in fluorescently stained actin just behind leading margins, and compare the centripetal movement of the bleached areas at times just before and just after the advancing margin touches another cell. Vasiliev and co-workers [21] showed an inhibition of the rearward movement of fluorescent markers on the outside of cells (i.e. capping of fluorescent concanavalin); that is exactly what ought to happen if the mechanism of contact inhibition is prevention of further actin assembly where cells touch one another. Comparable experiments need to be carried out inside the cell with fluorescently labelled actin.

Another set of related questions concerns changes in contractility and adhesiveness that occur during and following contact inhibition. Two alternative theories on this subject were proposed, one by myself and the other by Michael Abercrombie. Both theories concerned the tendency of cells to retract away from each other immediately following an instance of contact inhibition and contact paralysis. My proposal had also been intended to explain the formation of cell sheets in which only the cells at the margin of the sheet were adhering to the substratum, which is something that one often encounters, not just with epithelial cells but also with fibroblasts that are crowded together (especially at higher concentrations of serum in the medium). My proposal was that, besides inhibiting further forward protrusion and ruffling, cell–cell contact also inhibits the formation of new adhesions to the substratum in the areas just centripetal to where the ruffling has been inhibited. The idea was that the advancing cell forms new adhesions to the substratum as its margin advances, but that, when it contacts the other cell, it forms a cell–cell adhesion, but stops forming new cell–substratum adhesions in the contact area. Cell sheet formation would result in those cases where the cell–cell adhesion became sufficiently strong to support the cells' contractile tension before too many cell–substratum adhesions had been lost. Conversely, retraction of the cells away from each other would result from those cases in which the cell–substratum adhesions were lost too soon, before the cell–cell adhesion forces had become strong enough to support the tension. A further idea was that higher serum concentrations either delayed the loss of adhesions to the substratum or increased the relative strength of the cell–cell adhesions.

Michael Abercrombie's counter-theory was that contact induced a local increase in cell contractility just behind the contact, and that this pulled the cells apart. He and Dunn published a paper [22] in which they used interference reflection microscopy to look for the inhibition of cell–substratum adhesion that I had predicted to occur behind where the cell–cell contacts had formed; not finding such a loss of adhesions, they concluded that increased contractility must be the real cause of retraction. They also made the excellent point that recently contacted cells quite often retract at the trailing end, which could hardly result from decreased adhesion at the front.

My original paper in this exchange [23] had included still frames from a time-lapse sequence in which one cell's forward locomotion underwent contact inhibition after contacting the side of a second cell in a region of the latter that was not adhering to the substratum, instead being stretched between two adhesions some distance away. When the first cell retracted, it pulled the second cell sideways; this proved that a cell–cell adhesion had formed. In addition, the angle at which the second cell was flexed indicated that the tension exerted by the retracting cell was much less than that in the second cell (although it might have been more at the moment the retraction began). In subsequent observations with the silicone rubber substratum, I have looked at hundreds of instances of contact inhibition, seeking some case of increased contractility. Although there were many cases of retractions, and also cases in which cell contractility increased, I have yet to find a case where contact stimulated increased contraction. On the other hand, using interference reflection microscopy, I frequently record examples in which the broad grey-appearing adhesions cease to form just behind cell–cell contacts. To see this, it helps to illuminate with relatively dim light from a tungsten source; mercury arc illumination, probably because of its near-UV components, dramatically inhibits the ability of cells to let go of their adhesions to the substratum. For example, if you observe polymorphonuclear leucocytes with mercury arc illumination, the cells will be unable to traverse the illuminated area. Elsewhere the cells can crawl along, breaking their old adhesions at their uropod at their rear ends. But the arc illumination makes these rear adhesions unbreakable, in some way, so that the uropods expand and each cell stops forward movement, as if the cell were an animal whose tail had been nailed to the floor. Such inhibition of adhesion loss is itself unexplained, as far as I know, but might be a useful tool for studying whatever mechanism cells use to let go of their old adhesions; I suggest it might explain why Abercrombie and Dunn did not see loss of adhesions behind contact inhibition events. This question deserves further study.

Since we do not even understand the mechanisms by which actin assembly is concentrated at certain parts of the cell margins in preference to others (question 1, above), we are hardly in a good position even to guess intelligently as to why cell–cell contact should block this assembly. Any answers to our first question will automatically suggest possible explanations for this third one, and also for question 4 below.

Question 4: which properties of invasive cells allow continuation of forward locomotion at contacts, circumventing contact inhibition?

Perhaps the most important questions one could ask relating to the mechanics of contact inhibition are those concerning the invasiveness of cancerous cells and macrophages. As shown by Abercrombie, Heaysman and Karthauser [24], at least some cancerous cells are comparatively much less sensitive to contact inhibition than are equivalent normal cells. This makes much sense, in terms of explaining invasiveness. Critics of this interpretation have not been lacking, however, with much attention being drawn to the phenomenon of 'underlapping'; in which the lack of cell–substratum adhesions beneath cell bodies and along much of their margins allows neighbouring cells to crawl under them. The point here is that most cases of what had been interpreted as one cell crawling up on to its neighbour's surface are really cases in which one or the other cell has crawled under its neighbour. Is such underlapping to be regarded as a failure of contact inhibition? Many have thought it shouldn't be. On the other hand, one can certainly find cases in which one cell crawls under another, but nevertheless undergoes contact inhibition, including typical contact paralysis, once its upper surface touches the cell above it. Therefore, if a macrophage or cancerous cell is able to underlap a neighbouring cell freely, in the sense of continuing to ruffle and advance despite touching the overlying cell, then I would still call that a failure of contact inhibition, even though neither cell was actually using the other as its substratum, as had originally been assumed. Furthermore, I have seen many cases in which highly malignant sarcoma cells actually did use the upper surfaces of fibroblasts as substrata.

In the case of macrophages, contact with fibroblasts can actually encourage further locomotion in the direction of the contact [19]. Should this be regarded as some sort of 'reverse contact inhibition'? Perhaps in most cells actin synthesis is inhibited near cell–cell contacts, but in macrophages the reverse occurs, with contact actually stimulating actin synthesis. This is one of a whole series of cases in which macrophage motility responds to various stimuli in the exact opposite way to that of (nearly) all other cell types; for example, macrophages move towards the positive electrode in electric fields [25], whereas fibroblasts, epithelial cells etc. move towards the negative electrode. Another example is that macrophages accumulate on hydrophobic substrata, whereas fibroblasts etc. accumulate on the more hydrophilic substrata available to them; likewise, macrophages prefer roughened surfaces to smooth, whereas other cell types show the opposite behaviour [26]. No one knows the mechanism for any of these cases of opposite behaviour, much less whether they all share some common cause or function. An exciting new possibility is suggested by the very different behaviour of Rho-family GTP-binding proteins (see Chapter 8 in this volume) in macrophages as compared with other cell types.

There are many reasons to suspect that susceptibility to contact inhibition is closely linked to a cell's ability to form adhesions. Some of the most tantalizing findings have come from Bershadsky and his colleagues, who found that some transformed cells revert to forming more or less normal patterns of focal adhesions and stress fibres when they become multinucleate [27]. It is puzzling why such a quantitative change in nuclear content or size should result in a qualitative change in adhesion and cytoskeletal structure, but an explanatory theory has been proposed [28] involving the defective development of focal adhesions in transformed cells.

More attention should be paid to the many morphological and structural differences that recur when comparing freshly explanted fibroblasts and epithelial cells with cells of transformed lines of sarcoma and carcinoma lines. All such differences presumably must reflect mechanical abnormalities in the cytoskeleton and plasma membrane. Some of these differences are likely to be related to the cell's relative abilities to continue actin assembly, and thus forward locomotion, at sites of cell–cell contact. It is ironic that pathologists have found dozens of recurring morphological differences in cancer cells, many involving nuclear indentations, that are consistent enough to be the basis of judging whether biopsy specimens are malignant or benign, yet no-one knows the cytoskeletal basis of these morphological abnormalities. If their causes were known, there might be some hope of devising treatments that selectively kill cells having such structural abnormalities.

Although it is a myth that cancer cells grow more rapidly than normal cells, existing chemotherapy methods, and even searches for new anti-cancer drugs, continue to be based on this myth. The drugs used, and the new drugs sought, are designed to be selectively damaging to the fastest-growing cells. Even follicular lymphomas, the great majority of which are now known to result from overproduction of an intracellular protein that inhibits apoptosis, and thus have nothing at all to do with growth rates, are treated with drugs that selectively damage DNA or block DNA synthesis. Paradoxically, these drugs are nevertheless very effective in selectively killing the apoptosis-deficient lymphocytes, producing long-term remissions in most patients. This makes no logical sense whatsoever, except to tell us that we are still far from understanding how to kill cancer cells selectively, while damaging normal cells as little as possible. Much more thought needs to be given to targeting other abnormalities of malignant cells, rather than DNA synthesis and mitosis, which remain the (intended!) targets of nearly all anti-cancer drugs. What if we had a drug that would selectively kill cells based on abnormalities in actin treadmilling, say, or based on whatever causes reduced susceptibility to contact inhibition? In the case of follicular lymphomas, one of their diagnostic characteristics is something called a 'cleaved cell'. By this term, pathologists mean lymphocytes whose nuclear membranes have long grooves. These grooves must have some molecular and mechanical cause; they must result from some cytoskeletal abnormality. Any drug or other treatment that could selectively kill cells with this abnormality would help save many thousands of lives per year. Yet no-one even studies the causes of such morphological abnormalities, much less tries to find treatments that would selectively kill those cells having them. They are too busy looking for better ways to damage DNA and mitotic spindles. It is a preposterous situation.

Question 5: why do so many cells in culture detach their adhesions under their middle parts, what is the mechanism of this de-adhesion, and does it have any purpose, or equivalent, inside the body?

The individual shapes of tissue culture cells resemble those of water droplets on glass; this gives some people the idea that cells maximize their adhesions, but actually, cell adhesions to glass and plastic usually occupy only small fractions of their lower surface. A typical fibroblast's adhesions are concentrated just behind what I am calling the type A margins, with few or no adhesions in central areas, as has been shown both by micromanipulation [23] and by interference reflection microscopy [29]. Although this is no longer controversial, it has yet to be explained either how (mechanistically) or why (functionally) so much of the lower surface of cells does not adhere to the glass or plastic beneath it (Fig. 3). For this state to persist in an actively crawling cell, there must be continual detachment of adhesions, not only along the trailing margin, but also behind the leading margin just behind its band of newly forming adhesions.

The mechanisms of this detachment are not known. Some possibilities include: (a) removal of the adhesion molecules, such as integrins, from that part of the membrane back into the cell; (b) some kind of conformational changes in adhesion molecules that reduce their binding; (c) detachment of adhesive materials from the cell surface (such as the debris, and even cell fragments that many fibroblasts leave behind along their rear margins); or (d) secretion of materials (possibly including water itself) that could push the cell membrane away from the substratum. If interference reflection microscopy is used to scan through cells of established lines, it is not unusual to find occasional cells in which the contacts continue underneath the whole cell; it might be useful to select out cells with this peculiarity, and see to what extent this property is maintained by their progeny. Selection of sublines at the two extremes (those whose contacts continue everywhere under the cell, and those with the narrowest bands of adhesions behind their type A margins) could be the basis of comparisons of which other peculiarities of contents or behaviour are correlated with these types of adhesions.

Another approach would be to find conditions that either interfere with or favour the ability of the cells to let go beneath their middles. UV illumination with a mercury arc happens to inhibit breakage of adhesions to glass and plastic, as is seen when insufficiently filtered arc illumination is used for interference reflection microscopy of polymorphonuclear leucocytes. These cells normally have a dozen or so retraction fibres trailing behind their rear ends; these break their adhesions one after the other. Under arc illumination, however, many are unable to detach; more and more retraction fibres accumulate behind each cell, soon halting its forward progress. A dramatic consequence is that when a culture of leucocytes is arc-illuminated continuously at one spot, more and more of the cells will become trapped there, as they enter but become unable to leave the UV spot. The front ends continue the motility, tugging this way and that, a fact that might be useful in studies of chemotaxis or other directional responses.

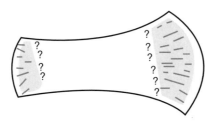

Fig. 3. Fibroblasts and epithelial cells tend to detach from culture substrata beneath each cell's central nuclear area.

A question which perhaps is too seldom asked is whether the black and grey areas seen in interference reflection microscopy always correspond to actual adhesions. After all, what this optical method actually 'sees' is the gap distance between the outer surface of the plasma membrane and the surface of the substratum, with small distances being seen as grey and the smallest distances as black. In principle, why couldn't there be places where the membrane comes very close to the substratum, but there are no adhesion molecules? A further consideration is that reflection depends on the sharpness of the increment of the refractive index at the cell surface; so maybe we should worry about places where the membrane comes very close to the substratum, but perhaps the cytoplasm just inside that part of the membrane has an unusually low refractive index (which is proportional to local concentrations of proteins or other materials). Yet another possible source of misleading images could be extracellular materials between cell and substratum that have a refractive index similar to that of the cytoplasm; this might reduce the amount of reflection at close contacts. Any secreted material with a refractive index higher than that of the cytoplasm could have even more confusing effects. With such possibilities in mind, I carried out extensive studies with a micromanipulator, pushing microneedles under various cultured fibroblasts while observing them by interference reflection, but found no exceptions to what everyone assumes: white-appearing areas did indeed lack adhesions and were easily pushed or pulled away from the glass; grey-appearing areas were always adhesions, but weak enough to be easily peeled away from the glass, usually without tearing the plasma membrane; and black-appearing areas were always strong adhesions that always tore when pulled away from the glass. Only when silicone rubber substrata were used in such studies was it possible to break black-appearing adhesions without ripping away the part of the cell membrane where the adhesion had been.

One further futile experiment using micromanipulation in combination with interference reflection was the following. The sides of microneedles were used to push straight down on the tops of fibroblasts in their central areas around the nucleus (Fig. 4) The question was whether such downward pressure would be able to create grey- and even black-appearing contacts in areas that had appeared white; and, if so, would these contacts then persist after release of the pressure? The answer to the first question is yes; pushing down can create contacts, sometimes even black-appearing ones, although cells are extremely sensitive to lysis whenever the needle undergoes any vibration, even from a lorry driving by outside. Although these experiments were done on a vibration-free table, they still had to be carried out late at night. The answer to the second question is no; grey and black contacts resulting from artificial pressure did not persist. Apparently, the membrane surface must have adhesion molecules in order not to spring apart from the glass.

Yet another contributory factor in the loss of adhesions beneath cell centres might be decreased pulling forces exerted in these areas. The strength of connections between the actin cytoskeleton and adhesion molecules has been found to change in response to pulling forces exerted by laser forceps [30], raising the possibility that reduced pulling under the cell centre might also result in

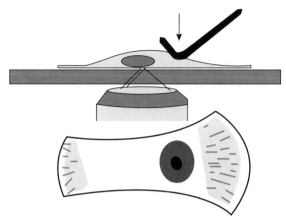

Fig. 4. Artificial formation of cell–substratum contacts. By pushing down on the top of a fibroblast with a micromanipulator needle, one can create cell–substratum contacts close enough to appear black by interference reflection microscopy. But these contacts do not become adhesions, even after a 1 min or longer in contact.

weakening of the connections between adhesion molecules and the substratum. We should also consider possible roles for GTPases in controlling the strength and location of adhesions, especially in view of the observed effects of non-hydrolysable GTP analogues [31], which inhibit detachment caused by RDG peptides, but not rounding-up in mitosis.

Question 6: what is the mechanism of formation of focal adhesions; specifically, are they crystallizations or aggregations?

Although an enormous amount is known about the molecular structure of focal adhesions [32], we cannot yet claim to understand the mechanisms by which the adhesion molecules become concentrated into their characteristic small elongated spots beneath the cell. Published descriptions of these adhesions often seem to me to be regarding their formation as analogous to crystallization, with the integrins, α-actinin, talin etc. clustering together simply by virtue of their mutual binding. At least one other possibility exists, which is that they are actively pulled or swept together by the cell's force-generating machinery, in particular by the mechanisms that actively transport adhering particles centripetally across the cell surface [33]. After all, when a particle is transported, so are the adhesion molecules by which it is attached, and so presumably are the proteins that link these to the actin cortex. In one attempt to test this possibility, I plated fibroblasts on to glass surfaces that had previously been strewn with polystyrene (so-called 'latex') beads of very small diameters (around 0.2 μm). Fibroblasts spread on top of these tiny beads (Fig. 5), and in some cases many of the beads became clustered together in a series of elongated bars behind the cells' leading margins, parallel to their direction of

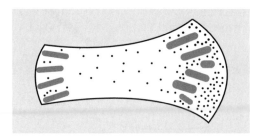

Fig. 5. Evidence of convergent surface flow as a possible mechanism of formation of focal adhesions. When fibroblasts are allowed to attach and flatten themselves on to a glass substratum that had previously been with strewn with 0.2 μm-diam. polystyrene spheres, sometimes the cells collect these tiny spheres into elongated 'footprints' behind the leading margins, with shapes and locations very similar to those of focal adhesions.

spreading, and generally resembling the typical black focal adhesions both in size and shape. Unfortunately, when these cells were then examined by interference reflection microscopy with the hope of finding that the lines of clustered styrene beads were located exactly at the focal adhesions, what was found instead was that the thickness of the beads lifted the cells' lower surfaces sufficiently far above the substratum so that no grey or black adhesions were visible at all. Perhaps equivalent experiments could be done with smaller markers, and probably a lower density of them.

A related aspect of cell adhesion that deserves more attention is the relationship between the grey-appearing and the black-appearing contacts. Clearly the gap distance between cell and substratum is larger in the grey adhesions than in the black ones. Likewise, micromanipulation indicates that the grey areas adhere, but less strongly than the black-appearing areas. So one suspects that the adhesion molecules, especially integrins, are more concentrated per unit area in the black contacts than in the grey ones. The formation of black adhesions could then occur by these adhesion molecules being swept together. But however tempting it is to think of the grey and black as if equivalent to amounts of ink, we may need to remind ourselves that these colours actually represent degrees of closeness. Why should it be possible to concentrate a lot of moderate closeness so as to create a smaller area of extreme closeness? Wouldn't that only be true if the adhesive molecules are stretched to a greater length when distributed more diffusely, and are only able to shorten to the 'black' distance when concentrated to some threshold density per area? An implication would be that some force pushes the membrane away from the substratum, stretching the adhesion molecules out to the 'grey' length, except where their density and/or longitudinal stiffness is sufficient to pull the membrane down to the 'black' distance.

Question 7: are stress fibres nucleated from focal adhesions, or are they gathered together from lamellipodial actin?

Apparently, all strongly contractile tissue culture cells form cytoplasmic strands of actin. These 'stress fibres', as they are called, also contain other muscle-related proteins, including type II myosin, tropomyosin and α-actinin. The ends of these fibres are mechanically attached to the focal adhesions that appear black by interference reflection. All cells that form focal adhesions can be expected to form stress fibres, and vice versa, while cells that have only the grey-appearing adhesions will not have stress fibres, and also will produce much less distortion of silicone rubber substrata.

How do the actin fibres become bunched together into stress fibres (Fig. 6)? One possibility is that the focal adhesions serve as nucleation sites, from which strands of actin grow out by progressive polymerization. This mode of formation seems to be widely assumed, although I am not aware of evidence specifically in its favour. A very different possibility would be that actin that is already polymerized becomes bunched together. In particular, the cortical actin that flows rearward from the lamellipodia might somehow become bunched together as it goes past the focal adhesions. Note how these two alternatives parallel the two possible explanations for the formation of the focal adhesions themselves, as discussed in the previous section. In both cases, i.e. the formation of focal adhesions as well as the formation of stress fibres, the question is whether to visualize the events as phenomena of nucleation and polymerization or as phenomena of aggregation of something that is already polymerized. Presumably, if stress fibres are formed by aggregation, then probably the focal adhesions are also, and conversely. Furthermore, forces exerted by the actin fibres could be what pulls the adhesion sites together.

A related question is how the other muscle-related proteins, such as type II myosin, become associated with the stress fibres. Once again, there seem to be two classes of possibility: aggregation or polymerization. For some unknown but important reason, these additional proteins do not co-polymerize into the networks of lamellipodial actin. It would be very interesting to know why they don't, and even more interesting to know what would happen

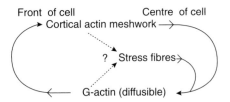

Fig. 6. Speculative diagram of the possible pathways of assembled and disassembled actin. A key question is the extent to which stress fibres obtain their actin from bundling together of cortical actin fibres, as opposed to the separate assembly of G-actin into stress fibres.

if they did co-polymerize there, or conversely what would happen if you could prevent myosin, tropomyosin etc. from becoming part of the stress fibres.

Question 8: what are the functions of the different myosins in cell locomotion?

Crawling tissue cells contain both type II (conventional) and type I myosins; both are believed to play roles in locomotion and force exertion, and both have their own characteristic patterns of localization and exclusion relative to cell polarity. The conventional, type II myosins become combined with actin in stress fibres, but are somehow excluded from lamellipodia. Some type I myosins become concentrated at the most distal tips of lamellipodia. Neither their precise functions nor the mechanisms of their geometric localizations and exclusions are adequately understood.

One first assumption would be that one or both types of myosins slide actively with respect to actin fibres, and that this is responsible for the cells' contractility and traction forces. In the case of the stress fibres, there is no reason to doubt this interpretation: each stress fibre seems to be equivalent to a tiny intracellular muscle. But what about the centripetal flow of cortical actin? Possibly this flow is driven, at least in part, by sliding-filament interactions with one or both classes of myosins. If so, the questions arise as to where these myosins are located, and to what other cytoplasmic components they are mechanically coupled, such that the actin flow is centripetal in the patterns observed. A rather different type of possibility would be that some of the type I myosins might function by travelling along cortical or other actin fibres, using the sliding-filament force for the purpose of their own propulsion out to the barbed ends, and perhaps carrying something out there that is important in controlling cell polarity.

Whole series of questions arise related to the role of myosins in controlling differences in the strength of cell contractility. Transformation and other cases of reduced contractility and reduced ability to form stress fibres seem quite likely to involve myosins. On the other hand, it is certainly not easy to think of possible reasons why microtubule poisons should promote the interactions of myosins with actin so as to produce the Danowski phenomenon discussed above.

Question 9: what is the cytoskeletal mechanism of cell alignment along fibres, ridges and grooves?

Since the earliest tissue culture observations, it has been noticed that cells tend to orient their shapes and their locomotion parallel to fibres, fibrin and even spiders' webs having been used in Harrison's initial studies [34,35]. These orientation responses can be so dramatic that they were once taken as evidence that the cell propulsion mechanism is a form of capillarity [31]. Now, of course, it is thought that these responses result from some kind of influence of surface curvature on the cytoplasmic polymerization of actin, microtubules or other cytoskeletal components. Two types of possibility have been considered: Dunn and Heath [36] proposed that spreading and actin fibre formation are inhibited in directions across sharp curvatures; and Curtis and co-workers (see [5]) has suggested that actin fibre formation is favoured at locations next to, and directions parallel to, sharp curvatures in the substratum. Although seemingly the opposite of one another, these two alternative types of mechanism need not be inconsistent, and could even be complementary. Perhaps sharp folds in the plasma membrane encourage the polymerization of cortical actin fibres next to them, and parallel to the long axis of the fold; once formed, these actin fibres could be stimulated to break or depolymerize wherever bent across their long axis. Another possibility is that directional differences in the mechanical resistance of substrata could result in corresponding directional changes in the strength of links between adhesion molecules and the actin cytoskeleton [30], so that spreading is favoured in whichever direction offers the strongest resistance to pulling forces.

A much lesser known phenomenon is the one that Rich and Harris [26] named 'rugophobia': if a fine-order roughness is created in a plastic or other solid surface, fibroblasts and epithelial cells refuse to crawl on to the roughened area. One possible explanation would be that this is a special case of the Dunn and Heath mechanism; in other words, the roughening of the substratum might be equivalent to many sharp grooves and ridges oriented perpendicular to the long axes of cortical actin fibres, except that tiny bumps play the part of ridges and tiny pits play the role of grooves, so that cells can't orient parallel to them and are inhibited from spreading in any axis. A further puzzle is that macrophages move preferentially on to these roughened areas (one of the many ways in which the behavioural responses of macrophages tend to be the exact opposite of those of fibroblasts). On the other hand, macrophages do orient (and very strongly, too) parallel to even very fine grooves or ridges. One might have thought that they would orient at right angles to grooves etc., but their orientation is parallel, just like fibroblasts.

Question 10: what different kinds of morphogenetic functions are served by these mechanisms that produce crawling in tissue culture cells?

Just as Michael Abercrombie himself began studying cell movements inside developing embryos, many of the students of tissue culture cell locomotion have been embryologists (including myself). Presumably, the main reason to study the locomotion of cells in tissue culture is in order to try to understand how this phenomenon is used inside the body, including during embryonic development. Many embryonic cells migrate actively from one place to another; skeletal myoblasts and all the derivatives of the neural crest are two important examples. It is also possible, however, that additional functions besides moving cells from place to place are served by the same basic mechanisms that cause locomotion in tissue culture. Another such function may be the displacement and alignment of collagen fibres. When fibroblasts are cultured on rigid substrata such as glass or polystyrene, one only sees the spreading and movement of the cells themselves; these artificial substrata are too rigid to be distorted appreciably by cellular forces. But if one uses sufficiently flexible substrata, such as fibrin gels, collagen gels or silicone rubber, then these same cellular forces that produce locomotion also have additional effects [37]. Instead of (or in addition to) pulling the cell past the substratum, these same forces will pull the flexible substratum centripetally past the cells' surfaces.

These mechanical distortions are so large that they convinced me that these additional effects correspond to the real functional purpose of the phenomenon that we have studied for so long as fibroblast 'locomotion'. For the more strongly contractile cell types, especially the ones called fibroblasts, substratum distortion and rearrangement may be the primary function of the actin-based forces that they exert. After all, these cells' normal substrata inside the body are not rigid plastic dishes, but flexible collagen gels. The strength of fibroblasts is such that they cannot help rearranging collagen around them; so we have to believe either that this rearrangement is part of their function, or that it is an unintentional and wasteful property, potentially even dangerous if it produced the wrong distortions. Evolution uses whatever is available, and research with artificial rubber substrata definitely shows that cell traction is available as an effective mechanism for constructing extracellular matrices, such as tendons, fascia or ligaments (Fig. 7). This possibility has not been widely accepted; nor is it much discussed in the cell motility literature.

In Harrison's earliest tissue culture studies [34], he noted that the clotted lymph substrata were somehow distorted around crawling cells. Paul Weiss later hypothesized [35] that this was a chemical effect, in which the cells induced the contraction of gels around them, perhaps by absorbing water from them. In other words, the cells supposedly didn't pull on the gels in any mechanical sense; instead they caused gels to shrink locally around where cells were concentrated. His main evidence for this was that nerve cells did not cause the distortion, whereas glial cells and fibroblasts did; the idea was that nerve cells didn't divide and therefore did not absorb water and cause shrinkage. For

Fig. 7. Diagram of a hypothetical mechanism for embryonic formation of tendons and skeletal muscles. Instead of muscles or tendons first forming and later finding their attachments to the skeleton, the hypothesis is that fibroblasts at certain paired locations on the skeletal surfaces exert especially strong traction, thereby aligning collagen fibres and muscle cells by stretching them between the sites of traction exertion. These sites thereby become the anatomical 'origins' and 'insertions' of the skeletal muscles being formed.

decades, Weiss's theory was taught as proven fact [38]. In one of my first conversations with Michael Abercrombie, he gave me his opinion that these gel distortion effects were really caused by the physical pull of cells on the gels, with the pulling forces being the same as those that cause cell locomotion. This conversation was what gave me the idea of making very flexible substrata, with carbon particles embedded in them, and using the centripetal distortion of the gels (seen by displacements of the particles) as a new method for mapping locations and directions of locomotory forces in individual cells. Abercrombie himself was doubtful if this approach could work, at least for single cells, because he expected that the forces would probably be too weak. Fortunately for me, fibroblasts are hundreds of times stronger than they would need to be if their own locomotion were the only function of these forces. Note that in the case of leucocytes, where the forces that produce locomotion are only about as strong as are needed for that function, I never did manage to create artificial substrata flexible enough to detect their contractility.

Initially, my flexible substrata were made from clotted blood plasma from roosters [39]; later I tried gels made of collagen, then acrylamide gels, then sheets of cellulose nitrate, then various denatured proteins, and eventually silicone rubber. Only by using silicone, a material inherently not subject to any dehydration shrinkage effects [40], was it finally possible to overcome the criticism of supporters of the Weiss theory. That criticism evaporated so abruptly that people now don't grasp how powerful it was. Recently, the silicone rubber method has been significantly improved by two other groups. Timothy Oliver found an alternative method of cross-linking the rubber, using plasma discharge, that produces a more sensitive, compliant and plastic surface, which he used to demonstrate the unexpected pattern of laterally convergent force exertion in fish keratocytes; he also developed a much improved technique for spreading marker particles over the rubber surface [41]. Burton has found that much more flexible substrata than mine can be made by using silicone fluids with benzene side chains, and irradiating them with UV light after cross-linking with heat [42]. The UV breaks many of the cross-linking bonds, thereby making the rubber progressively able to detect smaller and smaller forces.

The original silicone rubber method quickly revealed wide differences in contractile strength between one differentiated cell type and another: blood platelets are enormously strong, especially for their size; fibroblasts and glial cells are very strong; epithelia are weaker than fibroblasts; macrophages are much weaker than epithelial cells, and are right at the lower limit of the method's sensitivity; while neither polymorphonuclear leucocytes nor nerve growth cones exert enough force to distort the rubber detectably. Another surprise was that explanted fibroblasts become stronger over their first week or two in culture, which is apparently related to Gabbiani's phenomenon of conversion into 'myofibroblasts' in response to trauma [43]. Unexpectedly, cells from cancerous cell lines turned out to exert much weaker forces than non-cancerous cells, work that was extended by Barbara Danowski's demonstration that phorbol ester tumour promoters cause a sudden weakening of fibroblast contractility [44]. One would probably have predicted the opposite: that cancerous cells would exert significantly greater propulsive forces on the substratum, helping to account for their greater invasiveness. In fact, they are weaker. However, Danowski discovered how weakened contractility can contribute to increased invasiveness [10]: if you plate normal fibroblasts on to a substratum with sticky spots bounded by non-sticky areas, the cells will be unable to spread on the non-sticky parts of the surface, and those spread and crawling on the sticky parts will be unable to cross the boundaries on to the non-sticky ones. In fact, the cells tend to line up along the boundary.

Question 11: what force causes cell aggregates to round-up?

Embryologists try to understand how the collective behaviour of large groups of cells is caused by the properties of the individual cells making up the mass. This is comparable with the cell biologists' goal of understanding whole-cell behaviour in terms of molecular properties. Success in both tasks will eventually provide molecular explanations for the creation of anatomical structures by cell rearrangements.

One very common, seemingly simple, collective behaviour is the tendency of masses of cells to round-up. If you cut a fragment of just about any tissue out of an embryo, it will become approximately spherical within 1 h or less; likewise, if you use EDTA or enzymes to dissociate a tissue into individual cells, and then let these cells re-aggregate, they will form many multicellular spheres. Two or more such spheres put in contact will fuse to form one larger sphere. These events are especially rapid and dramatic in the case of sponge cells. Since cell masses would not exist without cell–cell adhesion, it is obvious that rounding-up has to depend on cell–cell adhesion; but, in addition, it is widely assumed that adhesion provides the actual physical force that drives rounding-up. The idea is that cells tend to maximize their net area of adhesion to one another, with the minimization of the net area of the cell surface exposed to the surrounding medium supposedly being a secondary effect, very much in the same way as the surface tension of water is caused by intermolecular attraction forces. This is the foundation of a popular theory intended to explain cell sorting as well as such morphogenetic cell movements as gastrulation and neurulation; I refer to Steinberg's Differential Adhesion Hypothesis [45]. Some textbooks call this the 'Thermodynamic Theory', reflecting the mistaken idea that systems only gravitate to one configuration in preference to others when (and because) this preferred state minimizes free energy. Steinberg demonstrated persuasively that the forces that cause cell sorting are the same as those that cause cell aggregates to round-up and resist flattening by external forces; he and Phillips were even able to measure the forces that resist aggregate flattening. From the postulate that rounding is driven by adhesion forces, they deduced that their measurements of resistance to flattening were measurements of adhesion, and that cell sorting, gastrulation etc. were also driven by changes in cell adhesiveness.

But is it possible that some quite different force actually drives rounding-up? In that case, Phillips would have been measuring this different force, and Steinberg's experiments would have proved that this different force causes gastrulation, neurulation and cell sorting. Cell contractility could be this different

Fig. 8. Rounding-up of cell aggregates. Cell aggregates behave as if they had contractile skins, not only in rounding-up and resisting flattening, but also in cell sorting and engulfment by aggregates of different cell types. One popular theory interprets this collective behaviour as resulting from a tendency to maximize cell–cell adhesions (by analogy with liquid surface tension). An alternative explanation is that the cells really do form contractile skins, in the sense of stronger contractility at those parts of each cell's surface that are exposed either to the fluid medium or to cells of other differentiated cell types.

force, as I have proposed previously [46]. The minimization of outside surface areas could itself be the primary cause of rounding, instead of just a side effect. After all, we know that tissue cells are actively contractile, so perhaps each cell's surface contractions vary according to whether that area is in contact with another cell. In particular, suppose that contraction is slightly stronger at the aggregate surface, in contact with the surrounding medium, as compared with that at areas where the cells contact each other: this would produce the effect of a contractile skin around the aggregate surface, causing rounding-up (Fig. 8). This rounding would not occur, nor would the aggregates exist, unless the cells adhered to one another; but this is a different question than whether the formation of adhesions produces the force that drives rounding, resists flattening and produces cell sorting, gastrulation, etc. Especially in the case of neurulation, there is strong evidence that an actin layer is formed along exposed cell surfaces, and that the contraction of this layer drives epithelial folding. Since neural tube formation can also occur by a sorting type of mechanism, it would make sense if sorting is caused by differences in cortical contractility.

Question 12: why do cells round-up during mitosis, and when treated with EDTA or very high hydrostatic pressures?

Any tissue culturist knows that cells round-up in mitosis; but no one knows why they round-up, either functionally or mechanistically. Embryologists also see this happening inside developing embryos, so it is not just some artifact of culture. The usual guess concerning mechanism is that the cells somehow lose or break their adhesions; but if you observe mitotic tissue culture cells at high magnifications you can see that they retain their adhesions via dozens of 'retraction fibres', which are long, thin strands of plasma membrane that are connected to the substratum (Fig. 9). Following division, the cytoplasms of the two daughter cells flow back down these retraction fibres, re-inflating them to approximate the same shape as the original cell. The same is

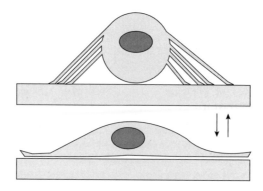

Fig. 9. Role of retraction fibres in adhesion. When tissue culture cells round-up in mitosis, they appear to have lost their adhesions to their substratum. Actually, the cytoplasm pulls away from the adhesions, leaving behind long strands of plasma membrane called retraction fibres, which the daughter cells will later re-inflate with cytoplasm after mitosis. Much the same occurs when cells are treated with chelators of bivalent cations, or are treated with 600 atm (approx. 6×10^7 Pa) of hydrostatic pressure. It is not known if the mechanisms are the same.

true when cells round-up as a result of treatment with EDTA solutions. Even if cells are adhering to the top coverslip in a perfusion chamber, they will not fall off when perfused with EDTA; instead, each cell hangs by its retraction fibres unless these are mechanically broken, such as by turbulence. If you then perfuse normal medium back in, these rounded cells will quickly re-spread by re-inflating their retraction fibres. Very high hydrostatic pressures [in the region of hundreds of atmospheres (1 atm = ~100 kPa)] produce a similar rounding-up, with continued adhesions mediated by large numbers of retraction fibres [47].

One way to think about this rounding phenomenon is that the inside of the plasma membrane becomes mechanically unlinked from the actomyosin cortex. Therefore the cytoplasm rounds up, leaving strands of plasma membrane behind, each still with an adhesion to the substratum at its distal end. Cytoskeletal detachment from the inside surface of the plasma membrane is followed by re-attachment when EDTA is removed, pressure is lowered or cleavage is completed. If this is the correct interpretation, what is the function of this process in relation to mitosis and cleavage? One possibility has to do with the problem of signalling from the mitotic spindle to the cell cortex to control where the actin and myosin need to accumulate so as to form the contractile ring whose contraction will form the cleavage furrow and eventually pinch the cell in two. Experiments on echinoderm eggs show that locations of furrow induction can be highly sensitive to distorting the egg into abnormal shapes, and the types of aster-cortex signalling mechanisms that have been proposed would be expected to have trouble inducing a furrow in cells as highly flattened as most tissue culture cells [48].

The contractile rings responsible for animal cell cleavage are made of actin and myosin, but it is unclear how this actin and myosin become concentrated and aligned in the correct location to form this ring. One theory is that it polymerizes there; another theory is that it flows away from the poles, probably pulled by imbalances of its own contractility, and piles up at the equator to form the contractile ring [49]. I mention this theory of lateral flow of cortical actin in order to contrast it with the postulated centripetal flow of actin inward from lamellipodia. In the case of the mitotic (rounded-up) cell, the purpose would be to accumulate and align the actin and myosin; the flow would be a means to an end, namely ring formation by accumulation. Note the contrast with the situation in the crawling cell, where the actin is not supposed to pile up in the centre, but merely to exert a lateral force through the membrane as it flows.

Consider a cell that crawls for a while, using centripetal actin flow to exert propulsive traction forces; then it goes into mitosis, and makes a contractile ring out of its same supply of cortical actin. The question is: why does actin flow have such different effects in these two cases? In the mitotic cell, the actin and myosin somehow accumulate to form the contractile ring. Contrast this with the crawling cell, whose centripetal flow of cortical actin is for the purpose of exerting traction forces through the plasma membrane, and in which the actin does not seem to accumulate, but instead disassembles and recycles back to the advancing margin. Note that Burton and Taylor [42] have shown that mitotic cells exert traction, converging towards the site of formation of the contractile ring, during the period of contractile ring formation. Could the difference between mitotic and crawling cells result from anything as simple as reduced disassembly of actin in mitotic cells? In other words, if disassembly were inhibited in a crawling cell, would it therefore accumulate its actin into something like a contractile ring around the nucleus? A related question is whether the detachment of the cytoskeleton from the membrane (i.e. apparently what causes rounding-up in mitotic cells) serves some function related to allowing the actin to accumulate into the ring.

Questions also arise about the role of microtubules in these events. We discussed the Danowski phenomenon above: how the disruption of microtubules in crawling cells will greatly strengthen their traction forces and promote the formation of stress fibres. In mitotic cells, also, it so happens that microtubules are believed to control where actin will accumulate to form the contractile ring in cleavage. More specifically, the contractile ring forms halfway between the two asters (asters being radiating arrays of microtubules). Researchers on cell cleavage debate whether the effect of the asters is to strengthen or to weaken cortical contractility. The advocates of strengthening say that the effects of the two asters overlap at the equator (thereby causing the contractile ring to form there); the advocates of weakening think that the poles are weakened more than the equator, so that the ring forms at the latter. Possibly students of crawling cells can help to settle some of those issues as well.

References

1. Abercrombie, M. and Johnson, M.L. (1946) J. Neurol. Neurosurg. Psychiatry 10, 89–92
2. Abercrombie, M., Flint, M.H. and James, D.W. (1954) J. Embryol. Exp. Morphol. 2, 264–274
3. Abercrombie, M. and James, D.W. (1957) Symp. Soc. Exp. Biol. 11, 235–254
4. Weiss, P.A. (1934) J. Exp. Zool. 68, 393–448
5. Clark, P., Connolly, P., Curtis, A.S.G., Dow, J.A.T. and Wilkinson, C.D.W. (1991) J. Cell Sci. 99, 73–79
6. Symons, M.H. and Mitchison T.J. (1991) J. Cell Biol. 114, 503–513
7. Kislauski, E.H., Zhu, X. and Singer, R.H. (1997) J. Cell Biol. 136, 1263–1270
8. Wessells, N.K. and Nuttall, R.P. (1978) Exp. Cell Res. 115, 111–122
9. Vasiliev, J.M., Gelfand, I.M., Domnina, L.V., Ivanova, O.Y., Komm, S.G. and Olshevkaja, L.V. (1970) J. Embryol. Exp. Morphol. 24, 625–640
10. Danowski, B.A. (1989) J. Cell Sci. 93, 255–266
11. Kolodney, M.S. and Wysolmerski, R.B. (1992) J. Cell Biol. 117, 73–82
12. Canman, J.C. and Bement, W.M. (1997) J. Cell Sci. 110, 1907–1917
13. Ingber, D.E. (1993) J. Cell Sci. 104, 613–627
14. Rodionov, V.I., Gyoeva, F.K., Tanaka, E., Bershadsky, A.D., Vasiliev, J.M. and Gelfand, V.I. (1993) J. Cell Biol. 123, 1811–1820
15. Abercrombie, M. and Heaysman, J.E.M. (1953) Exp. Cell Res. 5, 111–131
16. Abercrombie, M. and Heaysman, J.E.M. (1954) Exp. Cell Res. 6, 293–306
17. Abercrombie, M. and Ambrose, E.J. (1958) Exp. Cell Res. 15, 332–345
18. Abercrombie, M. (1967) Natl. Cancer Inst. Monogr. 26, 249–277
19. Harris, A.K. (1974) in Cell Communication (Cox, R.P., ed.), pp. 147–185, John Wiley & Sons, New York
20. Heaysman, J.E.M. (1978) Int. Rev. Cytol. 55, 49–66
21. Vasiliev, J.M., Gelfand, I.M., Domnina, L.V., Dorfman, N.A. and Pletyushkina, O.Y. (1976) Proc. Natl. Acad. Sci. U.S.A. 73, 4085–4089
22. Abercrombie, M. and Dunn, G.A. (1975) Exp. Cell Res. 92, 57–62
23. Harris, A.K. (1973) Dev. Biol. 35, 97–114
24. Abercrombie, M., Heaysman, J.E.M. and Karthauser, H.M. (1957) Exp. Cell Res. 13, 276–291
25. Orida, N. (1980) J. Cell Biol. 87, 92a
26. Rich, A.M. and Harris, A.K. (1981) J. Cell Sci. 50, 1–7
27. Bershadsky, A.D., Gelfand, V.I. and Vasiliev, J.M. (1981) Cell Biol. Int. Rep. 5, 143–150
28. Bershadsky, A.D., Tint, I.S., Neyfakh, Jr., A.A. and Vasiliev, J.M. (1985) Exp. Cell Res. 158, 433–444
29. Izzard, C.S. and Lochner, L.R. (1976) J. Cell Sci. 21, 129–159
30. Choquet, D., Felsenfeld, D.P. and Sheetz, M.P. (1997) Cell 88, 39–48
31. Symons, M.H. and Mitchison, T.J. (1992) J. Cell Biol. 118, 1235–1244
32. Burridge, K. and Fath, K. (1989) BioEssays 10, 104–108
33. Harris, A.K. and Dunn, G.A. (1972) Exp. Cell Res. 73, 519–523
34. Harrison, R.G. (1914) J. Exp. Zool. 17, 521–544
35. Weiss, P.A. (1961) Exp. Cell Res. Suppl. 8, 260–281
36. Dunn, G.A. and Heath, J.P. (1976) Exp. Cell Res. 101, 1–14
37. Harris, A.K., Stopak, D. and Wild, P. (1981) Nature (London) 290, 249–251
38. Weiss, P. (1955) in Analysis of Development (Willier, B.H., Weiss, P.A. and Hamburger, V., eds.), pp. 346–401, W.B. Saunders, Philadelphia
39. Harris, A. K. (1973) in Locomotion of Tissue Cells (Porter, R. and Fitzsimons, D.W., eds.), pp. 3–26, Association of Scientific Publishers, Amsterdam
40. Harris, A.K. Stopak, D. and Wild, P. (1980) Science 208, 177–179
41. Oliver, T., Dembo, M. and Jacobson, K. (1995) Cell Motil. Cytoskeleton 331, 225–240

42. Burton, K. and Taylor, D.L. (1997) Nature (London) **385**, 450–454

43. Gabbiani, G. (1994) Pathol. Res. Pract. **190**, 851–853

44. Danowski, B.A. and Harris, A.K. (1988) Exp. Cell Res. **177**, 47–59

45. Steinberg, M.S. (1970) J. Exp. Zool. **173**, 395–434

46. Harris, A.K. (1994) in Biomechanics of Active Movement and Division of Cells (Akkas, N., ed.), pp. 87–129, Springer, Berlin

47. Bourns, B., Franklin, S., Cassimeris, L. and Salmon, E.D. (1988) Cell Motil. Cytoskeleton **10**, 380–390

48. Rappaport, R. (1996) Cytokinesis in Animal Cells, Cambridge University Press, Cambridge

49. Cao, L.G. and Wang, Y.L. (1990) J. Cell Biol. **111**, 1905–1911

Subject index